Tomosynthesis Imaging

IMAGING IN MEDICAL DIAGNOSIS AND THERAPY

William R. Hendee, Series Editor

Published titles

Quality and Safety in Radiotherapy
Todd Pawlicki, Peter B. Dunscombe,
Arno J. Mundt, and Pierre Scalliet, Editors
ISBN: 978-1-4398-0436-0

Adaptive Radiation Therapy
X. Allen Li, Editor
ISBN: 978-1-4398-1634-9

Quantitative MRI in Cancer
Thomas E. Yankeelov, David R. Pickens,
and Ronald R. Price, Editors
ISBN: 978-1-4398-2057-5

Informatics in Medical Imaging
George C. Kagadis and Steve G. Langer,
Editors
ISBN: 978-1-4398-3124-3

Adaptive Motion Compensation in
Radiotherapy
Martin J. Murphy, Editor
ISBN: 978-1-4398-2193-0

Image-Guided Radiation Therapy
Daniel J. Bourland, Editor
ISBN: 978-1-4398-0273-1

Targeted Molecular Imaging
Michael J. Welch and William C. Eckelman,
Editors
ISBN: 978-1-4398-4195-0

Proton and Carbon Ion Therapy
C.-M. Charlie Ma and Tony Lomax, Editors
ISBN: 978-1-4398-1607-3

Comprehensive Brachytherapy:
Physical and Clinical Aspects
Jack Venselaar, Dimos Baltas, Peter J. Hoskin,
and Ali Soleimani-Meigooni, Editors
ISBN: 978-1-4398-4498-4

Physics of Mammographic Imaging
Mia K. Markey, Editor
ISBN: 978-1-4398-7544-5

Physics of Thermal Therapy:
Fundamentals and Clinical Applications
Eduardo Moros, Editor
ISBN: 978-1-4398-4890-6

Emerging Imaging Technologies in
Medicine
Mark A. Anastasio and Patrick La Riviere,
Editors
ISBN: 978-1-4398-8041-8

Cancer Nanotechnology: Principles and
Applications in Radiation Oncology
Sang Hyun Cho and Sunil Krishnan, Editors
ISBN: 978-1-4398-7875-0

Monte Carlo Techniques in Radiation
Therapy
Joao Seco and Frank Verhaegen, Editors
ISBN: 978-1-4665-0792-0

Image Processing in Radiation Therapy
Kristy Kay Brock, Editor
ISBN: 978-1-4398-3017-8

Informatics in Radiation Oncology
George Starkschall and R. Alfredo C. Siochi,
Editors
ISBN: 978-1-4398-2582-2

Cone Beam Computed Tomography
Chris C. Shaw, Editor
ISBN: 978-1-4398-4626-1

Tomosynthesis Imaging
Ingrid Reiser and Stephen Glick, Editors
ISBN: 978-1-4398-7870-5

Stereotactic Radiosurgery and
Radiotherapy
Stanley H. Benedict, Brian D. Kavanagh, and
David J. Schlesinger, Editors
ISBN: 978-1-4398-4197-6

IMAGING IN MEDICAL DIAGNOSIS AND THERAPY

William R. Hendee, Series Editor

Forthcoming titles

Computer-Aided Detection and Diagnosis in Medical Imaging
Qiang Li and Robert M. Nishikawa, Editors

Handbook of Small Animal Imaging: Preclinical Imaging, Therapy, and Applications
George Kagadis, Nancy L. Ford,
George K. Loudos, and Dimitrios Karnabatidis,
Editors

Physics of Cardiovascular and Neurovascular Imaging
Carlo Cavedon and Stephen Rudin, Editors

Ultrasound Imaging and Therapy
Aaron Fenster and James C. Lacefield, Editors

Physics of PET Imaging
Magnus Dahlbom, Editor

Hybrid Imaging in Cardiovascular Medicine
Yi-Hwa Liu and Albert Sinusas, Editors

Scintillation Dosimetry
Sam Beddar and Luc Beaulieu, Editors

Tomosynthesis Imaging

Edited by

Ingrid Reiser
Stephen Glick

CRC Press
Taylor & Francis Group
Boca Raton London New York

CRC Press is an imprint of the
Taylor & Francis Group, an **informa** business

A TAYLOR & FRANCIS BOOK

Cover image courtesy of Nikolaj Reiser.

Taylor & Francis
Taylor & Francis Group
6000 Broken Sound Parkway NW, Suite 300
Boca Raton, FL 33487-2742

First issued in paperback 2016

© 2014 by Taylor & Francis Group, LLC
Taylor & Francis is an Informa business

No claim to original U.S. Government works

Version Date: 20140108

ISBN 13: 978-1-138-19965-1 (pbk)
ISBN 13: 978-1-4398-7870-5 (hbk)

Library of Congress Cataloging-in-Publication Data

Tomosynthesis imaging / editors, Ingrid Reiser, Stephen Glick.
    p. ; cm. -- (Imaging in medical diagnosis and therapy)
    Includes bibliographical references and index.
    Summary: "Preface For much of the past century, projection radiography has been the workhorse in the diagnostic imaging clinic. Tomosynthesis, which introduces depth information to the x-ray radiographic image with little or no increase in radiation dose, could potentially replace projection radiography as we move into the twenty-first century. This book, Tomosynthesis Imaging, offers the most comprehensive resource to date for this new emerging imaging technology. Digital tomosynthesis imaging is a novel quasi-three dimensional x-ray imaging modality that has been primarily developed during the past two decades, owing to the availability of large-area digital x-ray detectors. The tomosynthesis image is reconstructed from a sequence of projection images that are acquired from a limited angle x-ray scan, therefore, conceptually, tomosynthesis might be considered as limited-angle CT. Because of the limited angle acquisition, resolution in the reconstructed volume is not isotropic. The resolution in image planes parallel to the detector surface is similar to the native detector resolution, but the resolution perpendicular to the detector surface direction is substantially worse, and depends on the scan arc length and on the size of the detail being imaged. Tomosynthesis imaging is being actively investigated for use in a variety of clinical tasks. Currently, tomosynthesis breast imaging is at the forefront, having received approval for clinical use in Europe and Canada in 2008, and FDA approval in the United States in 2011. Although conventional mammography has been very successful in reducing the breast cancer mortality rate, its sensitivity and specificity are less then desirable, especially for women with dense breast tissue"--Provided by publisher.
    ISBN 978-1-4398-7870-5 (hardcover : alk. paper)
    I. Reiser, Ingrid, editor of compilation. II. Glick, Stephen (Stephen Jeffrey), 1959- editor of compilation. III. Series: Imaging in medical diagnosis and therapy.
    [DNLM: 1. Tomography, X-Ray. 2. Breast Diseases--diagnosis. 3. Image Processing, Computer-Assisted--instrumentation. 4. Imaging, Three-Dimensional--methods. WN 206]

RC78.7.T6
616.07'572--dc23
    2013048168

Visit the Taylor & Francis Web site at
http://www.taylorandfrancis.com

and the CRC Press Web site at
http://www.crcpress.com

To my loving family.

Ingrid Reiser

To my wife Clare, for all your support, patience, and friendship.
I love you more than words can tell.

Stephen J. Glick

Contents

Series Preface

Since their inception over a century ago, advances in the science and technology of medical imaging and radiation therapy are more profound and rapid than ever before. Further, the disciplines are increasingly cross-linked as imaging methods become more widely used to plan, guide, monitor, and assess treatments in radiation therapy. Today, the technologies of medical imaging and radiation therapy are so complex and so computer-driven that it is difficult for the persons (physicians and technologists) responsible for their clinical use to know exactly what is happening at the point of care, when a patient is being examined or treated. The individuals best equipped to understand the technologies and their applications are medical physicists, and these individuals are assuming greater responsibilities in the clinical arena to ensure that what is intended for the patient is actually delivered in a safe and effective manner.

The growing responsibilities of medical physicists in the clinical arenas of medical imaging and radiation therapy are not without their challenges, however. Most medical physicists are knowledgeable in either radiation therapy or medical imaging, and expert in one or a small number of areas within their discipline. They sustain their expertise in these areas by reading scientific articles and attending scientific talks at meetings. In contrast, their responsibilities increasingly extend beyond their specific areas of expertise. To meet these responsibilities, medical physicists periodically must refresh their knowledge of advances in medical imaging or radiation therapy, and they must be prepared to function at the intersection of these two fields. How to accomplish these objectives is a challenge.

At the 2007 annual meeting of the American Association of Physicists in Medicine in Minneapolis, this challenge was the topic of conversation during a lunch hosted by Taylor & Francis Publishers and involving a group of senior medical physicists (Arthur L. Boyer, Joseph O. Deasy, C.-M. Charlie Ma, Todd A. Pawlicki, Ervin B. Podgorsak, Elke Reitzel, Anthony B. Wolbarst, and Ellen D. Yorke). The conclusion of this discussion was that a book series should be launched under the Taylor & Francis banner, with each volume in the series addressing a rapidly advancing area of medical imaging or radiation therapy of importance to medical physicists. The aim would be for each volume to provide medical physicists with the information needed to understand technologies driving a rapid advance and their applications to safe and effective delivery of patient care.

Each volume in the series is edited by one or more individuals with recognized expertise in the technological area encompassed by the book. The editors are responsible for selecting the authors of individual chapters and ensuring that the chapters are comprehensive and intelligible to someone without such expertise. The enthusiasm of volume editors and chapter authors has been gratifying and reinforces the conclusion of the Minneapolis luncheon that this series of books addresses a major need of medical physicists.

Imaging in Medical Diagnosis and Therapy would not have been possible without the encouragement and support of the series manager, Luna Han of Taylor & Francis Publishers. The editors and authors, and most of all I, are indebted to her steady guidance of the entire project.

SERIES EDITOR

William Hendee
Rochester, Minnesota

Preface

For much of the past century, projection radiography has been the workhorse in the diagnostic imaging clinic. Tomosynthesis, which introduces depth information to the x-ray radiographic image with little or no increase in radiation dose, could potentially replace projection radiography as we move further into the twenty-first century. This book, *Tomosynthesis Imaging*, offers the most comprehensive resource to date for this new emerging imaging technology.

Digital tomosynthesis imaging is a novel quasi-three-dimensional x-ray imaging modality that has been primarily developed during the past two decades, owing to the availability of large-area digital x-ray detectors. The tomosynthesis image is reconstructed from a sequence of projection images acquired from a limited angle x-ray scan; therefore, conceptually, tomosynthesis might be considered as limited-angle CT. Because of the limited angle acquisition, resolution in the reconstructed volume is not isotropic. The resolution in image planes parallel to the detector surface is similar to the native detector resolution, but the resolution perpendicular to the detector surface direction is substantially worse, and depends on the scan arc length and on the size of the detail being imaged.

Tomosynthesis imaging is being actively investigated for use in a variety of clinical tasks. Currently, tomosynthesis breast imaging is at the forefront, having received approval for clinical use in Europe and Canada in 2008, and FDA approval in the United States in 2011. Although conventional mammography has been very successful in reducing the breast cancer mortality rate, its sensitivity and specificity are less than desirable, especially for women with dense breast tissue. By providing tomographic information, breast tomosynthesis promises to greatly improve visualization of important diagnostic features. In addition to breast imaging, the detection of lung nodules by chest tomosynthesis and detection of hairline fractures with tomosynthesis skeletal imaging are also being investigated. The number of clinical applications for tomosynthesis imaging will, most likely, increase in the future.

This book provides an in-depth understanding of the tomosynthesis image formation process that will allow readers to tailor tomosynthesis systems for new clinical applications. The characteristics of the tomosynthesis reconstructed volume depend strongly on system design parameters; therefore, it is important to gain an understanding of the underlying factors and their effects on the reconstructed volume. This book provides an in-depth coverage of system design considerations, as well as image reconstruction strategies. It also describes the current state of clinical applications of tomosynthesis, including imaging of the breast and chest, as well as its use in radiotherapy. While use of tomosynthesis imaging for these clinical applications is at an early stage, they illustrate the merits of tomosynthesis imaging and its breadth of potential uses.

This book is written for clinicians and researchers. With breast tomosynthesis being approved for clinical use in several countries, medical institutions are faced with the decision to add tomosynthesis to their suite of imaging devices. Clinicians as well as other hospital personnel involved in the purchase and use of clinical tomosynthesis systems should find this book helpful in understanding the principles of tomosynthesis. This book may also be used as classroom teaching material or in workshops to improve understanding of tomosynthesis images, regardless of clinical application.

The book is divided into five sections. Section I introduces tomosynthesis imaging with a historical perspective (the principle of tomosynthesis dates back to work by Ziedsdes des Plantes in 1929). Section II discusses imaging system design considerations, including acquisition parameters, system components, optimization, and modeling schemes. The purpose of this section is to acquaint the reader with the flexibility of tomosynthesis acquisition and the impact of physical factors in the various schemes. Section III reviews image reconstruction algorithms that have been developed for tomosynthesis, including filtered back-projection methods that are used in most clinical systems, as well as advanced iterative methods that have the potential to reduce artifacts and improve image quality. Section IV describes system evaluation methodologies, including radiologist performance studies and assessment using mathematical model observer assessment. Finally, Section V is dedicated to current clinical applications, which include breast, chest and therapy applications, and concludes with a discussion of future directions for tomosynthesis.

The goal of this book is to cover the fundamentals of tomosynthesis imaging. As tomosynthesis and its applications are evolving rapidly, clinical trials to assess the clinical performance of tomosynthesis imaging are ongoing, and we hope that our readers will join us in eagerly awaiting the outcome of these trials.

EDITORS

Ingrid Reiser
Chicago, Illinois
Stephen J. Glick
Worcester, Massachusetts

Editors

Ingrid Reiser, PhD, is a research associate (assistant professor) in the Department of Radiology at the University of Chicago. After earning her PhD in physics from Kansas State University in 2002, she transitioned into medical physics research where she witnessed the presentation of the first breast tomosynthesis images at RSNA 2002 (Radiological Society of North America). Tomosynthesis captivated her interest and she has since investigated many aspects of tomosynthesis imaging, such as computer-aided detection, system modeling, and objective assessment. Her research interests further include image perception and observer performance, as well as tomosynthesis and CT image reconstruction.

Stephen J. Glick, PhD, is a professor of radiology at the University of Massachusetts Medical School and the director of the Tomographic Breast Imaging Research Laboratory. He earned his PhD from Worcester Polytechnic Institute (WPI) in 1991. Dr. Glick has published over 60 journal articles and over 100 conference proceedings papers. Over the past decade, his research has been focused on 3D breast imaging techniques including digital breast tomosynthesis and breast CT with an emphasis on radiation dose, imaging technique optimization, advanced iterative reconstruction methods, detection studies for lesions and microcalcifications, and photon counting detector CT.

Contributors

Magnus Båth
Department of Medical Physics and Biomedical Engineering
Sahlgrenska University Hospital
Gothenburg, Sweden

Ying (Ada) Chen
Department of Electrical and Computer Engineering and
    Biomedical Engineering Graduate Program
Southern Illinois University
Carbondale, Illinois

James T. Dobbins III
Departments of Radiology and Biomedical Engineering
Duke University
Durham, North Carolina

Stephen J. Glick
Department of Radiology
University of Massachusetts Medical School
Worcester, Massachusetts

Devon J. Godfrey
Department of Radiation Oncology
Duke University
Durham, North Carolina

Mitchell M. Goodsitt
Department of Radiology
University of Michigan Hospital
Ann Arbor, Michigan

Åse Allansdotter Johnsson
Department of Radiology
The Sahlgrenska Academy at the University of Gothenburg
Gothenburg, Sweden

Roberta Jong
Division of Breast Imaging
Sunnybrook Health Sciences Centre
University of Toronto
Toronto, Canada

Beverly Lau
Department of Radiology
Stony Brook Medicine
Stony Brook, New York

James G. Mainprize
Department of Physical Sciences
Sunnybrook Research Institute
University of Toronto
Toronto, Canada

Thomas Mertelmeier
Siemens AG
Erlangen, Germany

Robert M. Nishikawa
Department of Radiology
University of Pittsburgh
Pittsburgh, Pennsylvania

Subok Park
Division of Imaging and Applied Mathematics
Food and Drug Administration
White Oak, Maryland

Norbert J. Pelc
Departments of Bioengineering and Radiology
Stanford University
Stanford, California

Steven P. Poplack
Department of Radiology
Dartmouth-Hitchcock Medical Center
Lebanon, New Hampshire

Ingrid Reiser
Department of Radiology
University of Chicago
Chicago, Illinois

Lei Ren
Department of Radiation Oncology
Duke University
Durham, North Carolina

Ioannis Sechopoulos
Department of Radiology and Imaging Sciences
Emory University
Atlanta, Georgia

Emil Y. Sidky
Department of Radiology
University of Chicago
Chicago, Illinois

Grant M. Stevens
CT Advanced Science & Engineering
GE Healthcare
Waukesha, Wisconsin

Tony Martin Svahn
Department of Clinical Sciences
Lund University, Skåne University Hospital
Malmö, Sweden

Anders Tingberg
Department of Radiation Physics at Malmö
Lund University
and
Skåne University Hospital
Malmö, Sweden

Q. Jackie Wu
Department of Radiation Oncology
Duke University
Durham, North Carolina

Martin J. Yaffe
Department of Medical Biophysics
University of Toronto
Toronto, Canada

Fang-Fang Yin
Department of Radiation Oncology
Duke University
Durham, North Carolina

Wei Zhao
Department of Radiology
State University of New York at Stony Brook
Stony Brook, New York

Weihua Zhou
Department of Electrical and Computer Engineering
Southern Illinois University
Carbondale, Illinois

Introduction

1 The history of tomosynthesis

Mitchell M. Goodsitt

Contents

1.1 INVENTION OF TOMOSYNTHESIS IN THE LATE 1930S

1.1.1 CONCEPT: SHIFT AND ADD

Tomosynthesis addresses one of the primary weaknesses of conventional single-projection x-ray imaging, the superposition of objects in the image. This superposition may result in the obscuring of an object of interest and/or the production of pseudoobjects that mimic a disease (e.g., pseudomasses or a reduction in bone joint space). Tomosynthesis decreases superposition by generating slice images of the body from a series of projections taken at a variety of angles. In this sense, it is much like computed tomography (CT); however, instead of acquiring projections over 360° as in CT, the projections for tomosynthesis are acquired over a limited range of angles (e.g., 10° or 60°). The resulting tomosynthesis images have much better spatial resolution than CT within the slice, but poorer resolution in the depth direction, between the slices. The basic tomosynthesis principle is illustrated in Figure 1.1. A series of snapshot radiographs (films) are taken, each from a different viewing angle, as an x-ray tube translates across a patient. These snapshots are later combined by shifting the snapshots relative to one another and superimposing (adding) them to bring different planes within the patient in focus. Today, we use digital radiographs instead of films and more sophisticated reconstruction algorithms like those employed in CT instead of the "shift-and-add" technique.

1.1.2 INVENTORS ZIEDSES DES PLANTES AND KAUFMAN

A Dutch neuroradiologist and electrical engineer, Bernard George Ziedses des Plantes, is often credited with inventing tomosynthesis. Ziedses des Plantes published the first article about the technique, which he called "seriescopy," in 1935 (Ziedses des Plantes 1935). He subsequently patented seriescopy in 1936 (Ziedses des Plantes 1936). In a 1938 paper (Ziedses des Plantes 1938), he described seriescopy as a "radiographic method which makes it possible to view an infinite series of parallel planes in succession by means of a few exposures." Ziedses des Plantes built a prototype serioscope based on circular tomography in which the system took four stationary film exposures at equidistant locations along the circular path of the x-ray tube (e.g., at 0°, 90°, 180°, and 270°). Different planes of focus could be achieved by reciprocally moving the developed films above a single viewing screen. Ziedses des Plantes also described an optical method of reconstruction using mirrors to superimpose images of the films and the rocking of the mirrors to achieve the same effect as translating the films (Webb 1990). In addition, he described a device for recording the level of the plane of focus (Webb 1990).

A method very similar to seriescopy was invented by Julius Kaufman, MD, of Brooklyn, New York in 1936 (Kaufman 1936a). Kaufman called his method "planeography" and stated that with this method, "it is possible to demonstrate any plane in space, parallel to the plane of the plate from two (or more) roentgenograms (films) properly taken." Kaufman stressed the localization and depth measurement capabilities of his method. Depth measurements were achieved "by reference to a specially constructed curve, the 'standard depth curve,'" which related the height of a linear object in a plane parallel to the films to the spread of the projections of that object in the films. Although Kaufman is not recognized as the inventor of tomosynthesis, given the similarities between the concepts of Ziedses des Plantes and Kaufman, the fact that the inventions were made at about the same time (1935 for Ziedses des Plantes, 1936 for Kaufman) and neither was familiar with the other's work at that time, it seems fair that both Ziedses des Plantes and Kaufman should be credited with this invention. In addition to his original paper explaining the planeography method, Kaufman published two short papers on applications of the method (Kaufman 1936b, 1938). Also, after learning of a presentation on serioscopy by Cottenot in 1937 (described in the next section of this chapter),

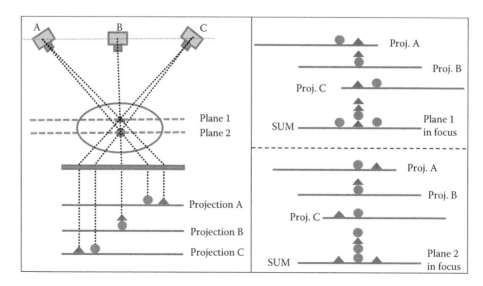

Figure 1.1 Illustration of the shift-and-add technique whereby the positions of images obtained at different angles (left) are shifted relative to one another and superimposed (added) to bring different planes (e.g., plane 1 and plane 2) within the object (patient) in focus (right).

Kaufman and a colleague, Harry Koster, MD, published a critical analysis comparing their planeography technique to Cottenot's serioscopy and Ziedses des Plante's serioscopy (Kaufman and Koster 1940). The latter are two different spellings for the same technique. Kaufman and Koster stated, "apparently Dr. Cottenot rediscovered planeography independently, terming the method 'serioscopy'" (Kaufman and Koster 1940). In their critical analysis, Kaufman and Koster illustrated how, with serioscopy, unequal magnification of parts of an object (e.g., the top and bottom of a vertically oriented cylinder) at different distances from the film detector results in an "erroneous impression of the true size, shape and relative position of the object" in the serioscope. They state that special measures suggested by Kaufman in his earlier planeography publications (Kaufman 1936b, 1938) are required to correct these problems.

1.1.3 COTTENOT'S EARLY CHEST TOMOSYNTHESIS SYSTEM

In 1937, Paul Cottenot presented a paper at the Fifth International Congress of Radiology in Chicago, Illinois, in which he described a tomosynthesis method he developed for the "study of pleuro-pulmonary lesions." He called this method "thoracic serioscopy" (Cottenot 1938). Cottenot noted that his method was based on the work of Ziedses des Plantes. In order to image the chest with tomosynthesis, it is critical that each of the projection radiographs be taken at the same level of inspiration. Cottenot accomplished this by developing a "respiratory trigger" consisting of a pneumatic belt that was wrapped around the patient. The belt sent a pressure wave to one side of a U-shaped mercury manometer, the other side of which contained a metal rod the position of which was adjusted so that the rod touched the mercury at a desired level of inspiration. When the mercury level on the side of the rod reached the rod, electrical contact was made, triggering the exposure. Four exposures were made, with the x-ray tube shifted 14 cm to the left, right, upward, and downward of the center. In order to limit motion blur, the exposure time for each film was set to about one-third the exposure time for a conventional radiograph. The thoracic serioscopy device is shown in Figure 1.2a.

The developed films were viewed on a special high-intensity light box with screws that shifted the relative positions of the

films (Figure 1.2b). Because the imaging geometry was known (focal distance of 1.4 m, and shifts of 14 cm, for a tomosynthesis angle of 11.4°), the serioscope could be calibrated, and the viewing system included a pointer and a dial that was graduated in centimeters (Figure 1.2b). Some of the successful clinical applications of serioscopy that Cottenot described included determining the dimensions and locations of a pneumothorax, abscesses of the lung, and tuberculosis foci (Cottenot 1938).

1.2 FILM-BASED SYSTEMS OF THE 1960S AND THE 1970S

1.2.1 GARRISON AND GRANT'S SYSTEM AT THE APPLIED PHYSICS LAB AT JOHNS HOPKINS

There were almost no developments in tomosynthesis between the late 1930s and the late 1960s when John Garrison and David Grant et al. of Johns Hopkins University in Baltimore, Maryland described their prototype system (Garrison et al. 1969). Their prototype, like Ziedses des Plante's, used a circular scan. Twenty individual uniformly spaced radiographs were recorded (Figure 1.3), photoreduced onto a single film, and reprojected using a sophisticated optical backprojection system with mirror assemblies (Figures 1.3b and 1.4).

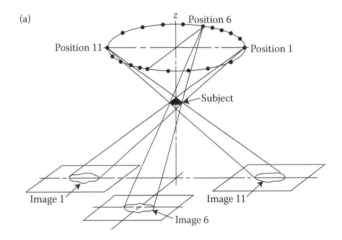

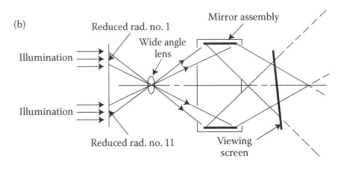

Figure 1.3 Garrison and Grant's image acquisition system (a). The 20 acquired images were each photoreduced onto a film strip using a custom camera recording system and then projected onto a single large film using the same 3D projector mirror system as was used for final viewing (see Figure 1.4). That film was then illuminated in the 3D projector mirror system (b) and the tomosynthesis slices were observed on a viewing screen. (Reproduced with permission from IEEE.)

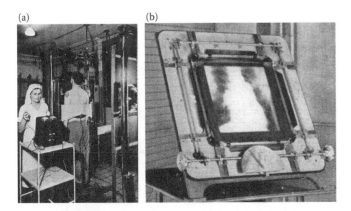

Figure 1.2 (a) Cottenot's thoracic serioscopy image acquisition system. (b) Cottenot's serioscope viewing system. (Reproduced with permission from *Radiology*.)

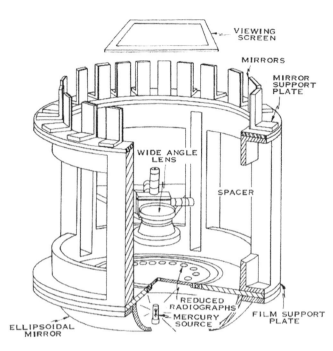

Figure 1.4 Optical display apparatus in Garrison and Grant's system. (Reproduced with permission from IEEE.)

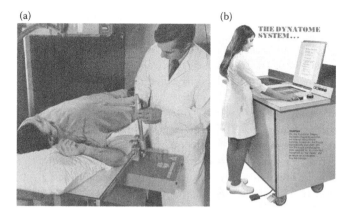

Figure 1.5 Calibrator (a) and film shaper (b) in Dynatome dynamic tomography system.

The projections were integrated on a viewing screen that was moved up and down to display different planes (slices) in the volume (Figure 1.4), and the image on the screen was captured with a TV camera and displayed on a TV monitor.

David Grant coined the term "tomosynthesis" in a paper he published in 1972 (Grant 1972).

1.2.2 RAPID FILM CHANGER TOMOSYNTHESIS SYSTEMS

1.2.2.1 Dynatome system of Albert Richards

The first commercial film-based tomosynthesis system, the Dynatome by CFC Products, Inc., was developed in the 1970s and the 1980s by Albert Richards, MS (Physics), a professor of dentistry at the University of Michigan Dental School. Ten radiographic films were acquired at different angles along a linear tomographic sweep of the patient. A rapid film changer was employed to acquire these films at a rate of ≥2 films/s. A special film developed by Eastman Kodak specifically for the low-density exposures associated with this system was employed. A calibrator was utilized (Figure 1.5a) for determining the angle for each film and the films were trimmed and shaped based on the projection angles (Figure 1.5b).

The developed films were arranged on a custom high-intensity light box viewer (Figure 1.6) and moved with a dial in a precise relationship to each other, so information from any level common to all films could be registered. By moving the dial, the equivalent of about 100 contiguous slices could be viewed, each with a thickness of 2 mm. The total radiation dose to the patient was about equal to that of two or three plain films.

Richards described the principles of dynamic tomography, his name for tomosynthesis, in a paper published in 1976 (Richards 1976). In that paper, he stated that he found moving the x-ray source in a circular path yielded "better results than can be

achieved with linear movement." Richards demonstrated the principle of "circular-movement" dynamic tomosynthesis by taking separate films of human cadaver skulls at eight locations along a circular path, each location separated by 45°, with manual orientation of each film cassette. For viewing, the developed films were superimposed on a bright viewbox; the long axis of each film was rotated 45° relative to the previous film, and 180° opposite film pairs (e.g., 0° and 180°, 45° and 225°) were shifted together along paths through their longitudinal axes. Richards stated that all that was needed to make circular movement dynamic tomography practical for live patient imaging was the construction of an apparatus that would provide rapid changing and proper orientation of the eight film cassettes. He also prophesized that a future application of dynamic tomography would be the examination of the beating heart, layer by layer, and this would be accomplished with EKG gating and very short exposures (Richards 1976).

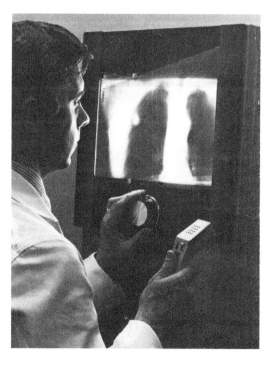

Figure 1.6 Dynatome tomosynthesis viewer.

1.2.2.2 Rapid film changer system of Miller

A rapid film changer tomosynthesis system similar to Richard's Dynatome was developed by Miller et al., but was never commercialized (Miller et al. 1971). Unique to Miller et al.'s rapid film changer system was the production of hardcopy films by illuminating the projection radiographs in a photographic camera, with shifting of either the camera or the radiographs to produce a film showing an image of the desired focal plane.

1.2.3 CODED APERTURE TOMOSYNTHESIS

One of the most intriguing and advanced tomosynthesis methods of the past, coded aperture imaging, was developed by Klotz and Weiss of Philips GmbH of Hamburg, Germany in the 1970s (Klotz and Weiss 1974, 1976, Weiss et al. 1977, 1979). This method could generate arbitrary tomosynthesis planes with imaging times of only milliseconds, essentially eliminating any problems associated with patient motion. They called their method "flashing tomosynthesis," and its application to coronary angiography was described in a paper that they coauthored with Woelke et al. (1982). The coding and decoding steps are illustrated in Figure 1.7. In general, for coding, many x-ray sources were either pulsed sequentially or simultaneously (Figure 1.7a). A single 60 × 60 cm film recorded the entire set of subimages, forming the coded image. For decoding, this coded image was illuminated with a light box and a three-dimensional (3D) image of the object in space was produced with an array of lenses that were arranged according to the distribution of x-ray tubes. A ground glass screen (Figure 1.7b) was positioned within the 3D image to display different layers within the object.

The system Woelke et al. employed consisted of 24 small stationary x-ray tubes that were fired simultaneously. The exposure time was about 50 ms, the radiation exposure to the skin was about 1 roentgen, and the slice thickness was about 1 mm. In addition to the multiple x-ray tubes, the hardware in the system included an optical postprocessing unit, a TV monitor on which reconstructed layers were displayed, and a film hardcopy unit (Woelke et al. 1982). In a study of 20 left coronary artery stenoses within 10 postmortem hearts that were placed within a thorax phantom, Woelke et al. found that the correlation between the degree of stenosis determined with flashing tomosynthesis and morphometry ($r = 0.92$, $p < 0.001$, SEE = 9%) was better than the correlation between the degree of stenosis determined with conventional 35-mm cine film cardiac imaging and morphometry ($r = 0.82$, $p < 0.001$, SEE = 16%). They also successfully employed their flashing tomosynthesis technique on five patients with coronary artery disease. Limitations of the flashing tomosynthesis method included restriction to a small 9-cm-diameter field of view, and the inability to evaluate blood flow dynamics due to the use of a single "flash" exposure (Woelke et al. 1982). Additional papers that have been published on flashing and coded aperture tomosynthesis include those of Groh (Groh 1977), Nadjmi et al. (1980), Stiel et al. (Stiel 1989, 1992, 1993), Haaker et al. (1985a,b, 1990), and Becher et al. (1983, 1985).

1.3 COMPARISON OF SERIOSCOPY, TOMOSYNTHESIS, AND CODED APERTURE TOMOGRAPHY

The distinctions between serioscopy, tomosynthesis, and coded-scan tomography are discussed in detail in a paper by Mandelkorn and Stark (1978).

1.4 FLUOROSCOPIC TOMOSYNTHESIS

Fluoroscopic applications of tomosynthesis were developed separately by Baily et al. of the University of California, San Diego (Lasser et al. 1971, Baily 1973, 1974) and Hoefer et al. of Philips in Hamburg, Germany (Hoefer 1974) in the early 1970s. The image intensifier–TV camera detector in fluoroscopic systems offered the advantages of high frame rates and good contrast for low radiation doses. Components of the early fluoroscopic tomosynthesis systems were similar. Those of Hoefer et al. included a linear tomography device with a pulsed x-ray source and an image intensifier–TV camera detector, a video disk recorder to sequentially save the tomographic projection video images, a minicomputer to store the positions of the x-ray tube and image intensifier at the time of each x-ray pulse and to compute the shift values for tomosynthesis, and a lithicon storage tube on which the projection images from the disk recorder were shifted and added to create tomosynthesized images. Those images (layers) were in turn recorded on the disk recorder and displayed on a video monitor.

The video signal from the image intensifier–TV camera chain of the fluoroscopic tomosynthesis system could also be digitized and processed to create digital projection images from which a set of tomosynthesis slices could be reconstructed. A set of slices generated before contrast injection could be subtracted from subsequent corresponding sets generated after intravenous or intra-arterial iodine contrast injection to produce digital subtraction angiography (DSA) tomosynthesis images that improved the perception of blood vessels. Pioneers in applying tomosynthesis to DSA included Kruger et al. of the University of Utah (Kruger et al. 1983, 1984, Anderson et al. 1984, deVries et al. 1985) and Maravilla et al. of the University of Texas (Maravilla et al. 1983a,b, 1984, Murry and Maravilla 1983).

Sone et al. employed an image intensifier detector digital video system for chest tomosynthesis (Sone et al. 1991, 1993, 1996).

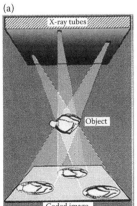
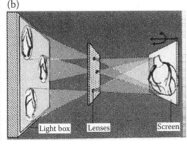

Figure 1.7 Coding step (a) and decoding step (b) of the flashing (coded aperture) tomosynthesis system described in Woelke et al. (Reproduced with permission from *Radiology*.)

Sklebitz and Haendle (1983) developed a real-time fluoroscopic tomography "tomoscopy" system that combined a fluoroscopic detector with a 16 x-ray tube source. A new type of x-ray tube was developed for this system. It had a large rotating anode (target) that was surrounded by 16 grid-controlled cathodes. The result was the equivalent of 16 x-ray tubes that were arranged in a circle and were pulsed sequentially, creating a source that electronically moved in a circle. The detected region of the image intensifier was magnetically adjusted in real time to electronically move in synchrony with the moving x-ray source. Each entire scan was accomplished in the time for one TV field (about 20 ms). The 16 projections were superimposed on a fluorescent screen for display, with electronic shifting of the projections to produce tomosynthesis slices at different heights.

A limitation of all of the image-intensifier-based tomosynthesis systems was geometric (e.g., pin cushion) distortion arising from poorer focusing of electrons within the image intensifier toward the outer edges of the image.

1.5 DIGITAL DETECTOR-BASED TOMOSYNTHESIS

There was a marked reduction in tomosynthesis research and development in the later 1980s because of the rising popularity of CT and because there were no suitable distortion-free, high-frame-rate digital detectors for tomosynthesis. There was a rebirth of tomosynthesis in the late 1990s with the development of high detective quantum efficiency (DQE) flat-panel and charge-coupled device (CCD) digital x-ray detectors with rapid readout, and the application of tomosynthesis to full-field breast imaging by Niklason and Kopans et al. of Massachusetts General Hospital (MGH) of Boston, MA, and scientists at General Electric (GE) Corporate Research and Development, Schenectady, New York (Niklason et al. 1997), the application of tomosynthesis to small (5 × 5 cm) field breast imaging by Webber of Bowman Gray School of Medicine, Winston-Salem, Chapel Hill, North Carolina and Instrumentarium Imaging, Inc. of Tuusula, Finland (Webber 1994–2000, Lehtimaki et al. 2003), and the application of tomosynthesis to chest imaging by Dobbins et al. of Duke University, Durham, North Carolina (Dobbins et al. 1998, 2008, Dobbins 2009).

1.5.1 DIGITAL DETECTOR BREAST IMAGING SYSTEMS

1.5.1.1 System developed by Niklason, Kopans, and GE

A picture of the system that was used by Loren Niklason and Daniel Kopans in their early breast tomosynthesis studies is shown in Figure 1.8.

It consisted of a GE DMR mammography system that was modified to incorporate a stationary GE-developed flat-panel digital detector made of a cesium iodine (CsI) phosphor backed by an amorphous silicon (a-Si) transistor–photodiode array. This detector had a pixel pitch of 100 microns and a readout time of 300 ms. The gantry was modified to permit manual positioning of the x-ray source along an arc. The tomosynthesis angle was 40° (±20°) with nine equally (5°) spaced views, and the radiation dose was about 1.4 times that of a standard film-screen mammogram. The projection images were transformed from the actual "x-ray source motion along an arc" geometry to "x-ray source motion in a horizontal plane parallel to the detector" geometry so

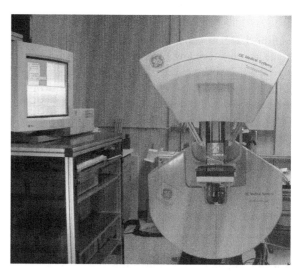

Figure 1.8 Digital mammography tomosynthesis system developed by GE and Niklason and Kopans. (With permission from GE.)

that linear shift-and-add reconstruction could be employed (Niklason et al. 1997). Based on preliminary phantom and breast specimen studies, Niklason et al. concluded that "tomosynthesis may improve the specificity of mammography with improved lesion margin visibility and may improve early breast cancer detection, especially in women with radiographically dense breasts" (Niklason et al. 1997). The research group at MGH also published early work comparing a variety of tomosynthesis reconstruction methods for breast imaging (Wu et al. 2004).

1.5.1.2 Tuned aperture CT digital spot tomosynthesis: Webber and instrumentarium

Tuned aperture computed tomography (TACT) is a tomosynthesis method that was originally developed by Richard Webber, DDS, PhD, of Bowman Gray School of Medicine, Winston-Salem, North Carolina for dental applications (Webber 1994, 1997, Webber et al. 1995, 1996, 1997, 1999), and later for breast imaging applications (Webber et al. 2000). It is based on optical aperture theory and requires the use of a fiduciary reference point. The method was licensed to Instrumentarium Imaging Inc., and they developed a commercial system called the Delta 32 TACT® 3D breast imaging system. The Delta 32 system debuted at the 1997 Radiological Society of North America meeting in Chicago, Illinois, and it received FDA approval in 2000. This was a spot imaging system that employed a 5 × 5 cm CCD detector. A fiducial marker was placed on the compression paddle and images were acquired at seven angles. According to a paper by Mari Lehtimaki et al. (Lehtimaki et al. 2003) of Instrumentarium Imaging, the TACT process includes (1) stacking the acquired images and recognizing the reference point corresponding to the fiducial marker in each image, (2) determining the center of gravity of the set of reference points, (3) drawing lines between each reference point and the center of gravity, (4) shifting each of the stacked images the same relative amount along its line between the reference point and center of gravity, and (5) averaging the superimposed pixels in the shifted stacked images to create a TACT image. Different TACT images were obtained by using different shift values. A unique feature of TACT is the use of the external fiducial marker, which defines the geometry, enabling the use of arbitrary acquisition projection angles.

1.5.1.3 Other early flat-panel digital tomosynthesis breast imaging systems

Others who pioneered the development of digital flat-panel tomosynthesis for breast imaging included Suryanarayanan and Karrelas et al. of the University of Massachusetts, Worcester, Massachusetts (Suryanarayanan et al. 2000, 2001). They employed a GE 2000D digital mammography system that was similar to the system employed by Niklason and Kopans at MGH; however, instead of a 40° tomosynthesis angle with nine 5° increments, Suryanarayanan and Karrelas et al. employed a 36° tomosynthesis angle with seven 6° increments. They also evaluated a variety of tomosynthesis reconstruction algorithms, including TACT backprojection, TACT maximization, TACT minimization, TACT-iterative restoration, and expectation maximization and Bayesian smoothing iterative methods (Suryanarayanan et al. 2000, 2001).

1.5.1.4 Optimization of acquisition parameters in breast tomosynthesis

Several research groups have been investigating the optimization of acquisition parameters (e.g., total sweep angle, angle increment, and radiation dose) for breast tomosynthesis (Wu et al. 2003, Godfrey et al. 2006b, Deller et al. 2007, Zhao et al. 2007, Zhou et al. 2007, Gifford et al. 2008, Reiser et al. 2008, Chan et al. 2008, Sechopoulos and Ghetti 2009, Chawla et al. 2009, Lu et al. 2010, 2011). This topic is discussed in great detail in Chapter 2 on geometry and systems design considerations in this textbook.

1.5.2 DIGITAL TOMOSYNTHESIS FOR BODY IMAGING

1.5.2.1 Chest imaging

Dobbins and his research group at Duke University, Durham, North Carolina constructed a prototype flat-panel detector tomosynthesis system for chest imaging (Godfrey et al. 2003). It employed a stationary commercial GE Revolution 41 × 41 cm CsI phosphor, a-Si detector flat-panel detector with 200 micron pixel pitch, a custom-made x-ray tube mover, and matrix inversion tomosynthesis (MITS) reconstruction. A linear actuator was used to move the x-ray tube vertically during the tomosynthesis scan, and a second actuator was employed to adjust the angle of the x-ray tube so the x-ray beam collimator was centered upon the detector for each projection. The digital detector could be operated at frame rates up to 5.8 frames/s. Typical acquisition parameters for chest tomosynthesis studies included 61 projections, ~10.5 s total scan time, 20° total tube motion, and 10 ms per exposure. The x-ray tube was moved continuously during the scan rather than use a step-and-shoot technique. A theoretical investigation indicated that for the chosen exposure time and scan rate, continuous motion of the x-ray tube would be satisfactory since the resulting blur at the detector was minimal even for objects in the most posterior thoracic plane of large subjects. One concern with digital detectors is image retention or ghosting in which faint versions of previous frames remain on the detector and corrupt successive frames. This would have a detrimental effect on tomosynthesis reconstructions. A solution is to employ scrub frames in which a blank frame is readout without x-ray exposure before each x-ray image is acquired. This scrub frame readout process cleans the detector of the majority of the information from prior images, but since twice as many frames are needed, it slows the total readout time by 50%. An experimental study was performed with and without scrub frames, and it was determined that for this detector and the chosen acquisition speed without scrub frames (5.8 frames/s), the difference between projections acquired with and without scrub frames was minimal, so scrub frames were not necessary (Godfrey 2003). The prototype system developed at Duke served as the basis for the commercial body tomosynthesis system presently marketed by GE Healthcare (VolumeRad). The GE system also uses continuous x-ray tube motion, but it uses a different reconstruction algorithm.

In 2009, Dobbins and McAdams published a review article titled "Chest Tomosynthesis: Technical Principles and Clinical Update" (Dobbins and McAdams 2009). In this article, Dobbins and McAdams described the initial results of a clinical trial for lung nodule detection that was conducted at Duke University. This trial compared tomosynthesis using their prototype unit to PA digital chest radiography using a GE Healthcare commercial unit (Dobbins et al. 2008). They found that when they "counted as true positives only those nodules that were scored as definitely visible, sensitivities for all nodules by tomosynthesis and PA radiography were 70% (±5%) and 22% (±4%), respectively ($p < 0.0001$)." They also described the results of a prospective human observer study of chest tomosynthesis that was conducted at the Sahlgrenska Academy of the University of Gothenburg, Sweden using the GE VolumeRad tomosynthesis system. In that study, it was found that on average, three times as many lung nodules were detected by tomosynthesis as by conventional chest radiography (Vikgren et al. 2008).

Other publications on the application of tomosynthesis to the evaluation of the chest, lung, and lung nodules include Rimkus et al. (1989), Sone et al. (1991, 1993, 1996), Matsuo et al. (1993), Dobbins et al. (1998), Godfrey et al. (2001a, b), Godfrey and Dobbins (2002), Fahrig et al. (2003), Yamada et al. (2011), Asplund et al. (2011), Gomi et al. (2011), Kim et al. (2010), Quaia et al. (2010, 2012), Johnsson et al. (2010a, b), Santoro et al. (2010), Zachrisson et al. (2009), and Jung et al. (2012).

1.5.2.2 Orthopedic, dental, radiotherapy, and other applications of tomosynthesis

Orthopedic applications of tomosynthesis such as the imaging of joints have been researched since 1989 (Rimkus et al. 1989, Sone et al. 1991, Kolitsi et al. 1996, Duryea and Dobbins 2001, Duryea 2003, Fahey et al. 2003, Gazaille et al. 2011). There have also been investigations of tomosynthesis imaging of fractures (Mermuys et al. 2008, Geijer et al. 2011), fracture reductions, and hip prostheses (Gomi et al. 2009). Tomosynthesis has been found to reduce metal artifacts from implanted hardware (Gomi et al. 2009). Other clinical applications of tomosynthesis include dental (Richards 1976, Groenhuis et al. 1983, Ruttimann et al. 1984, 1989, van der Stelt et al. 1986a,b, Engelke et al. 1992, Vandre et al. 1995, Webber 1995, 1996, 1997, Horton et al. 1996, Tyndall et al. 1997, Lauritsch and Harer 1998, Nair et al. 1998a,b, Abreu et al. 2001, Gomi et al. 2007, Ogawa et al. 2010, Katsumata et al. 2011), imaging of the inner ear (Chakraborty et al. 1984, Sone et al. 1991), abdomen (Nelson et al. 1985, Sone et al. 1991), kidney and gall stones (Mermuys et al. 2010), knee (Rimkus et al. 1989, Sone et al. 1991, Flynn et al. 2007), and radiotherapy (Zwicker and Atari 1997, Messaris et al. 1999, Persons 2001, Tutar et al. 2003, Godfrey et al. 2006a, Godfrey et al. 2007, Yan et al. 2007, Wu et al. 2007, Ren et al. 2008, Pang et al. 2008, Descovich et al. 2008, Maurer et al. 2008, 2010, Sarkar et al. 2009, Yoo et al. 2009, Zhang et al. 2009, Maltz et al. 2009, Mestrovic et al. 2009, Brunet-Benkhoucha et al. 2009, Lyatskaya et al. 2009, Winey et al. 2009, 2010, Park et al. 2011a,b, Wu et al. 2011).

1.5.2.3 Optimizing parameters in body tomosynthesis

Recently, Machida et al. published a tutorial in the *Radiographics* journal on optimizing parameters in flat-panel digital detector body tomosynthesis (Machida et al. 2010). The parameters that are discussed in this paper include "sweep angle, sweep direction, patient barrier–object distance, number of projections, and total radiation dose." In addition to these parameters, this paper also describes acquisition-related artifacts and ways to minimize those artifacts. Some of the artifacts include blurring from high-contrast objects that are oriented perpendicular to the sweep direction, ripple from high-contrast structures far from the plane of focus that are not sufficiently blurred, and ghosting or parasitic streaks from high-contrast objects located outside the plane of focus whose long axis is parallel to the sweep direction. Machida et al. demonstrate how blurring and ghosting artifacts can be minimized by appropriate selection of the sweep direction, and how ripple can be minimized by increasing the number of projections.

1.6 BRIEF HISTORY OF TOMOSYNTHESIS RECONSTRUCTION METHODS

The methods that have been employed to reconstruct tomosynthesis slices range from shift and add, which was employed in the first film-based systems, to iterative reconstruction, which is employed in modern digital detector-based systems. A brief description of these methods including their history follows.

1.6.1 SHIFT AND ADD

The shift-and-add method may be considered unfiltered backprojection. It has been used since 1935 (Ziedses des Plantes 1935). This method brings in-plane objects in focus while blurring out-of-plane features.

1.6.2 TUNED APERTURE CT

The TACT method is basically shift and add with fiducial markers. It allows images to be acquired at random angles and orientations and reconstructed in arbitrary planes. It has been employed since 1994 (Webber 1994).

1.6.3 MATRIX INVERSION

MITS employs linear algebra to solve and correct for out-of-plane blur using known blurring functions of all other planes when a given plane is reconstructed. It was developed by Dobbins et al. and has been used since 1987 (Dobbins et al. 1987, Dobbins 1990, Godfrey et al. 2001a,b, 2002, 2003, 2006b, Warp et al. 2000).

1.6.4 FILTERED BACKPROJECTION

Filtered backprojection is the most common CT reconstruction method. Low-pass filters are used in the spatial frequency domain to compensate for incomplete and/or nonuniform sampling of the tomography acquisition in the spatial domain to suppress

high frequencies. Filtered backprojection has been applied to tomosynthesis since 1998 (Lauritsch and Harer 1998, Badea et al. 2001, Stevens et al. 2001).

1.6.5 ALGEBRAIC RECONSTRUCTION TECHNIQUES

Algebraic reconstruction was the method employed in the first commercial CT scanners made by EMI Ltd (London, England). It involves an iterative solution to a set of linear equations, ray by ray. It has been employed in reconstruction since 1970 (Gordon et al. 1970, Meyer-Ebrecht and Wagner 1975). Several variants have also been employed including "simultaneous algebraic reconstruction" (SART) (Andersen and Kak 1984), "simultaneous iterative reconstruction technique" (SIRT) (Colsher 1977), and "iterative least squares technique" (ILST) (Colsher 1977, Bleuet et al. 2002).

1.6.6 STATISTICAL RECONSTRUCTION

Statistical reconstruction methods such as maximum likelihood (ML) determine the 3D model of x-ray attenuation coefficients, which maximize the probability of obtaining the measured projections. Some variants include "maximum likelihood–expectation-maximization" (ML-EM) (Lange and Fessler 1995) and "maximum likelihood with convex algorithm" (ML-convex) (Lange 1990, Lange and Fessler 1995, Wu et al. 2003, 2004).

1.7 METHODS TO REDUCE BLUR FROM OUT-OF-PLANE DETAILS

Although tomosynthesis reconstructions produce planes that are in focus, as illustrated in Figure 1.1, there are still contributions from out-of-plane objects to the focal planes. The blur from outside the focal plane reduces in-plane contrast. Various methods have been developed to address this problem. Edholm and Quiding created a photographic negative of the original reconstruction, blurred it in the tomosynthesis direction, and added the result to the original (Edholm and Quiding 1969, 1970). This unsharp masking method was in effect a high-pass frequency filter. Other high-pass frequency filter methods included those of Chakraborty et al. (1984), van der Stelt et al. (1986a,b), Liu et al. (1987), and Sone et al. (1996). Sone et al. (1991) also developed a band-pass frequency filter method. Other methods include the use of wavelets (Badea et al. 1998), and a method introduced by Ghosh Roy et al. (1985) that "used knowledge of the blurring functions to solve exactly for the distortion generated by a handful of planes immediately adjacent to the plane of interest" (Dobbins and Godfrey 2003). Methods similar to that of Ghosh Roy were developed by Kolitsi et al. (1993) and Sone et al. (1996), and Dobbins's MITS reconstruction technique is an extension of Ghosh Roy's method to the entire set of "conventionally reconstructed planes, enabling the exact solution of in-plane structures from a complete set of tomosynthesized planes" (Dobbins and Godfrey 2003).

1.8 PRESENT-DAY TOMOSYNTHESIS SYSTEMS

Many tomosynthesis systems have been developed, but only a few have received FDA approval and that approval was not obtained until very recently. GE Healthcare received FDA approval for their VolumeRad Tomosynthesis system for imaging the body

Table 1.1 Some characteristics of present-day breast tomosynthesis systems

UNIT	TOMO ANGLE	# VIEWS	PIXEL PITCH	2 × 2 BINNING[a]	DETECTOR	SCAN TIME (S)
GE Gen2	60°	21	100 micron	No	CsI-a-Si	7.5
GE DS	40°	15	100 micron	No	CsI-a-Si	11–20
GE Essential	25°	9	100 micron	No	CsI-a-Si	<10
IMS Giotto[b] (Dexela)	40°	13	85 micron	Yes and no	a-Se	12
Hologic	15°	15	70 micron	Yes	a-Se	4
Planmed	60°	15	85 micron	No	a-Se	14
Sectra (Philips)	11°	21	50 micron	No	Silicon	3–10
Siemens	50°	11–49	85 micron	Yes and no	a-Se	12–40
XCounter[c]	24°	48	60 micron	No	Gas	12–18

[a] Binning involves combining information from adjacent pixels. For example, the Hologic system combines 2 × 2 blocks of 70 × 70 micron pixels to create 140 × 140 micron "binned" pixels.

[b] The IMS Giotto system uses nonuniform angle increments—smaller increments at the center of the scan, larger increments toward the beginning and end of the scan, also different doses at different view angles.

[c] The XCounter system is no longer being manufactured.

(chest, knee, legs, etc.) in 2006, and Shimadzu received FDA approval for their Sonialvision Safire II body tomosynthesis system in 2008. On February 11, 2011, Hologic was the first company to receive FDA approval for their Selenia Dimensions digital breast tomosynthesis system (dimensions 3D). FDA approvals for the other manufacturers' breast tomosynthesis systems are anticipated in the near future.

Characteristics of the prototype and commercial breast and body tomosynthesis systems are described in detail in Chapters 2 and 13, respectively. A brief summary of some of the features of the breast imaging systems appears in Table 1.1. A brief summary of some of the features of the tomosynthesis systems that are currently available for body (e.g., chest, knee, and leg) imaging appears in Table 1.2.

1.9 PROMISING NEW APPLICATIONS AND DEVELOPMENTS IN TOMOSYNTHESIS IMAGING

Recently, researchers have been developing and investigating new applications of tomosynthesis imaging, including contrast-enhanced tomosynthesis of the breast and combinations of tomosynthesis with ultrasound, nuclear medicine, and optical imaging.

1.9.1 CONTRAST-ENHANCED (DSA) APPLICATIONS

As discussed previously, digital subtraction angiographic applications of tomosynthesis were developed many years ago

using image intensifier–TV camera detectors. The advent of flat-panel digital detectors makes it feasible to now employ similar techniques to image blood vessels and masses in the breast. The topic "contrast enhanced tomosynthesis of the breast" is discussed in detail in Chapter 16 of this book.

1.9.2 MULTIMODALITY BREAST IMAGING SYSTEMS

Information from tomosynthesis breast imaging can be combined with that from other imaging modalities for improved detection and characterization of masses. Three systems of this type have been developed. All involve imaging of the breast with two modalities in the same mammographic geometry, thereby insuring that there is a one-to-one correspondence between the masses observed with each modality. This solves the noncorresponding mass problem that sometimes occurs when the imaging geometry is different for the modalities. For example, it has been estimated that in at least 10% of cases, lesions found with free-hand ultrasound scanning (in the supine geometry) are different from lesions found in a mammogram (Conway et al. 1991).

1.9.2.1 Combined tomosynthesis and automated ultrasound imaging

Scientists at the University of Michigan and at GE global research teamed up to develop a combined x-ray tomosynthesis automated 3D ultrasound breast imaging system (Kapur et al. 2004, Carson et al. 2004, Booi et al. 2007, Sinha et al. 2007). Ultrasound imaging supplements x-ray tomosynthesis by enabling the distinction between cysts and tumors and by providing additional information

Table 1.2 Some characteristics of present-day body tomosynthesis systems

UNIT	TOMO ANGLE	# VIEWS	PIXEL PITCH	2 × 2 BINNING	DETECTOR	SCAN TIME
Knee Study						
GE	40°	40	200 micron	No	CsI-a-Si	8 s
Shimadzu	40°	75	150 micron	Yes	a-Se	2.5 s
Chest Study						
GE	30°	60	200 micron	No	CsI-a-Si	11.3 s
Shimadzu	40°	75	150 micron	Yes	a-Se	5 s

(a)

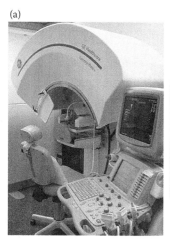

(b)

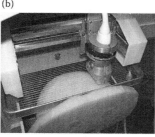

Figure 1.9 (a) Combined x-ray tomosynthesis automated ultrasound breast imaging system shown in the ultrasound scanning configuration. The ultrasound transducer is in its down position, where it scans just above the dual-modality mesh paddle (b). (The mesh paddle is shown compressing a breast simulating phantom.) The ultrasound scanning system is flipped up and back out of the x-ray field for the x-ray tomosynthesis acquisition.

for characterizing lesions. The combined x-ray/ultrasound system consists of a GE GEN II prototype tomosynthesis unit with a dual-modality (x-ray–ultrasound) mesh compression paddle and an ultrasound transducer translator (see Figure 1.9).

The Gen II tomosynthesis system has a stationary high-DQE CsI-aSi flat-panel x-ray detector similar to the one on the commercial GE Essential digital mammography system. The x-ray tube moves in an arc in a step-and-shoot mode, with a tomosynthesis angle of 60° and 21 angle increments (pulsed exposures/projection views) in 7.5 s. The patient is seated in a chair throughout the dual-modality procedure. The breast compression paddle is made of a material that is both x-ray and ultrasound compatible. The paddle is flipped up out of view for tomosynthesis acquisition and it is flipped down for ultrasound acquisition. For ultrasound imaging, acoustic coupling gel is used between the transducer and the paddle and the breast. A GE Logiq 9 ultrasound system is employed with an M12 L transducer operated at 10 MHz. The transducer is translated above the breast in an x–y raster mode and it is externally triggered to produce an image every 0.8 mm. Software was written to automatically register volumes of interest in the tomosynthesis and ultrasound images (Goodsitt et al. 2008). Recently, the group at the University of Michigan has been developing a photoacoustic tomography system (Wang et al. 2010) that can also be combined with the x-ray tomography and automated ultrasound unit. In photoacoustic tomography, short pulses (e.g., <25 ns) of near infrared laser light (e.g., 720–900 nm) are directed at the breast, the light heats up the inner tissues, and the resulting thermoexpansion leads to the emission of ultrasonic waves. The ultrasonic waves are detected with a large two-dimensional receiving ultrasound transducer array and a backprojection reconstruction algorithm is employed to produce an image of the optical absorption in the breast. For implementation with the combined system, the laser light is coupled to the compression paddle above the breast with a fiber optic array, and the ultrasound transducer receiving array is placed beneath the breast.

By using two different wavelengths of laser light in a technique called spectroscopic photoacoustic tomography, oxygenated and deoxygenated hemoglobin can be distinguished, allowing for functional imaging.

1.9.2.2 Combined x-ray tomosynthesis and nuclear medicine imaging

Mark Williams and his research group at the University of Virginia, Charlotte, Virginia developed a combined x-ray tomosynthesis and molecular breast imaging tomosynthesis system to provide coregistered anatomic and functional images of the breast in three dimensions (Williams et al. 2010).

The x-ray tomosynthesis part employs a full isocentric scanning motion in which both the x-ray tube and the digital detector are attached to a gantry that rotates about an axis. The breast support and compression paddle devices are independent of the rotating gantry, and the breast is positioned near the axis of rotation. The x-ray detector is a CCD device with a 20 × 30 cm field of view. It was not optimized for tomosynthesis and has a readout time of 2 s. Thirteen projections are acquired over a total angular range of 24° (±12°) and the total x-ray tomosynthesis scanning time is about 30 s. The gamma camera of the molecular breast imaging tomosynthesis system can be moved out of the way for x-ray imaging and into the field for gamma-ray imaging. When in use, a motor drive continuously adjusts the position of the camera to minimize the distance between the camera and the breast as the camera rotates, thereby maximizing spatial resolution and signal-to-noise ratio. The gamma camera has a 15 × 20 cm field of view. Five evenly spaced gamma camera views over a 40° angular range are employed for molecular breast imaging tomosynthesis with an imaging time of 2 min per view and a total scan time of 11 min. A pilot study was performed involving 17 women who were scheduled to undergo breast biopsy. Each was intravenously injected with 30 mCi (1110 MBq) of 99 mTc sestamibi for the molecular breast imaging portion of the study. It was found that adding "molecular breast imaging tomosynthesis notably improves specificity and positive predictive value compared with the specificity and positive predictive value of x-ray tomosynthesis alone." Close scrutiny of the published results (Williams et al. 2010) indicate the performance metrics (sensitivity, specificity, positive predictive value, negative predictive value, and accuracy) for x-ray tomosynthesis were worse than those for molecular breast imaging, and those for dual modality (x-ray/molecular) were identical to those for molecular breast imaging alone. Thus, for this study, the primary advantage of dual-modality over molecular breast imaging alone was the coregistration and improved localization. Future improvements in the x-ray and molecular imaging devices such as larger fields of view and shorter imaging times may result in additional advantages for the dual-modality mode.

1.9.2.3 Combined tomosynthesis and optical imaging

Qiangian Fang and his colleagues from MGH, Tufts University, and Northeastern University have developed a combined x-ray tomography and diffuse optical tomography system for imaging the breast (Fang et al. 2009, 2011). Diffuse optical tomography is a form of functional imaging. It employs near-infrared lasers to probe tissues and the resulting optical absorption and scattering of the laser light "are related to tissue physiological parameters, such as the concentrations of hemoglobin, deoxygenated hemoglobin, water, and lipids" (Fang et al. 2011). A disadvantage of diffuse optical tomography by itself is its poor spatial resolution. Fang et al. employed the anatomical information from coregistered x-ray tomosynthesis images to improve the optical tomography reconstructions.

The x-ray tomosynthesis system is a GE DS clinical prototype. It acquires 15 projections over a 45° tomosynthesis angle. The pixel size is 0.1 mm, and the reconstructed slice thickness is 1 mm. The optical system is described in detail in Fang et al. (2009). Two continuous wave (CW) frequency encoded laser systems are employed. One system has three lasers with wavelengths of 685, 810, and 830 nm. A fast Galvo scanner is used to direct these lasers to a maximum of 110 locations with a dwell time of 200 ms per location. The second system consists of 26 lasers, 13 at a wavelength of 685 nm and 13 at a wavelength of 830 nm, and these are used to continuously monitor tissue changes at 26 locations. Thirty-two avalanche photodiode (APD) detectors are employed and the CW signals are demodulated for each channel and wavelength. The laser source probes are placed in a cassette just above the x-ray detector and the detector probes are placed in an optically transparent dual-modality (x-ray and light) compression paddle. The optical data are employed to generate 3D maps of total hemoglobin concentration, oxygen saturation, and tissue-reduced scattering coefficients, and these are interpreted by relating them to coregistered x-ray tomosynthesis images. In the imaging protocol, the optical source and detector probes are initially attached to the tomosynthesis unit and the patient's breast is compressed. Optical data is acquired for 45 s. Then the optical probes are removed while the breast is maintained in compression, and the x-ray tomosynthesis image is acquired, which takes 23 s. In a study of 189 breasts in 125 subjects, Fang et al. found that "in 26 malignant tumors of 0.6–2.5 cm in size, the total hemoglobin concentration was significantly greater than that in the fibroglandular tissue of the same breast ($P = 0.0062$)" (Fang et al. 2011). Also, "solid benign lesions ($n = 17$) and cysts ($n = 8$) had significantly lower total hemoglobin concentration contrast than did the malignant lesions ($P = 0.025$ and $P = 0.0333$, respectively)" (Fang et al. 2011).

1.10 CONCLUSION

Tomosynthesis has had an exciting history, starting in 1935 and continuing to the present day. Some parts of this chapter on the history of tomosynthesis is based on two excellent, meticulously documented reviews of tomosynthesis imaging (Dobbins and Godfrey 2003, Dobbins 2009) and an excellent book on the origins of radiological tomography (Webb 1990). The reader is referred to those references and others in the reference list for additional information about the history and development of tomosynthesis. Finally, early this year, after this chapter on the history of tomosynthesis was finalized, Ioannis Sechopoulos published two excellent reviews of breast tomosynthesis (Sechopoulos 2013), Part I: The image Acquisition Process, and Part II: Image Reconstruction, Processing and Analysis, and Advanced Applications. These reviews are available online only at http://dx.doi.org/10.1118/1.4770279 and http://dx.doi.org/10.1118/1.4770281.

REFERENCES

Abreu, M. Jr, D.A. Tyndall, J.B. Ludlow. 2001. Effect of angular disparity of basis images and projection geometry on caries detection using tuned-aperture computed tomography. *Oral Surg. Oral Med. Oral Pathol. Oral Radiol. Endod.* 92:353–360.

Andersen, A.H., A.C. Kak. 1984. Simultaneous algebraic reconstruction (SART): A new implementation of the ART algorithm. *Ultrason. Imaging* 6:81–94.

Anderson, R.E., R.A. Kruger, R.G. Sherry et al. 1984. Tomographic DSA using temporal filtration: Initial neurovascular application. *Am. J. Neuroradiol.* 5:277–280.

Asplund, S.A., J.V. Johnsson et al. 2011. Learning aspects and potential pitfalls regarding detection of pulmonary nodules in chest tomosynthesis and proposed related quality criteria. *Acta Radiol.* 52:503–512.

Badea, C., Z. Kolitsi, N. Pallikarakis. 1998. A wavelet-based method for removal of out-of-plane structures in digital tomosynthesis. *Comput. Med. Imaging Graph.* 22:309–315.

Badea, C., Z. Kolitsi, N. Pallikarakis. 2001. Image quality in extended arc filtered digital tomosynthesis. *Acta Radiol.* 42:244–248.

Baily, N.A., E.C. Lasser, R.L. Crepeau. 1973. Electrofluoroplanigraphy. *Radiology* 107:669–671.

Baily, N.A., R.L. Crepeau, E.C. Lasser. 1974. Fluoroscopic tomography. *Invest. Radiol.* 9:94–103.

Becher, H., P. Hanrath, M. Schluter et al. 1983. Selective coronary angiography with flashing tomosynthesis in patients with coronary artery disease (abstract). *Circulation* 68 (Suppl III): III–178.

Becher, H., M. Schluter, D. Mathey et al. 1985. Coronary angiography with flashing tomosynthesis. *Eur. Heart J.* 6:399–408.

Bleuet, P., R. Guillemaud, I.E. Magnin. 2002. Resolution improvement in linear tomosynthesis with an adapted 3D regularization scheme. *Proc. SPIE* 4682:117–125.

Booi, R.C., J.F. Krücker, M.M. Goodsitt, M.O. Donnell, A. Kapur et al. 2007. Evaluation of thin compression paddles for mammographically compatible ultrasound. *J. Ultrasound Med. Biol.* 33:472–482.

Brunet-Benkhoucha, M., F. Verhaegen, S. Lassalle et al. 2009. Clinical implementation of a digital tomosynthesis-based seed reconstruction algorithm for intraoperative postimplant dose evaluation in low dose rate prostate brachytherapy. *Med. Phys.* 36:5235–5244.

Carson, P.L., G.L. LeCarpentier, M.A. Roubidoux, R.Q. Erkamp, J.B. Fowlkes, M.M. Goodsitt. 2004. Physics and technology of breast US imaging including automated three-dimensional US, in RSNA categorical course. In *Diagnostic Radiology Physics: Advances in Breast Imaging—Physics Technology and Clinical Applications*. A. Karellas and M. Geiger editors. Radiological Society of North America, Inc, Oak Brook, IL, 223–232.

Chan, H.P., J. Wei, Y. Zhang, B. Sahiner et al. 2008. Detection of masses in digital breast tomosynthesis mammography: Effects of the number of projection views and dose. In *IWDM 2008*. E.A. Krupinski editor. Springer-Verlag, Berlin. Vol. LNSC 5116, 279–285.

Chawla, A.S., J.Y. Lo, J.A. Baker et al. 2009. Optimized image acquisition for breast tomosynthesis in projection and reconstruction space. *Med. Phys.* 36:4859–4869.

Colsher, J.G. 1977. Iterative three-dimensional image reconstruction from tomographic projections. *Comput. Graph. Image* 6:513–537.

Conway, W.F., C.W. Hayes, W.H. Brewer. 1991. Occult breast masses: Use of mammographic localizing grid for US evaluation. *Radiology* 181:143–146.

Cottenot, P. 1938. Thoracic serioscopy: Method of study for pleuro-pulmonary lesions. *Radiology* 31:1–7.

Deller, T., K.N. Jabri, J.M. Sabol, X. Ni, G. Avinash, R. Saunders, R. Uppaluri. 2007. Effect of acquisition parameters on image quality in digital tomosynthesis. *Proc. SPIE* 6510:1L1–1L11.

Descovich, M., O. Morin, J.F. Aubry et al. 2008. Characteristics of megavoltage cone-beam digital tomosynthesis. *Med. Phys.* 35(4):1310–1316.

deVries, N., F.J. Miller, M.M. Wojtowycz, P.R. Brown, D.R. Yandow, J.A. Nelson, R.A. Kruger. 1985. Tomographic digital subtraction angiography: Initial clinical studies using tomosynthesis. *Radiology* 157:239–241.

Dobbins, J.T. III, 1990. Matrix inversion tomosynthesis improvements in longitudinal x-ray slice imaging. Patent #4,903,204 (United States).

Dobbins, J.T. III, A.O. Powell, Y.K. Weaver. 1987. Matrix inversion tomosynthesis: Initial image reconstruction. RSNA 73rd Scientific Assembly (Chicago, IL).

Dobbins, J.T. III, R.L. Webber, S.M. Hames. 1998. Tomosynthesis for improved pulmonary nodule detection (abstract). *Radiology* 209(P):280.

Dobbins, J.T. III, D.J. Godfrey. 2003. Digital x-ray tomosynthesis: Current state of the art and clinical potential. *Phys. Med. Biol.* 48:R65–R106.

Dobbins, J.T. III, H.P. McAdams, J-W Song et al. 2008. Digital tomosynthesis of the chest for lung nodule detection: Interim sensitivity results form an ongoing NIH sponsored trial. *Med. Phys.* 35:2554–2557.

Dobbins, J.T. III. 2009. Tomosynthesis imaging: At a translational crossroads. *Med. Phys.* 36:1956–1967.

Dobbins, J.T. III, H.P. McAdams. 2009. Chest tomosynthesis: Technical principles and clinical update. *Eur. J. Radiol.* 72:244–251.

Duryea, J., J.T. Dobbins III. 2001. Application of digital tomosynthesis to hand radiography for arthritis assessment. *Proc. SPIE* 4320:688–695.

Duryea, J., J.T. Dobbins III, J.A. Lynch. 2003. Digital tomosynthesis of hand joints for arthritis assessment. *Med. Phys.* 30:325–333.

Edholm, P.R., L. Quiding. 1969. Reduction of linear blurring in tomography. *Radiology* 92:1115–1118.

Edholm, P., L. Quiding. 1970. Elimination of blur in linear tomography. *Acta Radiol.* 10:441–447.

Engelke, W., U.E. Ruttimann, M. Tsuchimochi, J.D. Bacher. 1992. An experimental study of new diagnostic methods for the examination of osseous lesions in the temporomandibular joint. *Oral Surg. Oral Med. Oral Pathol.* 73:348–359.

Fahey, F.H., R.L. Webber, F-S-K. Chew, B.A. Dickerson. 2003. Application of TACT to the evaluation of total joint arthroplasty. *Med. Phys.* 30:454–460.

Fahrig, R., A.R. Pineda, E.G. Solomon, A.N. Leung, N.J. Pelc. 2003. Fast tomosynthesis for lung cancer detection using the SBDX geometry. *Proc. SPIE* 5030:371–378.

Fang, Q., S. Carp, J. Selb, G. Boverman, Q. Zhang, D. Kopans, E. Rafferty, R. Moore et al. 2009. Massachusetts General Hospital. *IEEE Trans. Med. Imaging* 28:30–42.

Fang, Q., J. Selb, S.A. Carp et al. Combined optical and X-ray tomosynthesis breast imaging. January 2011. *Radiology* 258:1 89–97. DOI: 10.1148/radiol.10082176.

Flynn, M.J., R. McGee, J. Blechinger. 2007. Spatial resolution of x-ray tomosynthesis in relation to computed tomography for coronal/sagittal images of the knee. In *Medical Imaging 2007: Physics of Medical Imaging*. J. Hsieh and M.J. Flynn editors. *Proceedings of the SPIE*. Vol. 6510, 65100D-1–65100D-9.

Garrison, J.B, D.G. Grant, W.H. Guier, R.J. Johns. 1969. Three dimensional roentgenography. *Am. J. Roentgenol.* 105:903–908.

Gazaille, R.E. III, M.J. Flynn, P. Walter III et al. November 2011. Technical innovation: Digital tomosynthesis of the hip following intra-articular administration of contrast. *Skeletal Radiol.* 40(11):1467–1471.

Geijer, M., A.M. Borjesson, J.H. Gothlin. 2011. Clinical utility of tomosynthesis in suspected scaphoid fracture. A pilot study. *Skeletal Radiol.* 40(7):863–867.

Ghosh Roy, D.N., R.A. Kruger, B. Yih, P. Del Rio. 1985. Selective plane removal in limited angle tomographic imaging. *Med. Phys.* 12:65–70.

Gifford, H.C., C.S. Didier, M. Das, S.J. Glick. 2008. Optimizing breast tomosynthesis acquisition parameters with scanning model observers. *Proc. SPIE* 6917:0S1–0S9.

Godfrey, D.J., J.T. Dobbins III. 2002. Optimization of matrix inversion tomosynthesis via impulse response simulations. RSNA 88th Scientific Assembly (Chicago, IL) (abstract).

Godfrey, D.J., R.J. Warp, J.T. Dobbins III. 2001a. Optimization of matrix inverse tomosynthesis. *Proc. SPIE* 4320:696–704.

Godfrey, D.J., R.J. Warp, J.T. Dobbins III. 2001b. Optimization of noise attributes in matrix inversion tomosynthesis. RSNA 87th Scientific Program (Chicago, IL) R104 Topical Review.

Godfrey, D.J., A. Rader, J.T. Dobbins III. 2003. Practical strategies for the clinical implementation of matrix inversion tomosynthesis (MITS). *Proc. SPIE* 5030:379–390.

Godfrey, D.J, F.F. Yin, M. Oldham et al. 2006a. Digital tomosynthesis with an on-board kilovoltage imaging device. *Int. J. Radiat. Oncol.* 65(1):8–15.

Godfrey, D.J, H.P. McAdams, J.T. Dobbins III. 2006b. Optimization of the matrix inversion tomosynthesis (MITS) impulse response and modulation transfer function characteristics for chest imaging. *Med. Phys.* 33:655–667.

Godfrey, D.J., L. Ren, H. Yan et al. 2007. Evaluation of three types of reference image data for external beam radiotherapy target localization using digital tomosynthesis (DTS). *Med. Phys.* 34(8):3374–3384.

Gomi, T., N. Yokoi, H. Hirano. 2007. Evaluation of digital linear tomosynthesis imaging of the temporomandibular joint: Initial clinical experience and evaluation. *Dentomaxillofac. Radiol.* 36(8):514–521.

Gomi, T., H. Hirano, T. Umeda. 2009. Evaluation of the X-ray digital linear tomosynthesis reconstruction processing method for metal artifact reduction. *Comput. Med. Imaging Graph.* 33(4):267–274.

Gomi, T., M. Nakajima, H. Fujiwara et al. 2011. Comparison of chest dual-energy subtraction digital tomosynthesis imaging and dual-energy subtraction radiography to detect simulated pulmonary nodules with and without calcifications: A phantom study. *Acta Rad.* 18(2):191–196.

Goodsitt, M.M., H.P. Chan, L. Hadjiiski, G.L. LeCarpentier, P.L. 2008. Carson. Automated registration of volumes of interest for a combined x-ray tomosynthesis and ultrasound breast imaging system. In *Digital Mammography Proceedings, IWDM*. E.A. Krupinski editor. Springer-Verlag, Berlin. Vol. 5116, 463–468.

Gordon, R., B. Bender, G.T. Herman. 1970. Algebraic reconstruction techniques (ART) for three dimensional electron microscopy and x-ray photography. *J. Theor. Biol.* 29:471–481.

Grant, D.G. 1972. Tomosynthesis: A three-dimensional radiographic imaging technique. *IEEE Trans. Biomed. Eng.* 19:20–28.

Groenhuis, R., J. Webber, U.E. Ruttimann. 1983. Computerized tomosynthesis of dental tissues. *Oral Surg. Oral Med. Oral Pathol.* 56:206–214.

Groh, G. 1977. Tomosynthesis and coded aperture imaging: New approaches to three-dimensional imaging in diagnostic radiography. *Proc. R. Soc. B* 195:299–306.

Haaker, P., E. Klotz, R. Koppe, R. Linde. 1990. Real-time distortion correction of digital x-ray II/TV-systems: An application example for digital flashing tomosynthesis (DFTS). *Int. J. Card. Imaging* 6:39–45.

Haaker, P., E. Klotz, R. Koppe, R. Linde, D.G. Mathey. 1985a. First clinical results with digital flashing tomosynthesis in coronary angiography. *Eur. Heart J.* 6:913–920.

Haaker, P., E. Klotz, R. Koppe, R. Linde, H. Moller. 1985b. A new digital tomosynthesis method with less artifacts for angiography. *Med. Phys.* 12:431–436.

Hoefer, E., H. Grimmert, B. Kieslich. 1974. Computer-controlled synthesis of tomograms by means of a TV storage tube. *IEEE T. Biomed. Eng. BME-21* 3:243–244.

Horton, R.A., J.B. Ludlow, R.L. Webber, W. Gates, R.H. Nason Jr, D. Reboussin. 1996. Detection of peri-implant bone changes with axial tomosynthesis. *Oral Surg. Oral Med. Oral Pathol.* 81:124–129.

Johnsson, A., J. Vikgren, A. Svalkvist et al. 2010a. Overview of two years of clinical experience of chest tomosynthesis at Sahlgrenska University Hospital. *Radiat. Prot. Dosim.* 139(1–3):124–129.

Johnsson, A., A. Svalkvist, J. Vikgren et al. 2010b. A phantom study of nodule size evaluation with chest tomosynthesis and computed tomography. *Radiat. Prot. Dosim.* 139(1–3):140–143.

Jung, H.N., M.J. Chung, J.H. Koo et al. 2012. Digital tomosynthesis of the chest: Utility for detection of lung metastasis in patients with colorectal cancer. *Clin. Radiol.* 67(3):232.

Kapur, A., P.L. Carson, J. Eberhard et al. 2004. Combination of digital mammography with semi-automated 3D breast ultrasound. *Tech. Can. Res. Treat.* 3(4):325–334.

Katsumata, A., K. Ogawa, K. Inukai et al. 2011. Initial evaluation of linear and spatially oriented planar images from a new dental panoramic system based on tomosynthesis. *Oral Surg. Oral Med. Oral Pathol. Oral Radiol. Endod.* 112(3):375–382.

Kaufman, J. 1936a. Planeogrpahy, localization, and mensuration: "Standard Depth curves". *Radiology* 27:168–174.

Kaufman, J. 1936b. The planeogram: Analysis and practical application with especial reference to mensuration of the pelvic inlet. *Radiology* 27:732–735.

Kaufman, J. 1938. Object reconstruction by planigraphy: Reconstruction and localization of planes. *Radiology* 30:763–765.

Kaufman, J, H. Koster. 1940. Serial planeography (serioscopy) and serial planigraphy: A critical analysis. *Radiology* 34:626–629.

Kim, E.Y., M.J. Chung, H.Y. Lee et al. 2010. Pulmonary mycobacterial disease: Diagnostic performance of low-dose digital tomosynthesis as compared with chest radiography. *Radiology* 257(1):269–277.

Klotz, E., H. Weiss. 1974. Three-dimensional coded aperture imaging using nonredundant point distributions. *Opt. Commun.* 11:368–372.

Klotz, E., Weiss H. 1976. Short-time tomosynthesis: The new image in tomography. In *Proceedings of the Symposium Actualitatis Tomographiae* (Genoa, Italy, 11–13 September 1975) L. Oliva and R.J. Berry editors. Excerpta Medica, Amsterdam, 65–70.

Kolitsi, Z., G. Panayiotakis, N. Pallikarakis. 1993. A method for selective removal of out-of-plane structures in digital tomosynthesis. *Med. Phys.* 20:47–50.

Kolitsi, Z., N. Yoldassis, T. Siozos, N. Pallikarakis. 1996. Volume imaging in fluoroscopy: A clinical prototype system based on a generalized digital tomosynthesis technique. *Acta Radiol.* 37:741–748.

Kruger, R.A., J.A. Nelson, R.D. Ghosh, F.J. Miller, R.E. Anderson, P. Liu. 1983. Dynamic tomographic digital subtraction angiography using temporal filtration. *Radiology* 147:863–867.

Kruger, R.A., M. Sedaghati, R. Ghosh, P. Liu, J.A. Nelson, W. Kubal, P. Del Rio. 1984. Tomosynthesis applied to digital subtraction angiography. *Radiology* 152:805–808.

Lange, K. 1990. An overview of Bayesian methods in image reconstruction. In *Digital Image Synthesis and Inverse Optics*. A.F. Gmitro, P.S. Idell, and I.S. La Haie editors. Society of Photo-Optical Engineering (SPIE), Bellingham, WA. 1351:270–287.

Lange, K., J.A. Fessler. 1995. Globally convergent algorithms for maximum a posteriori transmission tomography. *IEEE Trans. Image Process.* 4:1430–1438.

Lasser, E.C., N.A. Baily, R.L. Crepeau. 1971. A fluoroplanigraphy system for rapid presentation of single plane body sections. *Am. J. Roentgenol.* 113:574–577.

Lauritsch, G., W.H. Harer. 1998. A theoretical framework for filtered backprojection in tomosynthesis. *Proc. SPIE* 3338:1127–1137.

Lehtimaki, M., M. Pamilo, L. Raulisto et al. 2003. Diagnostic clinical benefits of digital spot and digital 3D mammography following analysis of screening findings. *Proc. SPIE* 5029:698.

Liu, J., D. Nishimura, A. Macovski. 1987. Vessel imaging using dual-energy tomosynthesis. *Med. Phys.* 14:950–955.

Lu, Y., H.P. Chan, M. Goodsitt et al. 2010. Effects of projection-view distributions on image quality of calcifications in digital breast tomosynthesis (DBT) Reconstruction. In *Proc. SPIE. Medical Imaging 2010: Physics of Medical Imaging*. E. Samei and N.J. Pelc editors. Society of Photo-Optical Engineering (SPIE), Bellingham, WA, 7622:76220D.

Lu, Y., H.P. Chan, J. Wei et al. 2011. Image quality of microcalcifications in digital breast tomosynthesis: Effects of projection-view distributions. *Med. Phys.* 38(10):5703–5712.

Lyatskaya, Y., A. Buehler, S.K. Ng et al. 2009. Optimal gantry angles and field sizes in kilovoltage cone-beam tomosynthesis for set-up of women with breast cancer undergoing radiotherapy treatment. *Radiother. Oncol.* 93(3):633–638.

Machida, H, T. Yuhara, T. Mori et al. 2010. Optimizing parameters for flat-panel detector digital tomosynthesis. *Radiographics* 30(2):549–562.

Maltz, J.S., F. Sprenger, F. Juerst et al. 2009. Fixed gantry tomosynthesis system for radiation therapy image guidance based on a multiple source x-ray tube with carbon nanotube cathodes. *Med. Phys.* 36(5):1624–1636.

Mandelkorn, F., H. Stark. 1978. Computerized tomosynthesis, serioscopy, and coded-scan tomography. *Appl. Opt.* 17:175–180.

Maravilla, K.R., R.C. Murry Jr, M. Deck, S. Horner. 1983a. Clinical application of digital tomosynthesis: A preliminary report. *Am. J. Neuroradiol* 4:277–280.

Maravilla, K.R., R.C. Murry Jr, S. Horner. 1983b. Digital tomosynthesis: Technique for electronic reconstructive tomography. *Am. J. Roentgenol.* 141:497–502.

Maravilla, K.R., R.C. Murry Jr, J. Diehl et al. 1984. Digital tomosynthesis: Technique modifications and clinical applications for neurovascular anatomy. *Radiology* 152:719–724.

Matsuo, H., A. Iwata, I. Horiba, N. Suzumura. 1993. Three-dimensional image reconstruction by digital tomo-synthesis using inverse filtering. *IEEE Trans. Med. Imaging* 12:307–313.

Maurer, J., D. Godfrey, Z. Wang et al. 2008. On-board four-dimensional digital tomosynthesis: First experimental results. *Med. Phys.* 35(8):3574–3583.

Maurer, J., T. Pan, F.F. Yin. 2010. Slow gantry rotation acquisition technique for on-board four-dimensional digital tomosynthesis. *Med. Phys.* 37(2):921–933.

Mermuys, K., K. Vanslambrouck, J. Goubau et al. 2008. Use of digital tomosynthesis: Case report of a suspected scaphoid fracture and technique. *Skeletal Radiol.* 37(6):569–572.

Mermuys, K., F. De Geeter, K. Bacher et al. 2010. Digital tomosynthesis in the detection of urolithiasis: Diagnostic performance and dosimetry compared with digital radiography with MDCT as the reference standard. *Am. J. Roent.* 195(1):161–167.

Messaris, G., Z. Kolitsi, C. Badea et al. 1999. Three-dimensional localisation based on projectional and tomographic image correlation: An application for digital tomosynthesis. *Med. Eng. Phys.* 21(2):101–109.

Mestrovic, A., A. Nichol, B.G. Clark et al. 2009. Integration of on-line imaging, plan adaptation and radiation delivery: Proof of concept using digital tomosynthesis. *Phys. Med. Biol.* 54(12):3803–3819.

Meyer-Ebrecht, D., W. Wagner. 1975. On the application of ART to conventional x-ray tomography. In *Digest from Topical Meeting on Image Processing for 2-D and 3-D Reconstruction from Projections: Theory and Practice in Medicine and the Physical Sciences*. Stanford University, CA. TuC3-1–TuC3-3.

Miller, E.R., E.M. McCurry, B. Hruska. 1971. An infinite number of laminagrams from a finite number of radiographs. *Radiology* 98:249–255 Topical Review R105.

Murry, R.C. Jr, K.R. Maravilla. 1983. Description of a digital tomosynthesis (DTS) system. *Proc. SPI.* 419:175–183.

Nadjmi, M., H. Weiss, E. Klotz, R. Linde. 1980. Flashing tomosynthesis—A new tomographic method. *Neuroradiology* 19:113–117.

Nair, M.K., D.A. Tyndall, J.B. Ludlow, K. May. 1998a. Tuned-aperture computed tomography and detection of recurrent caries. *Caries Res.* 32:23–30.

Nair, M.K., D.A. Tyndall, J.B. Ludlow, K. May, F. Ye. 1998b. The effects of restorative material and location on the detection of simulated recurrent caries: A comparison of dental film, direct digital radiography and tuned aperture computed tomography. *Dentomaxillofac. Radiol.* 27:80–84.

Nelson, J.A., F. Mann, F.J. Miller et al. 1985. Digital tomosynthesis in abdominal diagnosis (abstract). *Gastrointest. Radiol.* 10:302–303.

Niklason, L.T, B.T. Christian, L.E. Niklason et al. 1997. Digital tomosynthesis in breast imaging. *Radiology* 205:399–406.

Ogawa, K., R.P. Langlais, W.D. McDavid et al. 2010. Development of a new dental panoramic radiographic system based on a tomosynthesis method. *Dentomaxillofac. Radiol.* 29(1):47–53.

Park, J.C., S.H. Park, J.S. Kim et al. 2011a. Ultra-fast digital tomosynthesis reconstruction using general-purpose GPU programming for image-guided radiation therapy. *Tech. Can. Res. Treat.* 10(4):1028–1036.

Park, J.C., S.H. Park, J.H. Kim et al. 2011b. Four-dimensional cone-beam computed tomography and digital tomosynthesis reconstructions using respiratory signals extracted from transcutaneously inserted metal markers for liver. *SBRT Med. Phys.* 38(2):1028–1036.

Pang, G., A. Bani-Hashemi, P. Au et al. 2008. Megavoltage cone beam digital tomosynthesis (MV-CBDT) for image-guided radiotherapy: A clinical investigational system. *Phys. Med. Biol.* 53(4):999–1013.

Persons, T.M. 2001. Three-dimensional tomosynthetic image restoration for brachytherapy source localization. PhD Dissertation Wake Forest University.

Quaia, E., E. Baratella, V. Cioffi et al. 2010. The value of digital tomosynthesis in the diagnosis of suspected pulmonary lesions on chest radiography: Analysis of diagnostic accuracy and confidence. *Acad. Rad.* 17(10):1267–1274.

Quaia, E., E. Baratella, S. Cernic et al. 2012. Analysis of the impact of digital tomosynthesis on the radiological investigation of patients with suspected pulmonary lesions on chest radiography. *Eur. Rad.* 22(9):1912–1922.

Ren, L, D.J. Godfrey, H. Yan et al. 2008. Automatic registration between reference and on-board digital tomosynthesis images for positioning verification. *Med. Phys.* 35(2):664–672.

Reiser, I., B.A. Lau, R.M. Nishikawa. 2008. Effect of scan angle and reconstruction algorithm on model observer performance in tomosynthesis. In *Digital Mammography Proceedings, IWDM 2008.* E.A. Krupinski editor. Springer-Verlag, Berlin. Vol. LNCS 5116, 606–611.

Richards, A.G. 1976. Dynamic tomography. *Oral Surg.* 42(5):685–692.

Rimkus, D.S., B.M. Gill, N.A. Baily et al. 1989. Digital tomosynthesis: Phantom and patient studies with a prototype unit. *Comput. Med. Imaging Graph.* 13:307–318.

Ruttimann, U.E., R.A. Groenhuis, R.L. Webber. 1984. Restoration of digital multiplane tomosynthesis by a constrained iteration method. *IEEE Trans. Med. Imaging* 3:141–148.

Ruttimann, U.E., X-L Qi, R.L Webber. 1989. An optimal synthetic aperture for circular tomosynthesis. *Med. Phys.* 16:398–405.

Santoro, J., S. Kriminski, D.M. Lovelock et al. 2010. Evaluation of respiration-correlated digital tomosynthesis in lung. *Med. Phys.* 37(3):1237–1245.

Sarkar, V., C. Shi, P. Rassiah-Szegedi et al. 2009. The effect of a limited number of projections and reconstruction algorithms on the image quality of megavoltage digital tomosynthesis. *J. Appl. Clin. Med. Phys.* 10(3):155–172.

Sechopoulos, I., C. Ghetti. 2009. Optimization of the acquisition geometry in digital tomosynthesis of the breast. *Med. Phys.* 36(4):1199–1207.

Sechopoulos, I. 2013. A review of breast tomosynthesis. Part I. The image acquisition process. *Med. Phys.* 40:1. Online only: http://dx.doi.org/10.1118/1.4770279

Sechopoulos, I. 2013. A review of breast tomosynthesis. Part II. Image reconstruction, processing and analysis, and advanced applications. *Med. Phys.* 40:1. Online only: http://dx.doi.org/10.1118/1.4770281.

Sinha, S.P., M.M. Goodsitt, M.A. Roubidoux et al. 2007. Automated ultrasound scanning on a dual-modality breast imaging system—Coverage and motion issues and solutions. *J. Ultrasound Med.* 26(5):645–655.

Sklebitz, H., J. Haendle. 1983. Tomoscopy: Dynamic layer imaging without mechanical movements. *Am. J. Roentgenol.* 140:1247–1252.

Sone, S., T. Kasuga, F. Sakai et al. 1991. Development of a high resolution digital tomosynthesis system and its clinical application. *Radiographics* 11:807–822.

Sone, S., T. Kasuga, F. Sakai et al. 1993. Chest imaging with dual-energy subtraction digital tomosynthesis. *Acta Radiol.* 34:346–350.

Sone S, T. Kasuga, F. Sakai et al. Digital tomosynthesis imaging of the lung. *Radiat. Med.* 14:53–63.

Stevens, G.M., R. Fahrig, N.J. Pelc. 2001. Filtered backprojection for modifying the impulse response of circular tomosynthesis. *Med. Phys.* 28:372–380.

Stiel, G.M., L.S. Stiel, K. Donath et al. 1989. Digital flashing tomosynthesis (DFTS)—A technique for three-dimensional coronary angiography. *Int. J. Card. Imaging* 5:53–61.

Stiel, G.M., L.S. Stiel, C.A. Nienaber. 1992. Potential of digital flashing tomosynthesis for angiocardiographic evaluation. *J. Digit. Imaging* 5:194–205.

Stiel, G.M., L.S. Stiel, E. Klotz, C.A. Nienaber. 1993. Digital flashing tomosynthesis: A promising technique for angiocardiographic screening. *IEEE Trans. Med. Imaging* 12:314–321.

Suryanarayanan, S., A. Karellas, S. Vedantham et al. 2000 Comparison of tomosynthesis methods used with digital mammography. *Acad. Radiol.* 7:1085–1097.

Suryanarayanan, S., A. Karellas, S. Vedantham et al. 2001. Evaluation of linear and nonlinear tomosynthetic reconstruction methods in digital mammography. *Acad. Radiol.* 8:219–224.

Tutar, I.B., R. Managuli, V. Shamdasani et al. 2003. Tomosynthesis-based localization of radioactive seeds in prostate brachytherapy. *Med. Phys.* 30(12):3135–3142.

Tyndall, D.A., T.L. Clifton, R.L. Webber, J.B. Ludlow, R.A. Horton. 1997. TACT imaging of primary caries. *Oral Surg. Oral Med. Oral Pathol. Oral Radiol. Endod.* 84:214–225.

van der Stelt, P.F., U.E. Ruttimann, R. L. Webber. 1986a. Enhancement of tomosynthetic images in dental radiology. *J. Dent. Res.* 65:967–973.

van der Stelt, P.F., R.L. Webber, U.E. Ruttimann, R.A. Groenhuis. 1986b. A procedure for reconstruction and enhancement of tomosynthetic images. *Dentomaxillofac. Radiol.* 15:11–18.

Vandre, R., R. Webber, R. Horton, C. Cruz, J. Pajak. 1995. Comparison of TACT with film for detecting osseous jaw lesions. *J. Dent. Res.* 74:18.

Vikgren, J., S. Zachrisson, A. Svalkvist et al. 2008. Comparison of chest tomosynthesis and chest radiography for detection of pulmonary nodules: Human observer study of clinical cases. *Radiology* 249:1034–1041.

Wang, X., D. DeBusschere, C.H. Hu, J.B. Fowlkes, J.M. Cannata, G. McLaughlin, P. Carson. 2010. A high-speed photoacoustic tomography system based on a commercial ultrasound and a custom transducer array. 2010. In *Photons Plus Ultrasound: Imaging and Sensing 2010. Proceedings of the SPIE*. A.A. Oraevsky and L.V. Wang editors. 7564, 75642-1, 75642-9.

Warp, R.J., D.J. Godfrey, J.T. Dobbins JT III. 2000. Applications of matrix inverse tomosynthesis. *Proc. SPIE* 3977:376–383.

Webb, S. 1990. *From the Watching of Shadows, the Origins of Radiological Tomography*. Adam Hilger, Bristol, England.

Webber, R.L. 1994. Self-calibrated tomosynthetic, radiographic-imaging system, method and device. Patent #5,539,637 (United States).

Webber, R.L. 1997. Self-contained tomosynthetic, radiographic-imaging system, method and device. Patent #5,668,844 (United States).

Webber, R.L., R.A. Horton, D.A. Tyndall, J.B. Ludlow. 1997. Tuned-aperture computed tomography (TACT-TM): Theory and application for three-dimensional dentoalveolar imaging. *Dentomaxillofac. Radiol.* 26:53–62.

Webber, R.L., R.A. Horton, T.E. Underhill. 1995. Tuned-aperture computed tomography for diagnosis of crestal defects around implants. *J. Dent. Res.* 74:18.

Webber, R.L., R.A. Horton, T.E. Underhill, J.B. Ludlow, D.A. Tyndall. 1996. Comparison of film, direct digital, and tuned-aperture computed tomography images to identify the location of crestal defects around endosseous titanium implants. *Oral Surg. Oral Med. Oral Pathol. Oral Radiol. Endod.* 81:480–490.

Webber, R.L., H.R. Underhill, R.I. Freimanis. 2000. A controlled evaluation of tuned-aperture computed tomography applied to digital spot mammography. *J. Digit. Imaging* 13:90–97.

Webber, R.L., H.R. Underhill, P.F. Hemler, J. 1999. A nonlinear algorithm for task-specific tomosynthetic image reconstruction. *Proc. SPIE* 3659:258–265.

Weiss, H., E. Klotz, R. Linder. 1979. Flashing tomosynthesis—3-Dimensional x-ray imaging. *Acta Electron.* 22(1):41–50.

Weiss, H., E. Klotz, R. Linder et al. 1977. Coded aperture imaging with x-rays (flashing tomosynthesis). *Opt. Acta* 24(4):305–325.

Williams, M.B., P.G. Judy, S. Gunn, S. Majewski. 2010. Dual-modality breast tomosynthesis. *Radiology* 255(1):191–198.

Woelke, H., P. Hanrath, M. Schlueter, W. Bleifeld, E. Klotz, H. Weiss, D. Waller, J. Von Weltzien. 1982. Work in progress. Flashing tomosynthesis: A tomographic technique for quantitative coronary angiography. *Radiology* 145:357–360.

Winey, B., P. Zygmanski, Y. Lyatskaya. 2009. Evaluation of radiation dose delivered by cone beam CT and tomosynthesis employed for setup of external breast irradiation. *Med. Phys.* 36(1):164–173.

Winey, B.A., P. Zygmanski, R.A. Cormack et al. 2010. Balancing dose and image registration accuracy for cone beam tomosynthesis (CBTS) for breast patient setup. *Med. Phys.* 37(8):4414–4423.

Wu T, A. Stewart, M. Stanton et al. 2003. Tomographic mammography using a limited number of low-dose cone-beam projection images. *Med. Phys.* 30:365–380.

Wu, T., R.H. Moore, E.A. Rafferty, D.B. Kopans. 2004. A comparison of reconstruction algorithms for breast tomosynthesis. *Med. Phys.* 31:2636–2647.

Wu, Q.J., D.J. Godfrey, Z. Wang et al. 2007. On-board patient positioning for head-and-neck IMRT: Comparing digital tomosynthesis to kilovoltage radiography and cone-beam computed tomography. *Int. J. Radiat. Oncol.* 69(2):598–606.

Wu, Q.J., J. Meyer, J. Fuller et al. 2011. Digital tomosynthesis for respiratory gated liver treatment: Clinical feasibility for daily image guidance. *Int. J. Radiat. Oncol.* 79(1):289–296.

Yamada, Y., M. Jinzaki, I. Hasegawa et al. 2011. Fast scanning tomosynthesis for the detection of pulmonary nodules diagnostic performance compared with chest radiography, using multidetector-row computed tomography as the reference. *Invest. Radiol.* 46(8):471–477.

Yan, H., L. Ren, D.J. Godfrey et al. 2007. Accelerating reconstruction of reference digital tomosynthesis using graphics hardware. *Med. Phys.* 34(10):3768–3776.

Yoo, S., Q.J. Wu, D. Godfrey et al. 2009. Clinical evaluation of positioning verification using digital tomosynthesis and bony anatomy and soft tissues for prostate image-guided radiotherapy. *Int. J. Radiat. Oncol.* 73(1):296–305.

Zachrisson, S., J. Vikgren, A. Svalkvist et al. 2009. Effect of clinical experience of chest tomosynthesis on detection of pulmonary nodules. *Acta Radiol.* 50(8):884–891.

Zhao, W., B. Zhao, P.R. Fisher et al. 2007. Optimization of detector operation and imaging geometry for breast tomosynthesis. *Proc. SPIE* 6510:U547–U558.

Zhang, J., J. Wu, D.J. Godfrey et al. 2009. Comparing digital tomosynthesis to cone-beam CT for position verification in patients undergoing partial breast irradiation. *Int. J. Radiat. Oncol.* 73(3):952–957.

Zhou, J., B. Zhao, W. Zhao. 2007. A computer simulation platform for the optimization of a breast tomosynthesis system. *Med. Phys.* 34:1098–1109.

Ziedses des Plantes, B.G. 1935. Seriescopy, Een Rontgenographische method welke het mogelijk maakt achtereenvolgens een oneindig aantal evenwijdige vlakken van het te onderzoeken voorwerp afzonderlijk te beschouwen (Seriescopy, a Roentgenographic method which allows an infinite number of successive parallel planes of the test object to be considered separately) (English Translation). Ned. Tijdschr. *Geneesk* 51:5852–5856.

Ziedses des Plantes, B.G. 1936. Rontgenologic method and apparatus for consecutively observing a plurality of planes of an object. UK Patent 487389.

Ziedses des Plantes, B.G. 1938. Serisocopy: Ein röntgenographische Methode welke ermöglicht mit Hilfe einiger Aufnahmen eine unendlich Reihe paralleler Ebenen in Reichenfolge gesondert zu betrachten (A fading x-ray method which allows using some recordings to consider an infinite number of parallel planes in rich sequence separately) (English translation). Geb. *Röntgenstr.* 57:605–619.

Zwicker, R.D., N.A. Atari. 1997. Transverse tomosynthesis on a digital simulator. *Med. Phys.* 24:867–871.

Section II

System design

2

System design and acquisition parameters for breast tomosynthesis

Stephen J. Glick

Contents

This chapter reviews current variations on breast tomosynthesis (BT) system designs that have been investigated to date. In addition, the choices of acquisition parameters are also discussed. For a discussion on the design of tomosynthesis imaging systems dedicated for use in other applications such as imaging of the chest or orthopedic applications, please see Chapters 5 and 13.

2.1 BREAST TOMOSYNTHESIS SCANNER DESIGN

The concept of BT refers to any imaging system that uses limited-angle tomography to generate image slices through the three-dimensional (3D) breast. The basic continuous-to-discrete imaging equation for BT is given as

$$\Theta_m = \Theta'_m \exp\left(-\int_S f(r,E)\, h_m(r)\, dr \right), \qquad (2.1)$$

where Θ_m represents the number of x-rays detected at pixel m (detector pixels at all projection angles are lexicographically ordered to form one long vector), Θ'_m represents the blank scan x-ray flux at detector pixel m (i.e., x-ray flux with no object present), $f(r, E)$ is a continuous function representing the linear attenuation coefficient of the object (breast) as a function of spatial position r and energy E, and $h_m(r)$ represents the line joining the x-ray source and projection pixel m. Thus, the reconstruction problem is to best estimate a discretized version of the unknown object $f(r, E)$ given the measured projection data Θ_m.

There are a number of different system geometries that have been implemented for BT, basically all described by Equation 2.1 with differing $h_m(r)$. In this chapter, we will discuss the advantages and disadvantages of these different system geometries. Included will be a discussion comparing continuous versus step-and-shoot x-ray motion, and a stationary versus a moving detector. In addition, the use of novel stationary x-ray sources will be discussed. We will also discuss the effect of different acquisition parameters on image quality such as kVp setting, scan angular range, number of angular projections, detector pixel size, and choice of one or two BT views.

2.1.1 X-RAY TUBE MOTION

A number of different x-ray tube motions have been proposed for BT systems. Owing to geometric restrictions in the placement of the patient's head, most BT systems to date collect projection data from a source moving along a linear path. In the step-and-shoot geometry, the x-ray tube is moved in discrete steps, shooting an exposure at each step. The main advantage of this acquisition method is that unlike continuous motion acquisition, the source is stationary during the exposure. This eliminates source motion blur; however, it is possible that the microscopic vibrations present when the heavy x-ray tube comes to a stop may introduce a small motion blur. The second-generation General Electric (GE) prototype uses a step-and-shoot acquisition (Wu et al. 2006). In the continuous acquisition mode, the x-ray tube is pulsed at equal time intervals during the scan. The advantages of using a continuous x-ray tube motion is a shorter scan time that can possibly reduce the effect of patient motion. Of course, the trade-off with continuous motion systems is an increasing blur thereby degrading the system modulation transfer function (MTF) owing to the motion of the focal spot during the acquisition. In principle, this blur can be decreased by using a shorter x-ray pulse; however, this is problematic due to an increased x-ray tube power requirement.

Ren et al. (2005) have modeled a Hologic (Bedford, MA) prototype BT system and examined the effect of focal spot blurring with continuous tube motion. For this system, they modeled a BT scan with 11 exposures over 30° acquired in 18 s. The system used a direct conversion detector with a 70 μm pixel size; however, 2 × 2 binning was used so the effective pixel size was 140 μm. The authors showed that the blurring increases with distance away from the rotational center (iso-center), with blurring of a half-pixel at 3 cm distance away from the iso-center. This result suggests that focal spot blurring due to continuous motion is minor.

Zhao et al. (2007) have studied how the focal spot motion affects the MTF of each projection. They report that the MTF is degraded by a factor sinc $(k f_r)$, where k is the projected distance traveled by the x-ray source, and f_r is the radial frequency perpendicular to the ray of incidence. Since the focal spot motion does not affect the noise power spectrum, the degradation in the detective quantum efficiency (DQE) (which is defined as MTF^2/ (noise power spectrum) NPS) can be affected greatly by the degradation in MTF due to focal spot motion. In fact, Zhao et al. have suggested that focal spot motion can reduce the MTF and DQE by 30% and 50%, respectively.

Previous studies have suggested that focal spot motion can provide moderate degradation in both the MTF and DQE of BT. One approach for minimizing this degradation is to increase the acquisition time. Bissonnette et al. have suggested that a 39-s BT scan can eliminate image degradation due to focal spot motion. Of course the problem here is that longer acquisition times mean more degradation due to patient motion. Poplack et al. (2007) recently reported a comparison study between digital mammography (DM) and BT, and concluded that the accuracy in detecting microcalcifications was inferior on BT. They hypothesized that this result was possibly due to patient motion, and the relatively long acquisition times of BT (18 s). In fact, since this study was published, acquisition times of Hologic BT systems have been reduced to less than 5 s, further increasing the effect of focal spot blur due to continuous motion.

To date, there has been only one study that has investigated the trade-offs between increased blur due to focal spot motion and patient motion. Acciavatti and Maidment (2012) have recently presented an elegant analysis modeling both continuous tube motion and simple patient motion in BT. They suggested that continuous tube motion and patient motion have competing influences and that one can predict an optimal scan time for a given set of acquisition parameters. Through this analytical model, they suggest that it is possible to operate BT in continuous acquisition mode at a scan time that will provide improved image quality compared to step-and-shoot acquisition. This work also raises caution about applying less breast compression with BT because of increased patient motion.

2.1.2 STATIONARY X-RAY SOURCES

Most clinical BT systems use a single x-ray source that rotates along a linear path collecting projections at different source positions. Whether the source is moved continuously or whether the source is stopped during exposure, both strategies can cause blur due to focal spot motion (for the step-and-shoot acquisition, small vibrations can be presented when the heavy x-ray tube is stopped).

One promising new approach is the use multibeam field emission x-ray (MBFEX) technology using carbon nanotube (CNT) field emitters (Tucker et al. 2012). This system is composed of a field emission cathode with a linear array of 31 CNT emitting pixels, focusing electrodes, and tungsten anodes. The MBFEX system allows for the generation of x-rays from multiple source positions by electronically activating the corresponding CNT field emission cathodes. This activation of x-ray sources can be performed instantaneously either sequentially to emulate conventional BT or in parallel using a specified multiplexing approach. The advantage of the MBFEX system is that it can create a complete set of BT projections without any x-ray tube motion. Tucker et al. (2012) have recently reported on the use of the MBFEX system for BT using a Hologic Selenia Dimensions direct conversion detector. Results showed that the projection MTF under different imaging configurations did not fluctuate (whereas with a continuous source motion, the MTF would be degraded).

2.1.3 DETECTOR MOTION

Most BT systems use a full-field, flat-panel detector to acquire projection images. During acquisition, this detector can either remain stationary or rotate in some fashion opposite to the x-ray tube motion.

The acquisition geometry for the current GE (GE Healthcare, Buckinghamshire) BT system uses a detector that remains stationary throughout the scan (see Figure 2.1). The advantage of this geometry is that there is no motion blur due to detector movement during acquisition. In addition, keeping the detector stationary is mechanically simpler, and is a similar geometry to conventional mammography. Systems from other vendors (i.e., Hologic) use a different system geometry, where the x-ray tube and detector rotate around the stationary breast continuously over a specified angular range (see Figure 2.1). With this geometry, the breast is positioned on a separate holder at a short distance above the detector. One advantage of this approach is that the x-ray angle of incidence on the detector face generally spans a smaller range than that with a stationary detector. This is important in that x-rays entering into the detector at larger angles of incidence have a larger position uncertainty and thus cause more detector blurring. Badano et al. (2009) have studied the anisotropy in the point response due to oblique x-ray incidence in BT systems and reported that the point response can vary significantly across the detector face (using a stationary detector system).

2.1.4 ALTERNATIVE GEOMETRIES FOR BREAST TOMOSYNTHESIS

One unique BT system uses a scanning multislit setup with an x-ray tube, precollimator, and a photon-counting detector all mounted on a common arm (see Figure 2.2) (Fredenberg et al. 2009). The arm is then rotated around a central point located well below the detector, so that the x-ray source movement defines a linear arc above the compressed breast. With this rotation, points

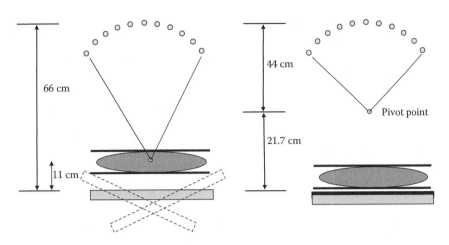

Figure 2.1 Illustration of two types of breast tomosynthesis geometries. The figure on the left shows a geometry in which both the source and the detector move during the acquisition, whereas the figure on the right shows a geometry where the detector is stationary during acquisition. (Reprinted from Glick, S. and X. Gong 2006. *Proc SPIE* **6142**: 569–577. With permission.)

in the object are "seen" by a range of source-to-detector angles, thus providing adequate tomographic sampling. The detector in this system consists of silicon-strip detectors with a pitch of 50 μm in both the readout and scan directions. Silicon has very low quantum efficiency (QE) for mammographic energies; thus, the silicon strips are positioned edge-on to increase the probability of x-ray absorption.

The detector of this system is operated in photon-counting mode; thus, unlike conventional mammography detectors that integrate energy of incident x-rays, this detector counts each x-ray individually. There are a number of possible advantages of

performing BT with photon-counting detectors, including (1) minimal electronic noise during readout, (2) reduced Swank noise, (3) improved dose efficiency, (4) possibility of improved signal-to-noise (SNR) ratio with energy weighting, and (5) the capability of performing dual-energy BT using electronic spectrum splitting for imaging of the contrast agent. Recently, Fredenberg et al. (2009) have studied this latter advantage using this silicon-based photon-counting detector. They report that dual-energy subtraction with this detector provided an improved detectability of iodine in phantom measurements.

2.2 BREAST TOMOSYNTHESIS ACQUISITION PARAMETERS

There are many design and acquisition parameters that must be decided to achieve maximum performance in BT. These parameters can be divided into three categories: system parameters, acquisition parameters, and reconstruction parameters. System parameters refer to aspects of the imaging system that would be of concern in designing a BT system. These include parameters such as x-ray tube target/filter, x-ray tube heat loading, CsI properties (thickness, light collection, etc.), pixel size (and fill factor), and detector electronic noise. Acquisition parameters of interest include kVp setting, number of projection angles, angular range, acquisition time, and imaging geometry. Finally, reconstruction parameters of importance include filtering method, number of iterations, and regularization scheme. In the remaining part of the chapter, we discuss each of these parameters and how they influence image quality.

2.2.1 kVp SETTING

The choice of kVp setting, as well as the target/filter material can have an important effect on BT image quality. Although there are many publications discussing the optimization of exposure technique for mammography, there are different factors that come into play with BT. One issue is the effect of electronic read-out noise. In BT, the x-ray fluence per view incident on the detector can be substantially lower than in mammography. This is because the x-ray fluence used in mammography is effectively split up

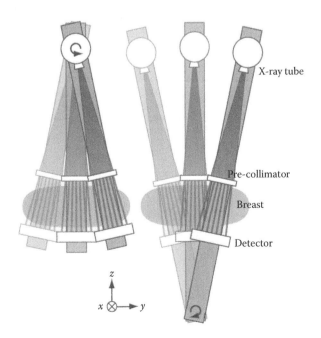

Figure 2.2 Illustration of a photon-counting mammography and breast tomosynthesis system using edge-on silicon-strip detectors. In the mammography system, shown on the left, the gantry arm is rotated around the center of the x-ray source to acquire the mammogram. In the BT system, the center of rotation is shifted to a point below the detector, as shown on the right. Thus, the x-ray source moves along a linear arc. (Figure originally published in Fredenberg, E., M. Lundqvist et al. 2009. *Proc SPIE* 7258. With permission.)

between the projections for BT. Each projection view then has lower incident x-ray fluence and the relative magnitude of the electronic noise becomes more important in comparison to the quantum counting noise. One way to decrease this importance of electronic noise is to increase the kVp setting and/or use a different target material.

Glick and Gong (2006) used an ideal observer SNR figure-of-merit (FOM) to evaluate different exposure techniques (i.e., kVp and target/filter) with an indirect-conversion detector-based BT system. The ideal observer SNR was formulated in the Fourier domain given as

$$\langle SNR^2 \rangle = N_{views} \iint_{u,v} \frac{G^2 \mid \Delta S(u,v)^2}{NPS_{tot}(u,v)} \, du dv, \qquad (2.2)$$

where u and v are two-dimensional (2D) Cartesian frequency coordinates, G represents the gain of the detector, ΔS represents the Fourier transform of the lesion projected onto the image plane, NPS_{tot} represents the 2D power spectrum of the noise (both quantum noise and electronic noise) in the detector, and N_{views} is the number of projection views acquired. Equation 2.2 was evaluated by modeling a CsI-based, indirect detector BT system as illustrated on the right of Figure 2.1. This system that was modeled acquires 11 projection views by rotating the x-ray tube over a 50° angular range, with the breast and detector remaining stationary. Three different x-ray tube anode/filter material combinations were studied, Mo/Mo (30 microns), Mo/Rh (25 microns), or W/Rh (50 microns) using the spectral models of Boone et al. (1997). Spectra were scaled to give a total of 2.4 mGy average glandular dose to the breast (i.e., approximately 50% greater than that for conventional mammography) by using DgN coefficients computed from Monte Carlo simulations (Thacker and Glick 2004). The imaging task here was defined as the detection of an isotropic lesion signal embedded in a compressed breast of varying composition and thickness. Two lesion signals were studied, a 5-mm-diameter sphere was used to model a mass, and a Gaussian function with sigma of 200 μm was used to model a microcalcification. Scatter for each projection view was modeled

by using previously reported scatter-to-primary ratios (SPRs) computed from Monte Carlo simulations (Boone 2000).

Figure 2.3 shows SNR results versus kVp setting for a 5-mm spherical tumor model and the three different target/filter combinations. Results from two different breast thicknesses, 4 and 6 cm, are shown. These data suggest that the W/Rh spectra provided the best performance, with up to 25% improvement over a standard Mo/Mo target/filter. Figure 2.4 demonstrates the effect of electronic noise on SNR for the detection of a 5-mm-diameter mass embedded in 100% glandular tissue. Shown are results for three different breast thicknesses (4, 6, and 8 cm) using an ideal detector with no electronic noise and a detector with electronic noise of 2500 electrons RMS. A moderate penalty in SNR is observed due to electronic noise in the detector, which decreases with increasing kVp setting. There appears to be very little effect of electronic noise for kVp settings over 36 kVp.

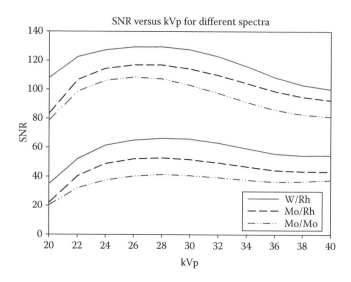

Figure 2.3 SNR versus kVp setting for three different target/filter combinations, Mo/Mo, Mo/Rh, and W/Rh. The top set of the three curves are for a 4-cm-thick breast, and the bottom set of three curves are for a 6-cm-thick breast. (Reprinted from Glick, S. and X. Gong 2006. *Proc SPIE* **6142**: 569–577. With permission.)

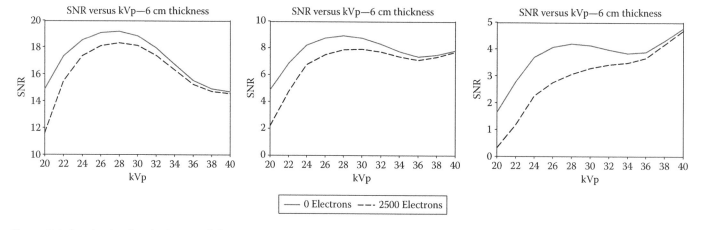

Figure 2.4 Graphs showing the impact of electronic noise on SNR. The SNR values versus kVp setting for a noise-free detector as well as a detector with 2500 electrons RMS electronic noise are shown. Each graph shows results for the detection of a 5-mm mass embedded into 100% fibroglandular tissue using a W/Rh spectra and 6 cm breast thickness. (Reprinted from Glick, S. and X. Gong 2006. *Proc SPIE* **6142**: 569–577. With permission.)

However, this result must be taken with some caution. The noise propagation through the detector is modeled here using a serial cascade model, which does not account for the increased noise due to characteristic K x-ray reabsorption. X-ray spectra with peak energy greater than 36 keV will have an increasingly higher probability of producing K-shell characteristic x-rays within the detector. Furthermore, the SNR model used here does not account for the normal breast structure, typically referred to as anatomical noise. Previous studies (Glick et al. 2007) in evaluating spectra for use in CT breast imaging have suggested that modeling normal anatomical noise can decrease the kVp setting at maximum SNR.

Zhao et al. (2007) have investigated optimal exposure techniques for BT when using an amorphous selenium detector-based system. They reported that the optimal kVp setting in BT was at least 2–3 kVp higher than the optimum for screening mammography. In addition, they reported that the W/Rh target/filter combination was optimal for all breast thicknesses studied.

2.2.2 SCAN ANGULAR RANGE AND NUMBER OF PROJECTION VIEWS

BT is limited-angle tomography where projections of the breast are acquired over an angular range somewhere between a few degrees and 90°. It is currently unclear what the optimal angular range should be, as well as the number of angular projection views. Not using enough projection views can result in "streak" artifacts in the reconstruction, whereas acquiring too many projection views will cause the x-ray fluence on the detector to be decreased, possibly increasing the influence of electronic noise on the reconstructed image. Even without factoring the effect of electronic noise, at some point increasing the number of projection angles will not provide improvements in image quality. Most likely, the decision on the optimal number of projection angles will depend on the task to be performed, for example, whether the goal of imaging is the detection of masses or microcalcifications.

A number of groups have been investigating this issue. Wu et al. (2003) performed a subjective analysis using a benchtop tomosynthesis system and a sterotactic needle-biopsy breast phantom to experimentally study how angular sampling effects reconstructed image quality. They noted that using a smaller angular range resulted in better spatial resolution within the tomosynthesis slice, but worse resolution in the z direction (i.e., in the direction perpendicular to the detector), whereas using a larger angular range improved the z resolution at the cost of reducing the in-slice resolution. This group also studied nonuniform angular sampling of the projections and reported that this type of sampling can provide moderate spatial resolution in the plane perpendicular to the detector face, while maintaining good resolution in the object plane parallel to the detector face.

Ren et al. (2006) have evaluated the number of projection views using a prototype benchtop system and a contrast-to-noise (CNR) FOM. They reported very little dependence on CNR when 11, 15, or 21 projection views were acquired using the same total dose. However, the authors point out that the CNR FOM might be suboptimal in that it does not account for shape distortion and other artifacts. This FOM also does not account for noise correlations in the reconstruction.

Sechopoulos and Ghetti (2009) used a computer simulation of a BT system to study 63 different combinations of angular range and number of projections. The simulation modeled a number of realistic effects, including the kVp spectrum, scatter, detector noise, and an angle-dependent MTF, and used a solid cube and a solid sphere to represent a mass and a calcification, respectively. The authors used CNR, the artifact spread function (ASF), and the ratio of these for the FOMs. They concluded that increasing the number of projection views beyond a "relatively low threshold" did not increasingly improve the z resolution, and that the minimum number of projection views for optimal performance was proportional to the angular range. Using the so-called quality factor, defined as the ratio of CNR to ASF, they concluded that the parameters with best performance were 13 projection angles over 60°.

Unlike the previous studies discussed, Reiser and Nishikawa (2010) used a task-based performance assessment of various BT scan parameters, using a range of signal sizes added to a structured background modeled with power law noise. Computer simulations were used to generate projections based on an ideal point x-ray source, and modeling Poisson quantum noise. Reconstruction was performed using an iterative maximum-likelihood expectation maximization algorithm (see Chapter 8). The defined task was signal-known-exactly (SKE) detectability of lesions modeled as designer nodules (Burgess et al. 2001). The FOM used was a prewhitening model observer, also referred to as d'. It was observed that increasing the angular range with constant angular sampling density resulted in increasing performance for all signal sizes. The detectability of larger-size lesion models was primarily affected by the scan angular range, whereas the detectability for smaller lesion models was limited by both quantum noise and the number of projection views. Reiser's study concurred with the Sechopoulos study in suggesting that increasing the number of views beyond some threshold does not improve the tomosynthesis image quality.

2.2.3 ONE-VIEW OR TWO-VIEW

As BT becomes more available in hospitals and clinics, one important question that needs to be addressed is whether acquisitions need to be performed using both mediolateral oblique (MLO) and craniocaudal (CC) views, or whether only one or the other view is needed. At first glance, one might think that since BT is a 3D imaging modality, only one-view would be required. However, owing to the limited angle acquisition, BT does not strictly provide a 3D isotropic reconstruction. The question is also affected by dose considerations. That is, is the extra radiation dose given by imaging two-views worth it? If one remembers the history of the movement to two-view mammography from single-view, the radiation dose was doubled with improvement in the prevalence of breast cancers detected by about 24% (Wald et al. 1995).

To date, studies comparing performance between two-view and one-view BT have been scant. Rafferty et al. (2009) investigated 34 malignant lesions imaged with BT using both CC and MLO views. They concluded that 65% were observed equally well on both views, 12% were more visible on the MLO view, 15% more visible on the CC view, and 9% were only observed on the CC view. Although a slight increase in performance was seen with the CC view, if only one-view BT is used, the MLO view might be preferred because it has been shown on mammography to provide more tissue coverage, especially near the axillary tail. Unfortunately, it appears as if maximum performance on BT would occur with two-view imaging. More studies are needed to answer the important question of whether the additional CC view is needed, and if so, would its inclusion in the diagnosis outweigh the increased examination and reading time, as well as increased radiation dose?

2.2.4 RADIATION DOSE: BT ALONE OR BT PLUS DM?

Another important question that needs to be addressed is whether BT screening should be used alone or in conjunction with DM. The biggest problem with routinely imaging women with both modalities is that it would require more radiation dose, as well as increased reading time. In Section 2.2.5, we discuss a variable-dose approach for imaging that would allow imaging with both BT and DM without increasing radiation dose. However, in general, screening with both modalities will result in increased patient dose. There are a number of advantages in screening patients with both DM and BT. First, BT is sufficiently different from DM that there will be some learning curve associated with reading the volume of the slices generated. For this reason alone, it is probably necessary to use both modalities until BT is more established. However, there might be other reasons for screening patients with both modalities. One currently unresolved issue is whether the accuracy in detecting microcalcifications on BT is good enough. Poplack et al. (2007) (see Chapter 12) have studied 99 women imaged with both BT and film-screen mammography and have concluded that BT was subjectively inferior for the visualization of microcalcifications. Other studies have reported a contrary conclusion. In particular, Kopans et al. (2011) studied 119 patients with relevant microcalcifications that were clearly not benign. Of these, two radiologists subjectively suggested that BT was superior on 41.6% of the cases, visibility was equivalent on 50.4% of the cases, and mammography provided better visualization in 8% of the cases. Clearly, a robust observer study evaluating microcalcification detection accuracy needs to be performed to fully understand this question. Before it is recommended that BT be used alone for screening, it is necessary that the technology provide equivalent or better accuracy in detecting microcalcifications than conventional mammography.

Rafferty et al. (2007) performed a reader study comparing performance with conventional mammography (2D) alone versus conventional mammography plus one-view BT (2D + 3D). The data included 310 image sets with 52 malignancies, 47 biopsy-proven benign abnormalities, 138 screening recalls, and 74 negative screening exams. Twelve experienced radiologists viewed each image set, and sensitivity, specificity, and ROC curves were computed. Results from comparing 2D to 2D + 3D showed that sensitivity was increased from 66% to 76% and specificity was increased from 81% to 89%. In addition, the recall rate was reduced by 43%. Unfortunately, the authors did not analyze how 3D imaging alone compared to 2D.

Gur et al. (2009) evaluated 125 women imaged with two-view BT and two-view DM. Their results suggested that the use of BT alone resulted in a 10% decrease in the recall rate as compared to DM. However, reading both BT and DM produced a 30% reduction in the recall rate.

Both these clinical studies suggest that the use of BT can provide improvements in detection accuracy, as well as a reduction in the recall rate; however, further clinical studies are needed to clarify whether screening with BT can or should be used alone or in conjunction with mammography.

2.2.5 VARIABLE-DOSE METHODOLOGY

A challenge in optimizing BT systems for maximum performance is that the optimal design is highly dependent on the task to be performed. In particular, the optimal system designed for the detection of masses is likely different from the system optimized for the detection of microcalcifications. As discussed above, there is some indication that the detection of masses might be superior with BT; however, for detecting microcalcifications, DM might provide superior performance. This suggests that a dual-modality screening approach might be beneficial; however, this paradigm requires increased radiation dose to the patient.

One approach for addressing this problem is the so-called variable-dose BT acquisition proposed by Nishikawa et al. (2007). In this acquisition method, one-half of the total dose of a BT scan is allotted to acquiring the central projection and the other half is distributed among the other projections acquired. The central projection would then be similar to a single-view mammography scan and could be used to search and localize microcalcifications. This higher dose central projection could also be combined with the other projection angles to generate a 3D BT reconstruction. The BT slices would then be used primarily to detect larger mass-like lesions.

Vecchio et al. (2011) have recently studied the variable-dose acquisition approach using BT and DM images from a breast phantom. The results suggested that the central BT projection view was similar to a standard view in mammography. In addition, they noted that BT reconstructions using uniform and variable-dose acquisitions were almost identical. Das et al. (2009) also studied the variable-dose acquisition method with somewhat inconclusive results. Their study used computer simulations with an anthropomorphic breast phantom. A comparison of variable-dose and uniform-dose acquisition was performed using location receiver operator characteristic analysis (LROC). In agreement with the Vecchio study, the LROC results did not show a statistically significant difference in mass detection between the uniform-dose and variable-dose acquisitions. However, the results also did not show a significant difference in detecting microcalcifications in the center projection view (with half of the total dose given in the BT study) and the reconstructed slices obtained with uniform-dose acquisition, although detection results were better when reading the uniform-dose BT study.

REFERENCES

Acciavatti, R. and A. Maidment 2012. Optimization of continuous tube motion and step-and-shoot motion in digital breast tomosynthesis systems with patient motion. *Proc SPIE* **8313**, *Medical Imaging 2012: Physics of Medical Imaging*, **831306**; doi:10.1117/12.911016

Badano, A., I. Kyprianou et al. 2009. Effect of oblique x-ray incidence in flat-panel computed tomography of the breast. *IEEE Trans Med Imag* **28**(5): 696–702.

Boone, J. 2000. Scatter/primary in mammography: Comprehensive results. *Med Phys* **27**: 2408–2416.

System design

Boone, J. M., T. R. Fewell et al. 1997. Molybdenum, rhodium, and tungsten anode spectral models using interpolating polynomials with application to mammography. *Med Phys* **24**(12): 1863–1874.

Burgess, A. E., F. L. Jacobson et al. 2001. Human observer detection experiments with mammograms and power-law noise. *Med Phys* **28**(4): 419–437.

Das, M., H. Gifford et al. 2009. Evaluation of a variable dose acquisition technique for microcalcification and mass detection in digital breast tomosynthesis. *Med Phys* **36**(6): 1976–1984.

Fredenberg, E., M. Lundqvist et al. 2009. A photon-counting detector for dual-energy breast tomosynthesis. *Proc* **7528**, *Medical Imaging 2009: Physics of Medical Imaging*, 72581J; doi: 10.1117/12.813037.

Glick, S. and X. Gong 2006. Optimal spectra for indirect detector breast tomosynthesis. *Proc SPIE* **6142**: 569–577.

Glick, S. J., S. Thacker et al. 2007. Evaluating the impact of x-ray spectral shape on image quality in flat-panel CT breast imaging. *Med Phys* **34**(1): 5–24.

Gur, D., G. Abrams et al. 2009. Digital breast tomosynthesis: Observer performance study. *AJR* **193**(2): 586–591.

Kopans, D., S. Gavenonis et al. 2011. Calcifications in the breast and digital breast tomosynthesis. *Breast J* **17**: 638–644.

Nishikawa, R. M., I. Reiser et al. 2007. A new approach to digital breast tomosynthesis for breast cancer screening. *Proc SPIE* **6510**, *Medical Imaging 2007: Physics of Medical Imaging*, 65103C; doi: 10.1117/12.713677.

Poplack, S. P., T. D. Tosteson et al. 2007. Digital breast tomosynthesis: Initial experience in 98 women with abnormal digital screening mammography. *AJR Am J Roentgenol* **189**(3): 616–623.

Rafferty, E., L. Niklason et al. 2007. Assessing radiologist performance using combined full-field digital mammogaphy and breast tomosynthesis versus full-field digital mammography alone: Results of a multi-center, multi-reader trial. *96th Annual Meeting of the Radiological Society of North America*, Chicago, IL.

Rafferty, E., L. Niklason et al. 2009. Breast tomosynthesis: One view or two? *95th Annual Meeting of Radiological Society of North America: Abstract SSG01-04*, Chicago, IL.

Reiser, I. and R. Nishikawa 2010. Task-based assessment of breast tomosynthesis: Effect of acquisition parameters and quantum noise. *Med Phys* **37**(4): 1591–1600.

Ren, B., C. Ruth et al. 2005. Design and performance of the prototype full field breast tomosynthesis system with selenium based flat panel detector. *Proc SPIE* **5745**, *Medical Imaging 2005: Physics of Medical Imaging*, 550; doi: 10.1117/12.595833.

Ren, B., T. Wu et al. 2006. The dependence of tomosynthesis imaging performance on the number of scan projections. In *Digital Mammography*. M. B. Susan, M. Astley, Chris Rose, Reyer Zwiggelaar, (eds) Springer. LNCS **4046**, Berlin.

Sechopoulos, I. and C. Ghetti 2009. Optimization of the acquisition geometry in digital tomosynthesis of the breast. *Med Phys* **36**(4): 1199–1207.

Thacker, S. and S. J. Glick 2004. Normalized glandular dose (DgN) coefficients for flat-panel CT breast imaging. *Phys Med Biol* **49**(24): 5433–5444.

Tucker, A., X. Qian et al. 2012. Optimizing configuration parameters of a stationary digital breast tomosynthesis system based on carbon nanotube X-ray sources. *Proc SPIE* **8313**, *Medical Imaging 2012: Physics of Medical Imaging* 831307; doi: 10.1117/12.911530.

Vecchio, S., A. Albanese et al. 2011. A novel approach to digital breast tomosynthesis for simultaneous acquisition of 2D and 3D images. *Eur Rad* **21**(6): 1207–1213.

Wald, N., P. Murphy et al. 1995. UKCCCR multicentre randomised controlled trial of one and two view mammography in breast cancer screening. *BMJ* **311**: 1189–1193.

Wu, T., A. Stewart et al. 2003. Tomographic mammography using a limited number of low-dose cone-beam projection images. *Med Phys* **30**(3): 365–380.

Wu, T., B. Liu et al. 2006. Optimal acquisition techniques for digital breast tomosynthesis screening. *Proc SPIE* **6142**, *Medical Imaging 2006: Physics of Medical Imaging*, 61425E; doi: 10.1117/12.652289.

Zhao, W., B. Zhao et al. 2007. Optimization of detector operation and imaging geometry for breast tomosynthesis. *Proc SPIE* **6510**, *Medical Imaging 2007: Physics of Medical Imaging*, 6510M; doi: 10.1117/12.713718.

3 Detectors for tomosynthesis

Wei Zhao

Contents

The renewed interest in, and rapid clinical translation of, digital tomosynthesis (DT) in recent years was enabled by the development and commercialization of flat-panel imagers (FPI). These detectors have a compact form factor, excellent image quality, and high image readout rate (frame rate). Naturally, all detectors for tomosynthesis have their root in digital radiography (DR) and full-field digital mammography (FFDM).

3.1 OVERVIEW OF DETECTOR TECHNOLOGIES USED IN TOMOSYNTHESIS

As discussed in other chapters throughout this book, tomosynthesis has made the transition into the clinic in both general radiographic and mammographic applications. While the detector technology shares common features, the design parameters vary substantially due to the different requirements in x-ray energy and image resolution.

3.1.1 GENERAL RADIOGRAPHIC APPLICATIONS

DT has been developed and investigated for several general radiographic clinical applications, such as chest, orthopedic, and head-and-neck imaging.[1–4] The majority of detectors used in radiographic DT systems are active matrix flat-panel imagers

(AMFPI) with rapid image readout (i.e., high frame rate). Here, we provide two examples of DT system implementations, as outlined in Table 3.1 with angular range, number of projection views, scan time, and detector parameters. Clinical investigations have shown that DT provides cross-sectional imaging information helpful for medical diagnosis, at a small fraction of the radiation dose in multidetector computed tomography (MDCT).[2,5]

The AMFPI used in DT systems are built using amorphous silicon (a-Si) semiconductor integrated circuit technology on large-area glass substrates. This is enabled by the rapid advancement in flat-panel active matrix liquid crystal displays (AMLCD), which use a two-dimensional array of a-Si thin-film transistors (TFT) to drive the display signal. Since the early 1990s, this technology has been incorporated into making large-area AMFPI. Depending on the x-ray detection materials used, AMFPI are divided into two main categories: *direct* and *indirect* conversion. As shown in Figure 3.1a, direct detection AMFPI employ a uniform layer of x-ray-sensitive photoconductor, for example, amorphous selenium (a-Se) to convert incident x-rays directly to charge, which is subsequently read out electronically by a TFT array.[6] Figure 3.1b shows that indirect AMFPI use an x-ray scintillator such as structured cesium iodide (CsI) to convert x-ray energy to optical photons, which are then converted to charge by integrated photodiodes (PDs) at each pixel of the TFT

Table 3.1 **Detector and image acquisition parameters for two examples of general radiographic tomosynthesis systems**

DETECTOR TYPE	INDIRECT[87]	DIRECT[3]
Maker/model	GE Healthcare/ Definium 8000	Shimadzu/ SonialVision Safire II
X-ray detection material	CsI (Tl)	a-Se
Thickness	0.6 mm	1.0 mm
Pixel pitch (micron)	200 × 200	150 × 150
Detector active area (cm × cm)	41 × 41	43.2 × 43.2
Detector matrix	2048 × 2048	2880 × 2880
Readout rate	~5 fps	30 fps max
Number of projections	25–60	75
Angular range (degrees)	20–40	40

Note: Parameters for both detectors were extracted from published data.

array.[7] Both direct and indirect AMFPI have been used in DR, and subsequently in DT systems.

3.1.2 DIGITAL BREAST TOMOSYNTHESIS

Digital breast tomosynthesis (DBT) was developed based on FFDM systems by increasing detector readout speed and incorporating x-ray tube gantry motion. Table 3.2 shows several examples of DBT systems incorporating different detector technologies: (1) a-Se based direct-conversion AMFPI used in several commercial DBT systems, with pixel size of either 70 or 85 microns; (2) indirect conversion AMFPI with pixel size of 100 microns; (3) indirect conversion FPI based on tiled, wafer-scale crystalline silicon (c-Si) complementary metal oxide semiconductor (CMOS) sensors; and (4) silicon strip detector with 50 micron pixel size, arranged in a multiple-slit geometry in a scanning FFDM/DBT system.

Although most DBT systems are implemented with a stationary detector to facilitate breast positioning, the scanning geometry (angular range and number of images), gantry motion (continuous or step-and-shoot), and detector pixel binning during a DBT scan vary substantially between different systems. As a result, the projection image quality not only depends on the inherent detector design but also on DBT system implementation.

3.2 DETECTOR PHYSICS AND PRINCIPLE OF OPERATION

In this section, the detector physics and principle of operation for three types of tomosynthesis detectors will be outlined: (1) AMFPI with either direct or indirect x-ray conversion; (2) CMOS FPI; and (3) silicon strip detectors with photon-counting capabilities.

3.2.1 ACTIVE MATRIX FLAT-PANEL IMAGERS

A large-area active matrix consists of a two-dimensional array of amorphous silicon (a-Si) TFTs, each of which functions as an electronic switch.[8] Both direct and indirect AMFPI use the same readout scheme: the scanning control circuit turns on the TFTs one row at a time, and transfer image charge from the pixels to external charge-sensitive amplifiers, which are shared by all the pixels in the same column. The readout rate is dictated by the pixel RC time constant, where R is the on-resistance of the TFT and C is the capacitance for the pixel-sensing element. For complete readout of charge, the TFT needs to be ON for at least $T_{on} = 5RC$. With nominal values of $R = 4$ MΩ and $C = 1$ pF, each row of the AMFPI would require ~20 µs to read out. In addition to the time required for charge transfer, the parallel to serial conversion of the digitized signal also necessitates an overhead. Hence, a detector with 1024 × 1024 pixels can be read out in real time (i.e., 30 frames per second (fps)). A faster readout rate, for example, 60 fps, may be possible by binning the pixels, that is, switching on more than one pixel at a time to reduce the image matrix. Both direct and indirect conversion AMFPI have been commercialized for a wide variety of clinical x-ray imaging applications. The direct method has the advantages of higher image resolution and simpler TFT array structure that can be

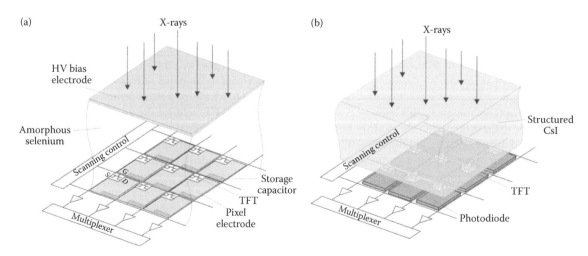

Figure 3.1 Diagram showing the concept of AMFPI with *direct* and *indirect* x-ray conversion: (a) direct detector uses an x-ray photoconductor (e.g., amorphous selenium) to convert x-rays directly charge; and (b) indirect detector uses a phosphor screen or structured scintillator to first convert x-rays to optical photons, which are then converted to charge by an integrated photodiode at each pixel of the detector.

Table 3.2 Examples of detectors used in different breast tomosynthesis systems

COMPANY	ANGULAR RANGE, TUBE MOTION	NUMBER OF VIEWS	DETECTOR	FRAME RATE	SCAN TIME (S)
GE[88]	±12.5°, step/shoot	9	CsI/a-Si, 100 µm pixel	~1 fps	<10
Hologic[89]	±7.5°, continuous	15	a-Se 70 µm pixel, 2 × 2 binning	3.5 fps, 2 × 2 binning	3.7
Siemens	±22°, continuous	25	a-Se, 85 µm pixel	1.2 fps, full resolution	20
Giotto[90]	±20°	13	a-Se, 85 µm pixel	1.2 fps, full resolution	10
Philips Microdose	±5.5°	21	Si-strip photon counting, 21 slits, 50 µm pixel	N/A (multiple slit)	3–8

manufactured in a standard facility for AMLCD. The advantage of the indirect method is the higher atomic number of Cs and I compared to that of a-Se. The factors affecting the x-ray detection process for direct and indirect AMFPI will be outlined.

3.2.1.1 Indirect detectors

3.2.1.1.1 X-ray scintillator

Indirect AMFPI have been manufactured by major medical imaging equipment vendors for different clinical applications. The most advanced form of detector construction uses thallium (Tl)-doped CsI with columnar (needle-like) structure as the x-ray scintillator. As shown in Figure 3.2a, the columns in CsI help channel light photons, which are generated by x-ray interaction, in the forward direction. Although the light guidance is not as perfect as in fiber optics with smooth walls, columnar CsI provides much better imaging performance than powder phosphor screens.[8,9] The CsI (Tl) used in AMFPI is less hygroscopic than the CsI (Na) layers used in x-ray image intensifiers (XRII) and its optical emission spectrum (green) is a better match to the spectral response of a-Si PDs, as shown in Figure 3.3.[8] For lower-cost detectors, which are mainly used for general radiography, powder phosphors such as gadolinium oxysulfide (GOS) have also been incorporated. There are also GOS screens optimized specifically for AMFPI, which differ from those used in screen-films in the sizes and spatial distribution of phosphor grains.

3.2.1.1.2 Design considerations for amorphous silicon optical sensing elements

An active matrix array designed specifically for medical imaging requires not only the TFT but also an image charge-sensing element. A number of factors can influence the choice of pixel design, some related to imaging performance and others not. Perhaps the most important nonimaging factor is the fabrication *yield* (the fraction of devices that are useful), which impacts directly the manufacturing cost of the AMFPI. By fabricating both the switching and sensing/storage elements at the same time, the number of lithographic steps can be minimized, allowing higher yield at the cost of suboptimal imaging performance.

The most common type of an a-Si optical sensing element is the a-Si PD. As shown in Figure 3.4a, each a-Si PD has a p-i-n multilayer structure, where the thin p- and n-doped layers are reverse biased to block injection of electrons and holes, respectively, from the bias electrodes into the intrinsic layer. The top transparent electrode is made of indium tin oxide (ITO), and connected to a negative bias potential, typically −5 V. Light photons enter the intrinsic (i) layer through the passivation layer, the ITO, and the p layer. Each of these layers absorbs some light, thereby reducing the absorption in the intrinsic layer, and hence quantum efficiency (QE) of the PD. Charge generated in the p- and n-layers does not contribute to the signal due to the very short drift lengths of minority carriers. Light absorbed in the intrinsic layer (typically 1–2 microns thick) generates

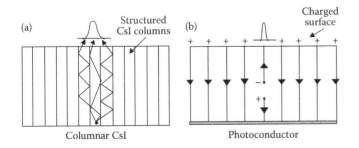

Figure 3.2 Image resolution of indirect and direct conversion x-ray detection materials: (a) columnar (needle) structures of CsI (Tl) help channel light in the forward direction, thus providing better resolution than powder phosphor screens; (b) the applied electric field in photoconductors draws x-ray generated image charge directly to surfaces without lateral spread.

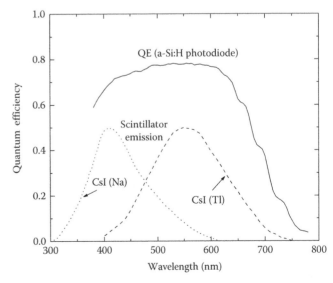

Figure 3.3 Optical quantum efficiency of a-Si photodiodes as a function of wavelength of light. Plotted in comparison are photon emission spectra for three types of x-ray scintillators: structured CsI (Na) and CsI (Tl). (Adapted from J. A. Rowlands and J. Yorkston, *Medical Imaging: Volume 1 Physics and Psychophysics, Psychophysics*, SPIE, Bellingham, 2000, pp. 223–328.)

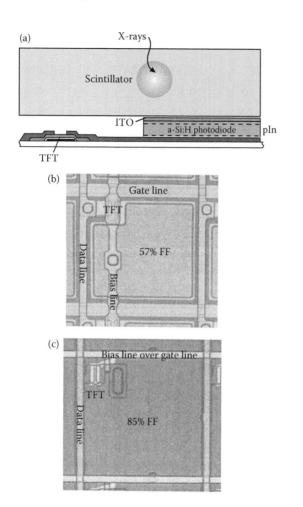

Figure 3.4 (a) Cross section of a single pixel of indirect AMFPI showing a-Si PD; (b) micrograph showing the top view of a pixel design with side-by-side TFT and photodiode; and (c) with photodiode on top of TFT. (Graphs b and c are reproduced from Figure 3.4b and 3.4c reproduced from R. L. Weisfield et al., *Proc SPIE* 5368, 338–348, 2004. With permission.)

electron–hole pairs, which are driven to the n and p contacts. The fabrication process for a-Si PD is different from that for a-Si TFT used in AMLCD. Flat-panel display manufacture does not require p+ doping or a thick (i.e., 2 micron) intrinsic a-Si layer, which is more than an order of magnitude thicker than that used for a-Si TFTs. These additional processes make the fabrication of a-Si PDs difficult at foundries designed specifically for AMLCD.

Figure 3.3 shows the QE as a function of photon wavelength in the visible range for a ~1.5-μm-thick p-i-n PD at −5 V reverse bias. The periodicity is due to the interference effects in the top passivation layer, and the material and thickness for this layer can be chosen to have antireflective properties for certain wavelengths. The lower QE at longer wavelengths is due to the decrease in absorption coefficient of intrinsic a-Si, and the decrease in QE at shorter wavelengths is due to increased absorption in the p-doped layer. Also shown in Figure 3.3 are the emission spectra for CsI scintillator with different dopants. It can be seen that the spectral response of a-Si:H PD is well matched to the emission from CsI:Tl.

To reduce manufacturing complexity, the PD is usually built side by side with the switching element. This leads to a suboptimal fill factor, which is the fraction of pixel area occupied by the sensing element. Figure 3.4b shows the micrograph of

the pixel designs for an indirect AMFPI with 127 μm pixel size.[10] The TFT is at one corner of the pixel. The gate lines, data lines, and the PD occupy the rest of the space. This is the typical design used in most commercial indirect AMFPI. Since the space taken by the TFT and the lines does not change as a function of the pixel pitch d, the fill factor f_p decreases rapidly as d decreases.[8] The value of f_p is 0.57 for $d = 127$ μm. Advanced sensor structures have been proposed to increase the fill factor. One method is shown in Figure 3.4c, where the PD is built on top of the other circuit elements (e.g., TFT, gate line, and data line) at each pixel.[10,11] With this design, the fill factor for the same pixel size of 127 μm is increased to $f_p = 0.85$. In principle, the fill factor can be increased to unity by building a continuous a-Si:H PD with common bias electrode on top of the entire TFT array, and some preliminary work has been conducted to demonstrate its feasibility. The main challenges for this approach are the increased line capacitance and the image blooming effect due to charge crosstalk between pixels.[10]

3.2.1.1.3 Pixel x-ray sensitivity for indirect AMFPI

The overall pixel x-ray sensitivity of an indirect AMFPI depends on four factors: (1) x-ray QE η of the scintillator; (2) inherent x-ray to optical photon conversion gain g_c, that is, the number of optical photons emitted from the scintillator for each absorbed x-ray; (3) optical QE of the a-Si PD, and (4) pixel fill factor. The thickness of CsI, d_{CsI}, used in indirect AMFPI varies depending on the clinical application. For the relatively high x-ray energy (>70 kVp) used in general radiographic DT, d_{CsI} is typically 600 μm.[12] Columnar-structured CsI layers have lower density compared with single crystals. The packing density could vary depending on the deposition procedures; however, the widely quoted value is ~75%, which results in a density ρ of 3.38 g/cm³. Shown in Figure 3.5a is the x-ray QE η of a 600-μm columnar CsI layer as a function of x-ray photon energy. The k-edge of Cs (35 keV) and I (33 keV) creates a boost for η over the energy range typically used for general radiography. With an RQA5 spectrum (70 kVp tungsten spectrum with 21 mm of added Al filtration), η with the above detector parameters is ~0.84. As shown in Figure 3.3, the optical QE of a-Si PDs is ~0.7 for the green light emitted from CsI (Tl). The reported conversion gain of CsI for indirect AMFPI is ~25 eV/photon,[8] which results in $g_c = 2000$ for a 50 keV x-ray photon. The fill factor f_p depends on the pixel pitch d, which ranges between 150 and 200 microns for general radiography, and the pixel design of the a-Si PD-TFT array. With all factors considered, the overall gain of an indirect AMFPI is ~1000 electrons per 50 keV x-ray photon with a nominal fill factor of $f_p = 0.7$.

The AMFPI used in breast tomosynthesis are made with much smaller pixel size, ranging from 83 to 100 μm.[13] Because of the higher spatial resolution requirement and lower x-ray energies, the thickness of the CsI is typically 150 μm to minimize image blur. However, thicker CsI layers (up to 200–300 μm) have also been proposed to improve the absorption of x-rays.[14] Shown in Figure 3.5b is the QE of a 150-μm-thick columnar CsI (with packing density of 75%) as a function of x-ray photon energy. Although the x-ray absorption does not benefit from the k-edge boost for typical mammographic x-ray energies (i.e., <30 keV), the Swank factor is also free from the effect of k-fluorescence escape. Shown in Figure 3.5b is the Swank factor A_S for a 150-μm-thick CsI

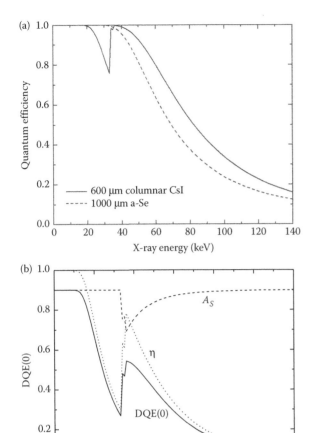

Figure 3.5 (a) Quantum efficiency (QE) η as a function of x-ray photon energy for two materials (a-Se and columnar CsI) used in x-ray imaging detectors: Thickness of layer is 1000 μm for a-Se and 600 μm for CsI. (b) Swank factor and DQE(0) for a 150 micron thick CsI layer.

layer as a function of x-ray energy. Below the k-edge, the value for A_S is mainly dependent on the optical factors of the CsI layer (e.g., with or without reflective backing, light absorption, light spreading between the columns). At the k-edge, A_S exhibits an abrupt decrease due to the effect of k-fluorescence escape, and slowly recovers as energy increases.[9,15] The resulting DQE(0), where DQE(0) = ηA_S, is also shown in the same graph.

The main disadvantage of AMFPI compared to crystalline Si detector technology, such as CMOS or CCD, is its higher electronic noise, which is dominated by the noise associated with the pixel reset and the charge amplifier connected to each data line.[16] The nominal value for the pixel electronic noise is ~1500 electrons, which is higher than the number of light photons captured by the PD for each absorbed x-ray in CsI. The excessive electronic noise would lead to degradation in imaging performance at low dose, which is the operational condition in DT. This degradation effect is severer at high spatial frequencies due to the image blur in CsI.

3.2.1.2 Direct detectors

In the direct detection approach, the energy absorbed in the photoconductor through x-ray interaction is converted directly to charge, as shown in the conceptual side view in Figure 3.6a.

Consequently, the sensing element is typically a pixel charge collection electrode, which is made using a thin metal layer and extended to cover as large an area as possible within a pixel. These pixel electrodes are connected electrically to a charge storage capacitor and the TFT, as shown in the schematic diagram of Figure 3.6b. A continuous electrode is applied to the upper surface of the photoconductor to allow the application of an external bias voltage. The fabrication of the storage capacitor only requires the standard metal and insulator layers with standard AMLCD manufacturing processes thus simplifying production and lowering cost.[17]

The operational requirements of the photoconductive layer are similar to those of the PD in the indirect approach in that a bias voltage must be applied across the thickness of the photoconductor to facilitate the separation and collection of the charge carriers produced by x-ray interaction. The most highly developed x-ray photoconductor is a-Se, which is currently used in all commercial direct AMFPI. To maintain an internal field of ~10 V/μm within a 1000-μm-thick layer of a-Se, a bias voltage of ~10,000 V must be applied. This high voltage necessitates blocking contacts to prevent charge injection. One type of thin blocking layers contains a large number of deep traps. These traps, when filled, permit a rapid decrease in the field at the interface between a-Se and metal contact, resulting in minimal charge injection.[18]

3.2.1.2.1 Design considerations for high-voltage protection of TFTs

Without careful design, it is possible for the high-voltage (HV) bias of a-Se, for example, 10,000 V, to appear across the TFT and storage capacitor of each pixel. This would result in permanent damage to the active matrix array, for example, through dielectric breakdown of storage capacitor and/or gate insulator of the TFT. When high voltage is first applied to the panel, the voltage drop is distributed between the pixel storage capacitance and the a-Se pixel capacitance, which are in series, as shown in Figure 3.6b. This configuration is a potential divider with the larger potential occurring across the smaller capacitance. Fortunately, the capacitance of an ~1000-μm-thick layer of a-Se is small (~2.5 pF cm^{-2}) so that an individual pixel of 200 × 200 μm has a capacitance of ~1 fF. If the storage capacitance is designed to be ~1 pF, the potential at the pixel electrode is only ~10 V with the other 9990 V dropped across the a-Se layer. Thus, the components on the active matrix panel can be protected in a purely passive manner from the application of the high potential.

However, if the TFT array is left without scanning, dark current or signal current from a-Se will cause the potential on the pixel electrode to rise toward the HV bias. Unless some preventive measures are taken, this voltage increase will eventually damage the active matrix circuit. Two general approaches have been implemented to prevent the voltage at the pixel electrode reaching damaging values, which is usually a few tens of volts with typical a-Si:H TFT designs. The first approach, as shown in Figure 3.6c, is to include an extra dielectric layer between a-Se and the HV bias electrode.[19,20] In this approach, the potential drop is redistributed across the dielectric layer rather than in the storage capacitor as the potential across the a-Se layer collapses. However, this protection results in increased

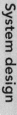

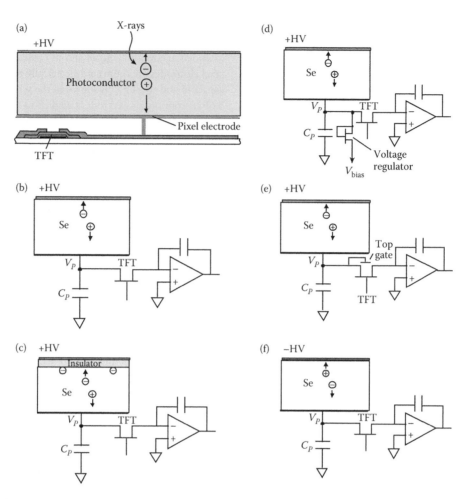

Figure 3.6 Schematic drawings showing the different detector configurations for direct conversion AMFPI: (a) side view of single pixel; (b) schematic diagram for the structure shown in (a); (c) detector implemented with an additional insulator layer on top of a-Se; (d) schematic diagram of a pixel with additional voltage regulation element to prevent HV damage; (e) pixel design using the top gate of the same TFT for HV protection; (f) using negative potential on the top bias electrode.

readout complexity, to eliminate the trapped image at the a-Se/ dielectric layer before subsequent exposures. This is achieved by the removal of the applied bias voltage and flooding the detector with light to generate charge in a-Se. The requirements of a refresh cycle make this approach incompatible with real-time readout. For DBT, the insulating layer has to be removed to allow rapid readout.[21]

The second approach is to maintain real-time imaging capability while providing high-voltage protection. The key pixel design is to include means to drain away charge on the pixel if the pixel potential exceeds a predetermined safe value. Three possible designs have been proposed: (1) To include an extra circuit element at each pixel in parallel with the storage capacitor for voltage regulation, as shown in Figure 3.6d. This device can be made using the standard TFT process, with a separate bias line to provide a path to bleed excess charge from the pixel electrode.[22] (2) To modify the TFT design by incorporating a second (top) gate, which is connected to the pixel electrode as shown in Figure 3.6e.[23,24] When the pixel potential approaches damaging levels, the top gate will increase channel current to drain away the excess charge. Since the excess charge is bled away along the readout (data) lines, the timing of the electronic scan needs special consideration to avoid corruption of image information on pixels sharing the same readout line with

overexposed pixels. (3) To apply a negative bias on the top electrode of a-Se to ensure that the ordinary TFT will start to conduct when the pixel potential reaches the threshold value, as shown in Figure 3.6f.[25] This approach requires the same consideration for proper timing of the electronic scan because excess charge from a single pixel could potentially corrupt the image data from all the pixels sharing the same data line. Although the simplicity of this design is attractive, the reversed a-Se structure makes it more susceptible to charge trapping.[26] This is because the electrons, which have much lower mobility than holes, have to travel a longer distance before reaching the pixel electrodes.

As can be seen, the basic design of the sensing/storage element for a direct detector is straightforward and compatible with the manufacturing process for TFT arrays in AMLCD. However, additional consideration for detector design is required to ensure the safe operation of TFT array in the presence of the high-voltage bias. With the rapid advancement in TFT manufacturing process and increased yield for AMLCD, the design complexity associated with direct AMFPI is no longer a concern.

3.2.1.2.2 Pixel x-ray sensitivity for direct AMFPI

The most highly developed x-ray photoconductor is a-Se, which is being used in all commercial direct AMFPI. Because of its lower atomic number than CsI, the thickness d_{Se} is 1000 μm for most clinical applications except mammography, where d_{Se} of 200 μm provides essentially complete absorption of mammographic spectral energies. The density of a-Se is 4.27 g/cm³, lower than that for crystalline selenium. As shown in Figure 3.5a, the absorption efficiency of a 1000-μm a-Se AMFPI is >50% for x-ray energies below 70 kVp. For an RQA 5 x-ray spectrum, η is 0.77 for d_{Se} = 1000 μm. For breast tomosynthesis, where tube voltage is usually less than 35 kVp, d_{Se} = 200 μm is sufficient to absorb more than 95% of the incident x-rays. The x-ray to charge conversion gain of a-Se is inversely proportional to the energy required to generate an electron–hole pair, W. The value for W in most single-crystal solid-state photoconductors, for example, Si, Ge, and CdTe, follows Klein's relation, that is, three times the band gap energy. In a-Se, W depends on the electric field E_{Se}. The nominal value is W = 50 eV at E_{Se} = 10 V/μm.[27] The geometric fill factor for direct AMFPI is high because the pixel electrode is built on top of the TFT and the gate and data lines. In addition, the image charge collection in a-Se is governed by the electric field. Because the field lines in the gap between pixels bend toward the pixel electrodes, the image charge created in this region can also be collected.[28] This leads to an effective fill factor of unity, which has been confirmed experimentally from direct AMFPI.[29] Compared to indirect AMFPI, a-Se direct detectors have approximately the same x-ray conversion gain (1000 electrons per incident 50 keV x-ray photon) and electronic noise. Hence they share the same advantages and limitations in low-dose imaging performance as the indirect AMFPI. One of the advantages of the direct AMFPI compared to the indirect is the ability to make smaller pixels because of its simpler array structure (no need for the PDs) and the unity fill-factor, which is independent of pixel size.

3.2.1.2.3 Other x-ray photoconductors

Despite the overwhelming success of a-Se in AMFPI, it has two shortcomings: one is the relatively low atomic number, Z = 34, which requires thick layers (e.g., 1000 μm) for high QE at general radiographic energies (>50 keV), and the other is the high electric field required to achieve a reasonable charge conversion gain, for example, W = 50 eV at 10 V/μm. Even at this field strength, the detector performance could be degraded in the dark part of the image in fluoroscopy,[30] or behind dense breast tissue in DBT.[31]

Other photoconductors with higher atomic numbers and/ or conversion gains have been investigated. Some of these photoconductors, for example, Cd(Zn)Te, HgI$_2$ and PbI$_2$, were first investigated in their single crystalline forms as nuclear radiation detectors. Cd(Zn)Te single crystals are currently under intensive investigation for single-photon-counting detectors in nuclear medicine (single photon emission computed tomography (SPECT) and positron emission tomography (PET)) imaging and computed tomography (CT).[32–34] For application in large-area AMFPI, polycrystalline thin films have been formed through physical vapor deposition (PVD), for example, Cd(Zn)Te, PbO, PbI$_2$, and HgI$_2$.[35–40] In addition to PVD, particle-in-binder (PIB) screen printing method has also been developed as a cost-effective alternative for depositing x-ray photoconductors.[41,42] Studies have shown that the trap-free limit of W of polycrystalline thick films approaches Klein's theoretical values.[43] However, under moderate bias field (~1 V/μm), which keeps dark current at a manageable

level, trapping of the slower charge carriers is a serious problem. This results in reduced sensitivity (i.e., higher effective W), decreased modulation transfer function (MTF) at high spatial frequencies, lower detective quantum efficiency (DQE), and ghosting artifacts.[44,45] This problem needs to be overcome before commercialization for regular clinical use.

3.2.2 CMOS-BASED FPI

With the advances in very large-scale integrated circuit (VLSI), recently there has been a steady increase in the effort of making wafer-scale c-Si CMOS image sensors for x-ray imaging.[46,47] Owing to the limited wafer size (mostly 6″ diameter due to cost considerations), the largest monolithic "tile" dimension is approximately 12 × 12 cm (or rectangular tiles with comparable total surface area).[48] The majority of wafer-scale CMOS sensors are amplified pixel sensors (APS), which have an amplification circuit at each pixel using three or more transistors.[49] The conceptual circuit diagram for a three-transistor (3T) pixel design is shown in Figure 3.7a. The sensing node for the PD is connected to the gate of the amplifying transistor operated as a source follower. When the readout transistor is turned on, the pixel voltage will be read out to the external (column) amplifier, followed by the reset of the pixel potential by the reset switch. They have the following advantages over the AMFPI with a single a-Si TFT per pixel: (1) pixel amplification (source follower) permitting nondestructive readout and lower electronic noise (100–300 e rms), (2) faster readout speed, and (3) smaller pixel size (~30–40 microns for breast imaging). To make large-area detectors, several CMOS tiles may be butted side by side with minimal dead zone (e.g., less than one pixel wide) between them. With each CMOS tile three-side buttable, tiled CMOS

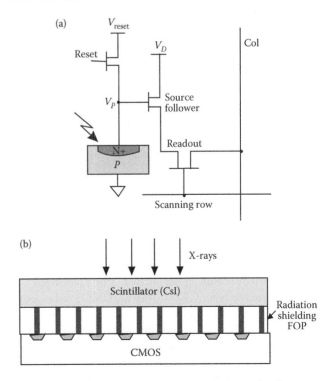

Figure 3.7 (a) Schematic diagram showing a single pixel of a 3-transistor design for CMOS APS with a photodiode structure of reverse biased N+/P; (b) side view showing the coupling of x-ray scintillator through a radiation shielding FOP to a CMOS sensor.

detectors with sizes up to 29 cm × 23 cm have been made,[50] and are currently being investigated for DBT.[51] With future cost reduction and increase in yield, tiled wafer-scale CMOS APS are expected to expand their applications.

Although CMOS APS with direct conversion has been investigated,[52,53] all existing commercial products use the indirect conversion approach, for which a PD is built at each pixel of the CMOS. The simplest PD structure, as shown in Figure 3.7a, is to form an N+ well on top of a p-type substrate. The PN junction is reverse-biased by applying a positive potential (up to 3.3 V) to the N+ well during reset of the pixel. As image charge begins to accumulate on the sensing node, the bias potential decreases and the pixel slowly reaches saturation. The full-well capacity, which is the maximum image charge before saturation, depends on the architecture of the PD and its bias condition. In order to increase dynamic range and improve sensitivity, more sophisticated PD structures have been developed, such as the pinned PD structure.[54–56] For medical imaging applications, some CMOS sensors incorporate selectable dynamic range, where a higher well capacity can be selected at the cost of increased electronic noise.[57] For x-ray detection, as shown in Figure 3.7b, a fiber optic faceplate (FOP) may be bonded to the CMOS APS before a scintillator is coupled. To protect the CMOS from radiation damage, the FOP can be doped with heavy element to further shield the CMOS from the radiation not attenuated by the scintillator. The scintillators used for CMOS APS are very similar to those used in indirect AMFPI, with high-end detectors using structured CsI:Tl and low-end detectors using GOS. The optical QE of Si PD depends on the PD design; it is usually ~45% for green light with the N+/p diode structure. The fill factor could reach 70–80%, even for small pixel sizes between 50 and 70 microns.[47,58] These result in a combined optical QE of 30–35%. The optical transmission of an FOP is typically 60%, which makes the total x-ray to charge conversion gain of CMOS APS in the range of 350–400 electrons per absorbed 50 keV x-ray photon. Although this gain is much smaller than that in indirect AMFPI, the low electronic noise of CMOS, which is 100–300 electrons, compensates for it. In addition, the low conversion gain also helps with the dynamic range of CMOS, which has limited well capacity in the range of 10^5–10^6 electrons.

3.2.3 PHOTON-COUNTING DETECTORS AND MULTISLIT GEOMETRY

In medical imaging, photon-counting detectors are routinely used in nuclear medicine, for example, SPECT and PET. Recently, a lot of effort has been devoted to developing photon-counting detectors for x-ray imaging, where the usage of small pixel sizes and count rate requirements are much more demanding. Commercial detectors with a pixel pitch of ~1 mm have been investigated extensively for CT.[34,59] They use CMOS technology to build a two-dimensional array of photon-counting circuitry, which includes amplifier, shaper, discriminator, and multiple counters to achieve photon counting with energy-resolving capability.

Photon counting could outperform energy integration detectors in the following aspects: (1) elimination of electronic noise and Swank's noise; (2) higher weighting of lower-energy x-ray photons, which helps improve low-contrast lesion signal-to-noise ratio (SNR); and (3) possibility of energy discrimination

(two or more energy bins), which permits variable energy-weighted subtraction techniques to provide additional contrast mechanisms. Existing photon-counting detectors for x-ray imaging mainly use direct conversion materials such as Si,[60] CdTe,[61] or Xe.[62] While the development of photon-counting detector for CT has been very promising, there has been no commercial 2D sensors for radiography or mammography due to the simultaneous requirement for both high resolution and large area. However, linear (1D) photon-counting detector combined with multislit system geometry has been applied to DBT.[63,64] Here, we will describe one example of such a system.

The Philips (formerly Sectra) MicroDose 2D Mammography system uses the multislit system geometry with 21 Si strip detectors. Through mechanical modification of this system, breast tomosynthesis prototypes have been built, and are currently under clinical investigation.[65] The concept of the system design is depicted in Figure 3.8a. The x-rays are divided into 21 narrow beams by a precollimator placed above the compressed breast. A postcollimator below is aligned with the precollimator to provide a slit-scan geometry, which has been shown to provide effective scatter rejection. The detector consists of silicon (Si)-strip detectors, which are aligned to the collimators. The x-ray tube, collimators, and detector are attached to a gantry, which is rotated around the x-ray source to complete data collection of 2D mammogram. In the tomosynthesis system, the center of rotation (COR) is located below the detector. By a scanning motion across

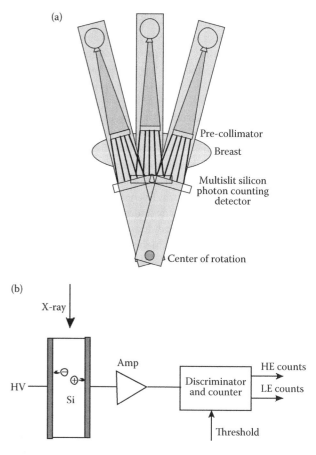

Figure 3.8 (a) Conceptual diagram showing digital breast tomosynthesis based on a multi-slit Si photon counting detector; (b) circuit block diagram showing photon counting readout. (Adapted from N. Dahlman et al. *Proc SPIE* 7961, **796114**, 2011.)

the breast, 21 projection views are acquired simultaneously with an angular range of 11°. The x-rays passing through the breast enter the Si strip detector from the side, as shown in Figure 3.8b, ensuring high absorption efficiency. The electron–hole pairs generated by x-rays are driven to the anode and cathode by the applied electric field. Each Si strip is connected to a preamplifier and shaper, which converts each x-ray interaction event into an electronic pulse. The pulse height is proportional to the energy of the absorbed photon. Because of the high x-ray to charge conversion gain of Si (~300 electrons/keV), pulses below a certain threshold can be regarded as noise and rejected. The detector has virtually no electronic noise or Swank's noise. There are two energy bins, each corresponding to a predetermined energy threshold. The counts from each energy bin can be used to perform dual-energy subtraction, that is, spectral imaging.[66]

3.3 DETECTOR PERFORMANCE IN TOMOSYNTHESIS

The most common clinical implementation of DT is for the x-ray tube to travel in a linear path, while keeping the detector either stationary or traveling in the opposite direction. In general radiographic tomosynthesis, the angular range used was between 20° and 60°, with 20–70 projection views. Both direct and indirect AMFPI have been used. Shown in Table 3.1 are the detector and tomosynthesis acquisition parameters for two commercial tomosynthesis systems. For breast tomosynthesis, a wider variety of system parameters have been implemented, with Table 3.2 showing a few different examples. Several factors related to tomosynthesis acquisition could impact the projection image quality: (1) additional focal spot blur due to continuous tube motion during x-ray exposure; (2) increased angle of oblique entry of x-rays, which causes additional image blur; (3) degradation effect of electronic noise at the reduced radiation dose used in each view; and (4) requirement of rapid image readout, which may lead to artifact due to image persistence. In this section, we will first review the inherent imaging performance of detectors, and then discuss the effect of specific factors related to tomosynthesis image acquisition. We will limit the discussion to direct and indirect AMFPI since they are the most widely used detector in commercial tomosynthesis systems. Tiled CMOS APS are being evaluated for DBT, and excellent image quality has been obtained at the low dose used in each DBT projection view.[67]

3.3.1 INHERENT IMAGING PERFORMANCE

Several international standards (IEC and AAPM task group) have adopted image quality metrics expressed in the spatial frequency domain to evaluate the image quality of projection x-ray images.[68] These image quality metrics, which include MTF, noise power spectrum (NPS), and DQE, will be reviewed here. Published measurements of commercial detectors will be used as examples for discussion.

AMFPI have been developed for a wide range of clinical imaging applications, which include general radiography, mammography, and radiography and fluoroscopy (R/F) mixed-mode operation. The clinical implementation of tomosynthesis is based on the modification of either general radiography or R/F systems; hence, the detector performance is inherited accordingly.

3.3.1.1 Image correction

Before quantitative evaluation of image quality can be performed, projection images acquired by FPI need to be corrected for imperfection due to detector nonuniformity and defects. Defect pixels are unavoidable during fabrication of the active matrix. Owing to the large number of pixels, even a 0.1% defect rate could result in 9000 bad pixels in a 3000×3000 pixel AMFPI. In addition, there is nonuniformity between pixels due to several reasons: (1) nonuniformity in the active matrix that results in variations in TFT characteristics; (2) variation in the thickness of x-ray detection material; and (3) gain nonuniformity between different charge amplifier channels. This necessitates image correction through postprocessing. The standard method is an offset and gain nonuniformity correction followed by a defect pixel replacement. An offset (or dark) image is obtained without x-ray exposure and subtracted from each x-ray image. To reduce the effect of electronic noise, the average of several dark images is usually used. Since there is temporal drift in offset due to device instability, offset images are constantly updated between x-ray examinations. The gain correction is performed by dividing the offset subtracted image by a gain table, which is obtained during a calibration procedure. During calibration, the detector is exposed to uniform radiation. By averaging several x-ray images, the gain of each pixel can be determined with minimal effect of x-ray quantum noise. Defect pixels are identified by setting a lower threshold of x-ray sensitivity based on the pixel statistics, and the result is stored in a defect map. After gain correction, the bad pixels are replaced by the average values of neighboring good pixels. The gain table and bad pixel map are much stabler compared to offset; hence, in commercial detectors, the calibration procedure only needs to be repeated once a month or even less frequently.

3.3.1.2 Spatial frequency domain image quality metrics
3.3.1.2.1 Spatial resolution: Modulation transfer function

The spatial resolution of projection images is quantified by MTF. MTF is defined as the Fourier transform (FT) of the point spread function (PSF). This concept applies to a linear system that is shift invariant. In practice, MTF is usually measured in two orthogonal directions using FT of the line spread function (LSF).

For AMFPI, the shift-invariance condition is violated because the digital detectors are undersampled, which makes the PSF and LSF position-dependent. For a digital detector with pixel-sensing element width $a = 140 \ \mu m$ and pixel pitch $d = 150 \ \mu m$, the detector aperture response is a sinc function with the first zero at $f = 1/a = 7.1$ cycles/mm. The Nyquist frequency of pixel sampling is $f_N = 1/(2d) = 3.3$ cycles/mm. This means that a digital detector is always undersampled except when the frequency response of the x-ray detection material is very poor. To measure the MTF of a digital detector, the concept of presampling MTF is usually used. It describes the frequency response of the detector before sampling occurs. The standard experimental technique adopted by IEC for measuring the presampling MTF is the slanted edge method. Shown in Figure 3.9a are the measured presampling MTF of the indirect AMFPI used in the GE general radiographic tomosynthesis (200 μm pixel size) and breast tomosynthesis (100 μm pixel size) systems. The MTF for the general radiographic AMFPI is much lower than that for mammography due to two factors: (1) the larger pixel size, and (2) a thicker CsI layer,

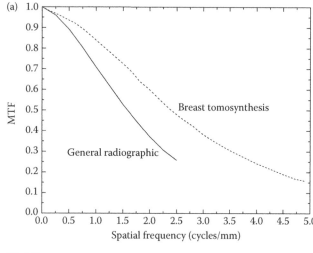

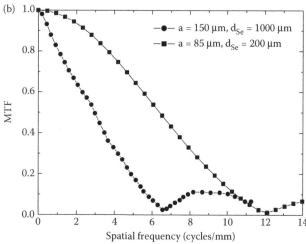

Figure 3.9 MTF of commercial AMFPI detectors used in digital radiographic and breast tomosynthesis systems. (a) Indirect FPI with pixel size of d = 200 μm for general radiographic (Adapted from P. R. Granfors and R. *Med Phys* **27**, 1324–1331, 2000.), and d = 100 μm for breast tomosynthesis. (Adapted from C. Ghetti et al. *Med Phys* **35**, 456–463, 2008.) (b) Direct FPI with d = 150 μm and 1000 μm thick a-Se for general radiographic (Reproduced from D. C. Hunt et al. *Med Phys* **31**, 1166–1175, 2004.), and d = 85 μm and 200 μm thick a-Se for breast tomosynthesis. (Adapted from B. Zhao and W. Zhao, *Med Phys* **35**, 1978–1987, PMCID: PMC2673645, 2008.)

which suffers from more light spreading between columns. The presampling MTF of direct AMFPI, on the other hand, is only limited by the pixel aperture function because there is essentially no image blur in the x-ray photoconductor. Shown in Figure 3.9b is the measured presampling MTF for a direct AMFPI with 150 μm pixel size and d_{Se} = 1000 μm (FPD14, Anrad Corp.),[30] which has similar detector parameters as the ones used in the Shimadzu tomosynthesis system. Shown in the same graph is the presampling MTF of a direct AMFPI used in DBT with 85 μm pixel size and d_{Se} = 200 μm.[31] The presampling MTF of the direct AMFPI have their first zeros at $1/d$, where d is the pixel pitch, with effective fill factor of unity. Image blur has been observed in some a-Se detectors due to charge trapping and recombination in the bulk of very thick a-Se layers or near the pixel electrode interface. This is a source of presampling blur, and could contribute to ~10–20% drop in presampling MTF at the Nyquist frequency depending on the material properties and thickness of a-Se.[29,30]

3.3.1.2.2 Noise power spectra

The noise in a projection image can be characterized in the spatial frequency domain using NPS, which is the Fourier transform of the autocorrelation of a flat-fielded x-ray image. The inherent stochastic (Poisson) noise of incident x-rays is white, that is, no spatial correlation. Image blur in an AMFPI detector could lead to spatial correlation of noise, which results in a high-frequency drop of NPS. When presampling NPS has frequency components above the Nyquist frequency of detector sampling, aliasing of NPS occurs, leading to an increase in NPS. Noise aliasing in direct conversion AMFPI results in an NPS that is essentially white. NPS of projection images can be measured experimentally using flat-fielded x-ray images under uniform x-ray exposures. For accurate measurement of NPS, the spatiotemporal behavior of detectors needs to be taken into account. Temporal performance of AMFPI (to be discussed later) such as image lag could lead to noise correlation between frames, resulting in a reduction in NPS if NPS analysis is performed using a temporal sequence of x-ray images. Two methods have been used to account for this factor: (1) measure the spatiotemporal NPS by adding time domain as a third-dimensional variable (in additional to the two spatial dimensions x and y), and determine the 2D NPS after correction of lag effect[69]; (2) measure 2D spatial domain NPS by eliminating the temporal effect, that is, at low frame rate where temporal correlation is negligible. For accurate measurement of 2D NPS, fixed patterns in projection images must be removed through offset and gain correction. Then a region of interest (ROI) $I(x, y)$ is selected from each flat-fielded image with its mean value subtracted before Fourier's transform to obtain NPS:

$$\text{NPS}(\mu, \nu) = \frac{d_x d_y}{N_x N_y} \langle |FT[I(x, y) - \bar{I}(x, y)]|^2 \rangle \qquad (3.1)$$

where $\langle \rangle$ represents the ensemble average, N_x and N_y are the number of elements in the x and y directions, respectively, and d_x and d_y are the pixel pitch in each direction.

3.3.1.2.3 Detective quantum efficiency

The overall imaging performance of an x-ray detector is best represented by its DQE. It is defined as the ratio between the SNR squared at the output of the detector and that at the input, which is equal to the number of x-ray photons per unit area q_0[68]:

$$\text{DQE} = \frac{\text{SNR}_{\text{out}}^2}{q_0} \qquad (3.2)$$

$\text{SNR}_{\text{out}}^2$ is also known as the number of noise equivalent quanta (NEQ). Hence, DQE describes the efficiency of the detector in utilizing the incident x-rays, and its upper limit is the QE η of the detection material. In order to describe the ability of the detector in transferring information with different frequency content, DQE is usually measured as a function of spatial frequency f using

$$\text{DQE}(f) = \frac{k_0 \text{MTF}(f)^2}{q_0 \text{NPS}(f)} \qquad (3.3)$$

where k_0 is the pixel x-ray response of the detector at a given x-ray exposure and NPS(f) is the NPS (Equation 3.1), which is the FT of the autocorrelation function of the detector at the same exposure level. Any additional noise source in an imaging system (e.g., detector electronic noise) increases NPS(f) from the x-ray quantum noise of ηq_0 and degrades the DQE. Because MTF(f) always decreases as a function of f, added noise (which is usually white) will cause DQE(f) to decrease with increasing f. Because it reflects both the signal and noise transfer of an imaging system, DQE(f) is regarded as the gold standard for performance comparison between different detectors.

Dose dependence of DQE(f) is another important imaging performance criteria. The DQE of AMFPI at very low exposures could be degraded due to the readout electronic noise, and it has been recognized as a major disadvantage compared to the more established x-ray detectors such as XRII, which has internal signal gain. Figure 3.10a shows the DQE(f) of the GE general radiographic tomosynthesis detector.[70] It shows that at a high radiographic exposure of 8.3 mR (72.5 μGy), DQE(0) of the detector for an RQA 5 spectrum is ~0.66, which is approaching the theoretical limit of DQE(0) = ηA_S. However, as the exposure decreases to 56 μR (0.49 μGy), which is ~25% of the mean detector exposure in radiography, DQE drops due to the degradation effect of added electronic noise. The exposure used in each view of tomosynthesis typically results in a total dose that is approximately twice that of a single general radiographic exposure.[71] Hence, a tomosynthesis scan with more than 10 views will suffer from image degradation due to electronic noise. The electronic noise can be reduced by increasing the system gain, so that the image quality of each view of tomosynthesis can be improved. Indirect AMFPI with the same detector parameters have been developed for R/F applications with lower noise electronics.[72] It is possible that these detectors may be used for tomosynthesis to further reduce the electronic noise at lower dose. Shown in Figure 3.10b is the measured DQE of a direct conversion AMFPI (FPD 14, Anrad) developed for R/F applications. The measurements were performed at fluoroscopic exposures, and it is x-ray quantum noise limited at 3.9 μR. Hence, it will be adequate for tomosynthesis, where the mean detector exposure for each view is ~10–30 μR.

Shown in Figure 3.11a are the measured DQE of an indirect AMFPI (GE Senographe Essential) used in DBT.[73] It shows that at the high end of screening mammography exposure of 26.5 mR (231 μGy), the DQE is x-ray quantum noise limited. As the exposure decreases to 2.87 mR (25 μGy), the DQE is degraded by electronic noise. The magnitude of this effect is more pronounced at high frequencies because the electronic noise is white, whereas the x-ray quantum noise decreases with frequency in indirect AMFPI. Shown in Figure 3.11b are the measured DQE of a direct AMFPI with 85-μm pixel pitch.[31] It shows that the detector is quantum noise limited for exposures greater than 1 mR. When exposure decreases, the effect of electronic noise can be seen.

3.3.2 TEMPORAL PERFORMANCE OF DIFFERENT X-RAY DETECTOR TECHNOLOGIES

Temporal performance of AMFPI is crucial for minimizing image artifact in tomosynthesis. Temporal imaging characteristics of AMFPI can be divided into two categories:

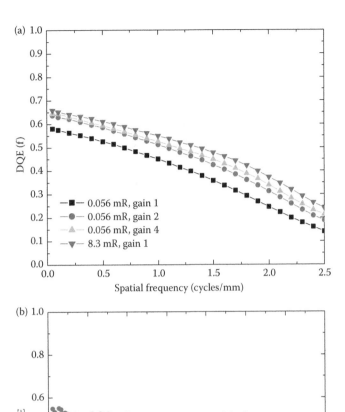

Figure 3.10 (a) DQE for an indirect AMFPI developed for radiographic applications with pixel size of 200 μm pixel size. DQE degradation due to electronic noise can be mitigated by higher gain of the readout electronics. (Adapted from P. R. Granfors and R. *Med Phys* **27**, 1324–1331, 2000.) (b) DQE for a direct AMFPI with pixel size of 150 μm pixel size and 1000 μm thick a-Se layer. The DQE shows x-ray quantum noise limited performance for exposures above 4 μR. (Adapted from O. Tousignant et al. *Proc SPIE* **5745**, 207–215, 2005.)

lag and ghosting. As shown in Figure 3.12, lag is the carryover of image charge generated by previous x-ray exposures into subsequent image frames. It is manifested as changes in dark images, that is, readout of the detector without an x-ray exposure. As shown in Figure 3.13, ghosting is the change of x-ray sensitivity, or gain, of the detector as a result of previous exposures to radiation. It can only be seen with subsequent x-ray exposures. Both lag and ghosting could lead to image artifacts in projection and reconstructed images in tomosynthesis.[74,75] An overview of the physical mechanism for and the measurement of lag and ghosting will be provided here for both indirect and direct AMFPI. Because of availability and completeness of measurement data, the examples used in this section are real-time AMFPI originally developed for R/F applications. They could be regarded as the upper limit of temporal performance in tomosynthesis.

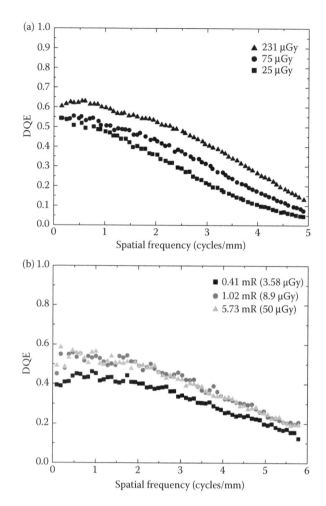

(a)

(b)

Figure 3.11 Measured DQE for AMFPI used in digital breast tomosynthesis systems. (a) Indirect AMFPI with 100 μm pixel size (GE Senographe Essential). (Adapted from C. Ghetti et al. *Med Phys* **35**, 456–463, 2008.) (b) Direct AMFPI with 85 μm pixel size (Anrad LMAM). (Adapted from B. Zhao and W. Zhao, *Med Phys* **35**, 1978–1987, PMCID: PMC2673645, 2008.) (*Note*: 1 mR = 8.73 μGy).

3.3.2.1 Temporal performance of indirect AMFPI

The lag and ghosting of indirect AMFPI can be attributed to three sources of mechanisms: (1) charge trapping and release in a-Si PD, (2) afterglow from the CsI scintillator, and (3) incomplete readout of charge from the pixel to the charge

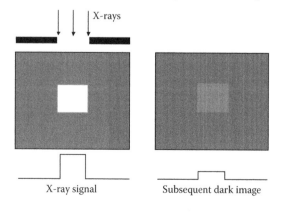

Figure 3.12 Conceptual images showing lag of an x-ray imaging system. Lag is defined as the residual signal from the detector's previous exposure to radiation. It is manifested as an enhanced signal in a subsequent dark image (acquired without x-rays).

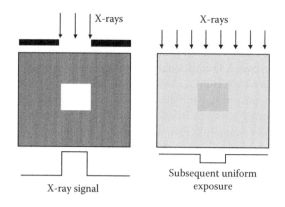

Figure 3.13 Conceptual images showing ghosting of an x-ray imaging detector. Ghosting is defined as the change in x-ray sensitivity as a result of the detector's exposure to radiation. It can only be seen with subsequent x-ray exposures.

amplifiers (when $T_{on} < 5RC$).[76] During x-ray exposure, the a-Si PD is biased with an electric field in order for image charge to be collected efficiently. Since electrons in a-Si have better charge transport properties, a-Si PDs are usually biased negatively at the light-entrance side. When electrons move toward pixel electrodes, they could be captured by localized states (traps) in the a-Si material, and released at a later time, for example, during the subsequent image frames. Lag has been investigated extensively under different imaging conditions, for example, detector exposure and frame rate.[77] Shown in Figure 3.14a is the relative signal intensity measured from an indirect AMFPI (Varian 2020) with the x-ray exposure delivered to frame zero, whose signal is set as the reference level (100%).[78] It shows that the first frame lag depends on the frame rate, and ranged between 2% and 10% depending on operational conditions and entrance exposure. The time required for a trapped charge to be released depends on the energy depth of the traps. Shallow traps are responsible for short-term lag, and deep traps for long-term residual signal, which could be visible tens of minutes after exposures, and the magnitude of long-term lag depends on the degree of pixel saturation and frame time.[77] Usually, lag is severer at higher exposures when the electric field across a-Si PDs nearly collapses due to pixel saturation. This is because charge is more likely to be trapped under low electric field.

Ghosting of indirect AMFPI has been observed as an increase in x-ray sensitivity after the detector is exposed to radiation.[76] Shown in Figure 3.14b is the relative x-ray sensitivity of indirect AMFPI as a function of exposure, which exhibits a 2% increase in x-ray sensitivity even at 10 s after x-ray exposure of 20 μGy. This is because the charge trapped in a-Si due to previous radiation exposure fills the traps and reduces the probability of further charge trapping in subsequent exposures, whereas a "rested" (i.e., no recent history of radiation exposure) detector experiences reduction in x-ray sensitivity due to charge trapping. To alleviate ghosting due to this mechanism, reset light exposure has been implemented, where short pulses (~100 μs) of light delivered between x-ray exposures would generate charge to fill the traps in a-Si PD, thereby minimize the probability of charge trapping during x-ray exposure.[76] As shown in Figure 3.14b, the longer the reset light duration (RLD), the lower the sensitivity ghost. This was also found to improve the x-ray sensitivity of AMFPI,

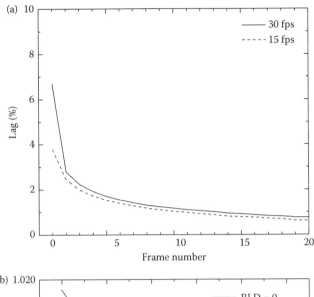

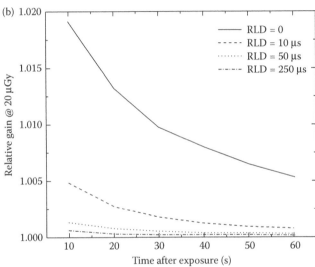

which is manifested as lag, that is, elevated dark signal, after radiation exposure. Shown in Figure 3.15a are lag measurements from a real-time a-Se AMFPI (FPD14, Anrad Corporation), as well as an a-Se layer identical to that used in AMFPI but without TFT readout.[82] The first frame lag (30 fps) of the AMFPI depends on the radiation exposure, and increases from 1.7% at 48 μR to 3.9% at 384 μR. Ghosting in a-Se detectors is manifested as a reduction in x-ray sensitivity. This is due to the recombination between previously trapped electrons in the bulk of a-Se and the x-ray-generated free holes.[83] It increases as a function of radiation dose and decreases with increasing electric field E_{Se}. Shown in Figure 3.15b are the quantitative measurements of ghosting in the same a-Se AMFPI (FPD14), where the relative x-ray sensitivity is measured as a function of time after the x-ray has been delivered at a rate of 33 mR/min. It shows that the x-ray sensitivity continues to decrease with accumulation of exposure. Sensitivity recovery, or ghosting erasure, can be achieved through charge

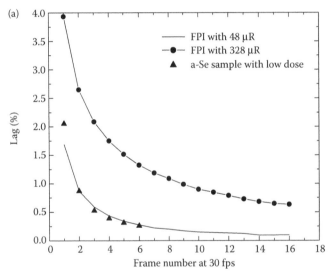

Figure 3.14 Measured lag for: (a) indirect AMFPI (Varian 2020) with frame rates of 15 and 30 fps (Reproduced from C. A. Tognina et al. *Proc SPIE* **5368**, 648–656, 2004.); (b) ghosting of indirect AMFPI (Trixel dynamic FPI) (Reproduced from M. Overdick et al., *Proc SPIE* **4320**, 47–58, 2001.).

compared to that without reset light, by ~8% for RLD > 100 μs. An alternative method developed to overcome lag and ghosting caused by charging trapping is to put a-Si PDs in the forward-bias condition for a short period between two subsequent exposures. Forward-bias causes a large number of charge carriers injected from the bias electrodes of a-Si PD, and fill the traps before the next x-ray exposure.[79]

3.3.2.2 Temporal performance of direct AMFPI

Lag and ghosting in a-Se AMFPI are due to charge trapping in the bulk of the a-Se layer or at the interface between a-Se and the pixel electrodes.[80,81] Comparison of temporal performance of complete AMFPI and a-Se samples (without TFT readout) showed that the dominant factors are the charge trapping and recombination in the a-Se layer.[82] The drift mobility of electrons in a-Se is only ~1/50 of that of holes, and there are a large number of deep electron traps, which could capture electrons for up to several hours. Trapped electrons enhance the electric field near the positive bias electrode, and increase the injection of holes,

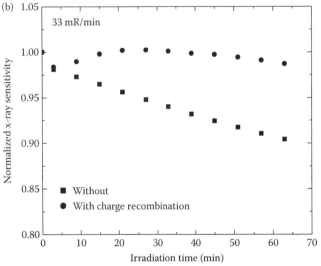

Figure 3.15 Temporal performance of a-Se-based AMFPI (a) lag for FPI at two different exposures, as well as for a matching a-Se layer (without TFT readout) at the lower exposure; (b) ghosting, i.e. measurement of relative x-ray sensitivity, as a function of radiation exposure. (Adapted from O. Tousignant et al. *Proc SPIE* **5745**, 207–215, 2005.)

recombination technique. It has been shown previously that the injection of holes into the bulk of a-Se between subsequent exposures accelerates the recovery of x-ray sensitivity because the trapped electrons are neutralized through recombination with holes.[81,84,85] This approach is different from the ghost erasure method used for indirect AMFPI, where saturation of electron traps through injection of charge carriers was shown to be the effective mechanism.

3.3.3 IMAGE BLUR DUE TO TOMOSYNTHESIS ACQUISITION

The effects of tomosynthesis acquisition on projection image quality have been a major concern for DBT because of the requirement for high spatial resolution for the visualization of microcalcifications. Since the majority of DBT systems use a stationary detector while the x-ray tube travels in an arc above the compressed breast, the following factors would cause additional image blur: (1) oblique entry of x-rays; (2) additional focal spot blur with continuous tube travel during radiation exposures; and (3) pixel binning, which increases readout speed at the cost of detector resolution. Figure 3.16 compares the presampling MTF due to several factors related to tomosynthesis acquisition,[31] and how they compare with the inherent detector resolution. It shows that with full-resolution readout, the MTF due to focal spot motion (FSM) of 0.65 mm is slightly higher than the inherent detector MTF. It will result in a moderate decrease of MTF of ~20% at the Nyquist frequency f_{NY} = 5.88 cycles/mm. In detector binning mode, the faster readout led to faster gantry travel speed, and hence worse MTF due to FSM of 1.15 mm. However, it still remains relatively higher than the inherent MTF in binning mode.

Oblique entry of x-rays in a DBT system with stationary detectors can result in additional blur, and this effect has been investigated for both direct and indirect AMFPI.[31,86] Shown in Figure 3.16 is the estimated MTF due to an oblique entry angle of 30°, which corresponds to the far side of the detector with the x-ray tube column at the angle of 22. This source of blur causes MTF to decrease by 28% at f_{NY} = 5.88 cycles/mm for a direct conversion a-Se AMFPI with 85 μm pixel size. It is seen from Figure 3.16 that although each acquisition-related factor alone is not the dominant source of blur compared to the inherent resolution of the detector, it degrades the overall MTF of the projection image. In addition, since the blur due to FSM and oblique entry of x-rays are for primary (x-ray) photons, the NPS remains white. This leads to degradation of DQE. Therefore, these factors need to be considered during optimization of DBT image acquisition.

REFERENCES

1. J. T. Dobbins, III and D. J. Godfrey, Digital x-ray tomosynthesis: Current state of the art and clinical potential, *Phys Med Biol* **48**, R65–R106, 2003.
2. H. Machida, T. Yuhara, M. Tamura, T. Numano, S. Abe, J. M. Sabol, S. Suzuki, and E. Ueno, Radiation dose of digital tomosynthesis for sinonasal examination: Comparison with multi-detector CT, *Eur J Radiol* **81**(6), 1140–1145, 2012.
3. M. J. Flynn, R. McGee, and J. Blechinger, Spatial resolution of x-ray tomosynthesis in relation to computed tomography for coronal/sagittal images of the knee, *Proc SPIE* **6510**, 65100D–65109D, 2007.
4. J. S. Ksar, J. Craig, M. Flynn, K. Lange, F. Nelson, and M. Van Holsbeeck, Digital tomosynthesis: A new radiographic technique for orthopedic imaging, *Am J Roentgenol* **188**, 2007.
5. J. T. Dobbins, 3rd and H. P. McAdams, Chest tomosynthesis: Technical principles and clinical update, *Eur J Radiol* **72**, 244–251, 2009.
6. W. Zhao and J. A. Rowlands, X-ray imaging using amorphous selenium: Feasibility of a flat panel self-scanned detector for digital radiology, *Med Phys* **22**, 1595–1604, 1995.
7. L. E. Antonuk, J. Boudry, W. Huang, D. L. McShan, E. J. Morton, J. Yorkston, M. J. Longo, and R. A. Street, Demonstration of megavoltage and diagnostic x-ray imaging with hydrogenated amorphous silicon arrays, *Med Phys* **19**, 1455–1466, 1992.
8. J. A. Rowlands and J. Yorkston, Flat panel detectors for digital radiography, in *Medical Imaging: Volume 1 Physics and Psychophysics, Psychophysics,* edited by H. K. a. R. V. M. J. Beutel, SPIE, Bellingham, 2000, pp. 223–328.
9. W. Zhao, G. Ristic, and J. A. Rowlands, X-ray imaging performance of structured cesium iodide scintillators, *Med Phys* **31**, 2594–2605, 2004.
10. R. L. Weisfield, W. Yao, T. Speaker, K. Zhou, R. E. Colbeth, and C. Proano, Performance analysis of a 127-micron pixel large-area TFT/photodiode array with boosted fill factor, *Proc SPIE* **5368**, 338–348, 2004.
11. A. Koch, J. M. Macherel, T. Wirth, P. De Groot, T. Ducourant, D. Couder, J. P. Moy, and E. Calais, Presented at the *Proc SPIE*, San Diego, CA, 2001 (unpublished).
12. D. A. Jaffray and J. H. Siewerdsen, Cone-beam computed tomography with a flat-panel imager: Initial performance characterization, *Med Phys* **27**, 1311–1323, 2000.
13. I. D. Job, N. Taie-Nobraie, R. E. Colbeth, I. Mollov, K. D. Gray, C. Webb, J. M. Pavkovich, F. Zoghi, C. A. Tognina, and P. G. Roos, Improved DQE by means of x-ray spectra and scintillator optimization for FFDM, *Proc SPIE* **8313**, 83135T–83138T, 2012.
14. D. Albagli, S. Han, A. Couture, H. Hudspeth, C. Collazo, and P. Granfors, Performance of optimized amorphous silicon, cesium-iodide based large field-of-view detector for mammography, *Proc SPIE*, **5745**, 1078–1086, 2005.
15. R. K. Swank, Measurement of absorption, and noise in an x-ray image intensifier, *J Appl Phys* **45**, 3673–3678, 1974.

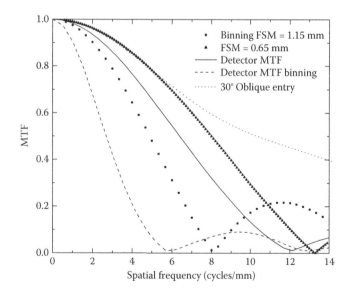

Figure 3.16 A comparison of MTF due to detector inherent resolution, focal spot motion, and oblique entry angle. Graph was adapted from Ref. 31 by Zhao et al., on the measurements of a Siemens prototype DBT system equipped with an a-Se direct conversion AMFPI with 85 μm pixel size and 200 μm thick a-Se.

16. R. L. Weisfield and N. R. Bennett, Electronic noise analysis of a 127-mu m pixel TFT/photodiode array, *Proc SPIE* **4320**, 209–218, 2001.

17. W. den Boer, S. Aggas, Y. H. Byun, T. Gu, J. Q. Zhong, S. V. Thomsen, L. S. Jeromin, and D. L. Y. Lee, Thin-film transistor array technology for high-performance direct-conversion x-ray sensors, *Proc SPIE* **3336**, 520–528, 1998.

18. M. Z. Kabir, F. Manouchehri, S. A. Mahmood, V. K. Devabhaktuni, O. Tousignant, H. Mani, J. Greenspan, and P. Botka, Modeling of dark current and ghosting in multilayer amorphous selenium X-ray detectors, *Proc SPIE* **6913**, 69133, 2008.

19. D. L. Y. Lee, L. K. Cheung, B. G. Rodricks, and G. F. Powell, Improved imaging performance of a 14″ × 17″ direct radiography system using a Se/TFT detector, *Proc SPIE* **3336**, 14–23, 1998.

20. E. L. Gingold, D. L. Y. Lee, L. S. Jeromin, B. G. Rodricks, M. G. Hoffberg, and C. L. Williams, Development of a novel high-resolution direct conversion x-ray detector, *Proc SPIE* **3977**, 185–193, 2000.

21. L. K. Cheung, Z. Jing, S. Bogdanovich, K. Golden, S. Robinson, E. Beliaevskaia, and S. Parikh, Image performance of a new amorphous selenium flat panel x-ray detector designed for digital breast tomosynthesis, *Proc SPIE* **5745**, 1282–1290, 2005.

22. A. Tsukamoto, S. Yamada, T. Tomisaki, M. Tanaka, T. Sakaguchi, H. Asahina, K. Suzuki, and M. Ikeda, Development and evaluation of a large-area selenium-based flat-panel detector for real-time radiography and fluoroscopy, *Proc SPIE* **3659**, 14–23, 1999.

23. J. Lehnert and W. Zhao, High voltage protection in active matrix flat-panel imagers, *Proc SPIE* **6142**, 61420, 2006.

24. W. Zhao, J. Law, D. Waechter, Z. Huang, and J. A. Rowlands, Digital radiology using active matrix readout of amorphous selenium: detectors with high voltage protection, *Med Phys* **25**, 539–549, 1998.

25. B. Polischuk, K. W. H. Rougeot, A. Debrie, E. Poliquin, J. P. Martin, T. T. Truong, M. Choquette, L, Laperri, and Z. Shuki, Direct conversion detector for digital mammography, *Proc SPIE* **3659**, 417–425, 1999.

26. F. Manouchehri, M. Z. Kabir, O. Tousignant, H. Mani, and V. K. Devabhaktuni, Time and exposure dependent x-ray sensitivity in multilayer amorphous selenium detectors, *J Phys D Appl Phys* **41**, 235106, 2008.

27. J. A. Rowlands, G. DeCrescenzo, and N. Araj, X-ray imaging using amorphous selenium: Determination of x-ray sensitivity by pulse height spectroscopy, *Med Phys* **19**, 1065–1069, 1992.

28. G. Pang, W. Zhao, and J. A. Rowlands, Digital radiology using active matrix readout of amorphous selenium: Geometrical and effective fill factors, *Med Phys* **25**, 1636–1646, 1998.

29. W. Zhao, W. G. Ji, A. Debrie, and J. A. Rowlands, Imaging performance of amorphous selenium based flat-panel detectors for digital mammography: Characterization of a small area prototype detector, *Med Phys* **30**, 254–263, 2003.

30. D. C. Hunt, O. Tousignant, and J. A. Rowlands, Evaluation of the imaging properties of an amorphous selenium-based flat panel detector for digital fluoroscopy, *Med Phys* **31**, 1166–1175, 2004.

31. B. Zhao and W. Zhao, Imaging performance of an amorphous selenium digital mammography detector in a breast tomosynthesis system, *Med Phys* **35**, 1978–1987, PMCID: PMC2673645, 2008.

32. M. E. Myronakis and D. G. Darambara, Monte Carlo investigation of charge-transport effects on energy resolution and detection efficiency of pixelated CZT detectors for SPECT/PET applications, *Med Phys* **38**, 455–467, 2011.

33. H. Peng and C. S. Levin, Design study of a high-resolution breast-dedicated PET system built from cadmium zinc telluride detectors, *Phys Med Biol* **55**, 2761–2788, 2010.

34. K. Taguchi, M. Zhang, E. C. Frey, X. Wang, J. S. Iwanczyk, E. Nygard, N. E. Hartsough, B. M. Tsui, and W. C. Barber, Modeling the performance of a photon counting x-ray detector for CT: Energy response and pulse pileup effects, *Med Phys* **38**, 1089–1102, 2011.

35. J. T. Rahn, F. Lemmi, J. P. Lu, P. Mei, R. B. Apte, R. A. Street, R. Lujan, R. L. Weisfield, and J. A. Heanue, High resolution x-ray imaging using amorphous silicon flat-panel arrays, *IEEE Trans Nucl Sci* **46**, 457–461, 1999.

36. R. A. Street, S. E. Ready, J. T. Rahn, M. Mulato, K. Shah, P. R. Bennett, P. Mei et al., High resolution, direct detection X-ray imagers, *Proc SPIE* **3977**, 418–428, 2000.

37. M. Mulato, F. Lemmi, R. Lau, J. P. Lu, J. Ho, S. E. Ready, J. T. Rahn, and R. A. Street, Charge collection and capacitance in continuous film flat panel detectors, *Proc SPIE* **3977**, 26–37, 2000.

38. R. A. Street, S. E. Ready, K. Van Schuylenbergh, J. Ho, J. B. Boyce, P. Nylen, K. Shah, L. Melekhov, and H. Hermon, Comparison of PbI_2 and HgI_2 for direct detection active matrix x-ray image sensors, *J Appl Phys* **91**, 3345, 2002.

39. Y. Kang, L. E. Antonuk, Y. El-Mohri, L. Hu, Y. Li, A. Sawant, Z. Su, Y. Wang, J. Yamamoto, and Q. Zhao, Examination of PbI_2 and HgI_2 photoconductive materials for direct detection, active matrix, flat-panel imagers for diagnostic X-ray imaging, *IEEE Trans Nucl Sci* **52**, 38–45, 2005.

40. M. Simon, R. A. Ford, A. R. Franklin, S. P. Grabowski, B. Menser, G. Much, A. Nascetti, M. Overdick, M. J. Powell, and D. U. Wiechert, PbO as direct conversion X-ray detector material, *Proc SPIE* **5368**, 188–199, 2004.

41. H. Du, L. E. Antonuk, Y. El-Mohri, Q. Zhao, Z. Su, J. Yamamoto, and Y. Wang, Investigation of the signal behavior at diagnostic energies of prototype, direct detection, active matrix, flat-panel imagers incorporating polycrystalline HgI2, *Phys Med Biol* **53**, 1325–1351, 2008.

42. S. Tokuda, S. Adachi, T. Sato, T. Yoshimuta, H. Nagata, K. Uehara, Y. Izumi, O. Teranuma, and S. Yamada, Experimental evaluation of a novel CdZnTe flat-panel x-ray detector for digital radiography and fluoroscopy, *Proc SPIE* **4320**, 140–147, 2001.

43. R. A. Street, S.E. Ready, L. Melkhov, J. Ho, A. Zuck, and B. Breen, Approaching the theoretical x-ray sensitivity with HgI_2 direction image sensors, *Proc SPIE* **4682**, 414–422, 2002.

44. M. Z. Kabir and S. O. Kasap, Charge collection and absorption-limited sensitivity of x-ray photoconductors: Applications to a-Se and HgI2, *Appl Phys Lett* **80**, 1664, 2002.

45. M. Z. Kabir, Effects of charge carrier trapping on polycrystalline PbO x-ray imaging detectors, *J Appl Phys* **104**, 074506, 2008.

46. D. Scheffer, A wafer scale active pixel CMOS image sensor for generic X-ray radiology, *Medical Imaging 2007: Physics of Medical Imaging, February 18, 2007–February 22, 2007* **6510**, SPIE, 2007.

47. S. K. Heo, J. Kosonen, S. H. Hwang, T. W. Kim, S. Yun, and H. K. Kim, 12-inch-wafer-scale CMOS active-pixel sensor for digital mammography, *Proc SPIE* **7961**, 79610O, 2011.

48. S. E. Bohndiek, A. Blue, J. Cabello, A. T. Clark, N. Guerrini, P. M. Evans, E. J. Harris et al., Characterization and testing of LAS: A prototype 'large area sensor' with performance characteristics suitable for medical imaging applications, *IEEE Trans Nucl Sci* **56**, 2938–2946, 2009.

49. M. Farrier, T. Graeve Achterkirchen, G. P. Weckler, and A. Mrozack, Very large area CMOS active-pixel sensor for digital radiography, *IEEE Trans Electron Devices* **56**, 2623–2631, 2009.

50. S. Naday, E. F. Bullard, S. Gunn, J. E. Brodrick, E. O. O'Tuairisg, A. McArthur, H. Amin, M. B. Williams, P. G. Judy, and A. Konstantinidis, Optimised breast tomosynthesis with a novel CMOS flat panel detector, in *10th International Workshop on Digital Mammography, IWDM 2010, June 16, 2010–June 18, 2010, Vol. 6136 LNCS*, Springer Verlag, Girona, Catalonia, Spain, 2010, pp. 428–435.

System design

51. T. Patel, K. Klanian, Z. Gong, and M. B. Williams, Detective quantum efficiency of a CsI-CMOS X-ray detector for breast tomosynthesis operating in high dynamic range and high sensitivity modes, in *11th International Workshop on Breast Imaging, IWDM 2012, July 8, 2012–July 11, 2012, Vol. 7361 LNCS,* Springer Verlag, Philadelphia, PA, United States, 2012, pp. 80–87.

52. M. P. Andre, B. A. Spivey, P. J. Martin, A. L. Morsell, E. Atlas, and T. Pellegrino, Integrated CMOS-selenium x-ray detector for digital mammography, *Proc SPIE* **3336**, 204–209, 1998.

53. M. Arques, S. Renet, A. Brambilla, G. Feuillet, A. Gasse, N. Billon-Pierron, M. Jolliot, L. Mathieu, and P. Rohr, Fluoroscopic x-ray demonstrator using a CdTe polycrystalline layer coupled to a CMOS readout chip, *Proc SPIE* **7622**, 76221 K–76221 K, 2010.

54. A. El Gamal and H. Eltoukhy, CMOS image sensors, *IEEE Circuits Devices Magaz* **21**, 6–20, 2005.

55. E. R. Fossum, CMOS image sensors: Electronic camera-on-a-chip, *IEEE Trans Electron Devices* **44**, 1689–1698, 1997.

56. A. J. P. Theuwissen, CMOS image sensors: State-of-the-art, *Solid-State Electron* **52**, 1401–1406, 2008.

57. S. Naday, E. F. Bullard, S. Gunn, J. E. Brodrick, E. O. O'Tuairisg, A. McArthur, H. Amin, M. B. Williams, P. G. Judy, and A. Konstantinidis, Optimised breast tomosynthesis with a novel CMOS flat panel detector, 10th International Workshop on Digital Mammography, IWDM 2010, June 16, 2010–June 18, 2010, 6136 LNCS, 428–435, 2010.

58. M. Esposito, T. Anaxagoras, A. Fant, K. Wells, A. Konstantinidis, J. P. F. Osmond, P. M. Evans, R. D. Speller, and N. M. Allinson, DynAMITe: A wafer scale sensor for biomedical applications, *J Instrum* **6**, C12064–C12064, 2011.

59. J. S. Iwanczyk, E. Nygard, O. Meirav, J. Arenson, W. C. Barber, N. E. Hartsough, N. Malakhov, and J. C. Wessel, Photon counting energy dispersive detector arrays for X-ray imaging, *IEEE Nuclear Science Symposium Conference Record,* 2741–2748, 2007.

60. H. Bornefalk and M. Lundqvist, Dual-energy imaging using a photon counting detector with electronic spectrum-splitting, *Proc SPIE* **6142**, 1–11, 2006, .

61. G. Blanchot, M. Chmeissani, A. Díaz, F. Díaz, J. Fernández, and E. García, Dear-Mama: A photon counting X-ray imaging project for medical applications, *Nucl Instrum Meth* **A569**, 136, 2006.

62. S. Thunberg, T. Franckep, J. Egerstroem, M. Eklund, and L. Ericsson, Evaluation of a photon counting mammography system, *Proc SPIE* **4682**, 202–208, 2006.

63. N. Dahlman, E. Fredenberg, M. Aslund, M. Lundqvist, F. Diekmann, and M. Danielsson, Evaluation of photon-counting spectral breast tomosynthesis, *Proc SPIE* **7961**, 796114, 2011.

64. A. D. A. Maidment, C. Ullberg, K. Lindman, L. Adelow, J. Egerstrom, M. Eklund, T. Francke et al., Evaluation of a photon-counting breast tomosynthesis imaging system, *Proc SPIE* **6142**, 61420B–61411, 2006.

65. F. F. Schmitzberger, E. M. Fallenberg, R. Lawaczeck, M. Hemmendorff, E. Moa, M. Danielsson, U. Bick et al., Development of low-dose photon-counting contrast-enhanced tomosynthesis with spectral imaging, *Radiology,* **259**(2), 558–564, 2011.

66. E. Fredenberg, M. Lundqvist, M. Aslund, M. Hemmendorff, B. Cederstrom, and M. Danielsson, A photon-counting detector for dual-energy breast tomosynthesis, *Proc SPIE* **7258**, 72581 J, 2009.

67. A. C. Konstantinidis, M. B. Szafraniec, R. D. Speller, and A. Olivo, The Dexela 2923 CMOS X-ray detector: A flat panel detector based on CMOS active pixel sensors for medical imaging applications, *NIMA* **689**, 12–21, 2012.

68. IEC62220–1:2004, *Medical Electrical Equipment—Characteristics of Digital X-Ray Imaging Devices—Part 1: Determination of the Detective Quantum Efficiency.*

69. S. N. Friedman and I. A. Cunningham, A spatio-temporal detective quantum efficiency and its application to fluoroscopic systems, *Med Phys* **37**, 6061–6069, 2010.

70. P. R. Granfors and R. Aufrichtig, Performance of a 41X41-cm2 amorphous silicon flat panel x-ray detector for radiographic imaging applications, *Med Phys* **27**, 1324–1331, 2000.

71. C. Canella, P. Philippe, V. Pansini, J. Salleron, R. M. Flipo, and A. Cotten, Use of tomosynthesis for erosion evaluation in rheumatoid arthritic hands and wrists, *Radiology* **258**, 199–205, 2011.

72. P. R. Granfors, R. Aufrichtig, G. E. Possin, B. W. Giambattista, Z. S. Huang, J. Liu, and B. Ma, Performance of a 41 × 41 cm^2 amorphous silicon flat panel x-ray detector designed for angiographic and R&F imaging applications, *Med Phys* **30**, 2715, 2003.

73. C. Ghetti, A. Borrini, O. Ortenzia, R. Rossi, and P. L. Ordonez, Physical characteristics of GE senographe essential and DS digital mammography detectors, *Med Phys* **35**, 456–463, 2008.

74. J. G. Mainprize, X. Wanga, and M. J. Yaffea, The effect of lag on image quality for a digital breast tomosynthesis system, *Proc SPIE* **7258**, 72580R, 2009.

75. A.-K. Carton, S. Puong, R. Iordache, and S. Muller, Effects of image lag, and scatter for dual-energy contrast-enhanced digital breast tomosynthesis using a CsI flat-panel based system, *Proc SPIE* **7961**, 79611D, 2011.

76. M. Overdick, T. Solf, and H. Wischmann, Temporal artifacts in flat-dynamic x-ray detectors, *Proc SPIE* **4320**, 47–58, 2001.

77. J. H. Siewerdsen and D. A. Jaffray, A ghost story: Spatio-temporal response characteristics of an indirect-detection flat-panel imager, *Med Phys* **26**, 1624–1641, 1999.

78. C. A. Tognina, I. Mollov, J. M. Yu, C. Webb, P. G. Roos, M. Batts, D. Trinh et al., Design and performance of a new a-Si flat-panel imager for use in cardiovascular and mobile C-arm imaging systems, *Proc SPIE* **5368**, 648–656, 2004.

79. I. Mollov, C. Tognina, and R. Colbeth, Photodiode forward bias to reduce temporal effects in a-Si based flat panel detectors, *Proc SPIE* **6913**, 69133S, 2008.

80. W. Zhao, G. DeCrescenzo, and J. A. Rowlands, Investigation of lag and ghosting in amorphous selenium flat-panel x-ray detectors, *Proc SPIE* **4682**, 9–20, 2002.

81. W. Zhao, G. DeCrescenzo, S. O. Kasap, and J. A. Rowlands, Ghosting caused by bulk charge trapping in direct conversion flat-panel detectors using amorphous selenium, *Med Phys* **32**, 488–500, 2005.

82. O. Tousignant, Y. Demers, L. Laperriere, H. Mani, P. Gauthier, and J. Leboeuf, Spatial and temporal image characteristics of a real-time large area a-Se x-ray detector, *Proc SPIE* **5745**, 207–215, 2005.

83. B. Fogal, M. Z. Kabir, S. K. O'Leary, R. E. Johanson, and S. O. Kasap, X-ray-induced recombination effects in a-Se-based x-ray photoconductors used in direct conversion x-ray sensors, *J Vac Sci Technol A* **22**, 1005–1009, 2004.

84. B. Zhao and W. Zhao, Temporal performance of amorphous selenium mammography detectors, *Med Phys* **32**, 128–136, 2005.

85. S. A. Mahmood, M. Z. Kabir, O. Tousignant, and H. Mani, Ghosting and its recovery mechanisms in multilayer selenium detectors for mammography, *Proc SPIE* **7258**, 725860–725811, 2009.

86. J. G. Mainprize, A. K. Bloomquist, M. P. Kempston, and M. J. Yaffe, Resolution at oblique incidence angles of a flat panel imager for breast tomosynthesis, *Med Phys* **33**, 3159–3164, 2006.

87. H. Machida, T. Yuhara, T. Mori, E. Ueno, Y. Moribe, and J. M. Sabol, Optimizing parameters for flat-panel detector digital tomosynthesis, *Radiographics* **30**, 549–562, 2010.

88. F. Diekmann, H. Meyer, S. Diekmann, S. Puong, S. Muller, U. Bick, and P. Rogalla, Thick slices from tomosynthesis data sets: Phantom study for the evaluation of different algorithms, *J Digit Imaging* **22**, 519–526, 2009.

89. B. Ren, C. Ruth, T. Wu, Y. Zhang, A. Smith, L. Niklason, C. Williams, E. Ingal, B. Polischuk, and Z. Jing, A new generation FFDM/tomosynthesis fusion system with selenium detector, *Proc SPIE* **7622**, 76220B, 2010.

90. S. Vecchio, A. Albanese, P. Vignoli, and A. Taibi, A novel approach to digital breast tomosynthesis for simultaneous acquisition of 2D and 3D images, *Eur Radiol* **21**, 1207–1213, 2011.

4 Patient dose

Ioannis Sechopoulos

Contents

4.1 INTRODUCTION

Before a new medical imaging technology involving ionizing radiation is introduced to clinical use, a comprehensive understanding of its dosimetric characteristics is necessary. This is especially true in imaging technologies that have the potential to be used for screening of the general population, as could be the case for breast tomosynthesis (Poplack et al. 2007). In addition, estimation tools and data on dose and how it varies with the different possible acquisition parameters are necessary to be able to optimize the technology so as to maximize image quality while minimizing radiation dose. Therefore, many studies have been performed on the dose involved in tomosynthesis imaging, and how changes in the values of various acquisition parameters impact it.

This chapter will introduce the basic concepts of radiation dosimetry metrics relevant to tomosynthesis imaging and how these can be measured and/or estimated. Then the ways in which image acquisition parameters involved in tomosynthesis imaging affect the resulting dose will be discussed. Finally, application-specific patient dosimetry will be reviewed.

4.2 RADIATION DOSE METRICS

Many different metrics, and their corresponding units, are used when discussing patient dose in medical imaging involving ionizing radiation. As we shall see, these metrics are not interchangeable as they represent different physical phenomena or attempt to represent different biological effects. The following are the relevant metrics for the study of patient dose in tomosynthesis imaging.

4.2.1 EXPOSURE

Exposure, denoted with the symbol X, is traditionally defined as the number of electron–ion pairs formed by incident x-rays per unit volume of dry air. The traditional unit of exposure is the *roentgen* (R). The SI definition of exposure, however, refers to the amount of charge created by incident x-rays per unit mass of air, and therefore has the units of coulomb per kilogram (C/kg). The conversion between roentgen and coulomb per kilogram is an exact one: $1\ R \equiv 2.58 \times 10^{-4}$ C/kg.

Given its definition, exposure has traditionally been a very popular metric for communicating the quantity and quality of an x-ray beam since it is easily and directly measured with an ionization chamber. However, the definition of exposure is limited to x-rays and air, and so its use has been phased out in favor of the *air kerma*.

4.2.2 KERMA AND AIR KERMA

Kerma, an acronym for kinetic energy released in matter, is defined as the amount of kinetic energy transferred to charged particles (in x-ray imaging, electrons) by the incident radiation per unit mass of matter. The unit of kerma is joule per kilogram (J/kg), which is given the special name *gray* (Gy). Of course, *air kerma* refers to the kerma in air, and, as mentioned above, it is the current metric used in place of exposure. Conversion between exposure (in R or C/kg) and air kerma is possible, using $1\ R \equiv 2.58 \times 10^{-4}$ C/kg = 8.76 mGy air kerma. Although this conversion value varies a small amount with incident x-ray energy, this conversion value is valid for x-ray energies in the diagnostic range.

4.2.3 ABSORBED DOSE

The primary metric that relates to biologic damage is the *absorbed dose*, defined as the amount of energy deposited in matter per unit mass. The absorbed dose is denoted by the letter D, and also has units of gray (Gy = J/kg). When the term "dose" is used with no qualifier, then this is usually a reference to absorbed dose.

Absorbed dose is related to both deterministic and stochastic risks. When referring to the same type of ionizing radiation (e.g., x-rays) and the same organ (e.g., lungs), then absorbed dose is directly related to biological damage. However, the same absorbed dose due to different incident radiation (e.g., x-rays vs. alpha particles) can result in different stochastic risk. This difference in impact due to different types of incident radiation is addressed by the introduction of the concept of *equivalent dose*. In addition, the organ in which the absorbed dose is deposited can also affect the resultant stochastic risk, since different organs have varying *radiosensitivity*. This variation is addressed by the *effective dose*.

As will be discussed later, equivalent and effective dose are nonphysical metrics, but arrived at by convention. Absorbed dose does not have this limitation, since it is based purely on physical phenomena. Of course, even though absorbed dose is a physical metric, the relationship between absorbed dose and actual increase in stochastic risk does have large uncertainties, and is a topic of extensive research and ongoing debate.

4.2.3.1 Estimating absorbed dose

Absorbed dose in tissue cannot be measured directly, but it is estimated either empirically or by simulation. To estimate the absorbed dose due to an imaging acquisition experimentally, a common procedure is to use an anthropomorphic phantom that includes the relevant sections of the body for the imaging application being tested and insert some form of dosimeter at various points in the phantom, as shown in Figure 4.1.

Various types of dosimeters are used for these types of studies, including thermoluminescent dosimeters (TLDs),

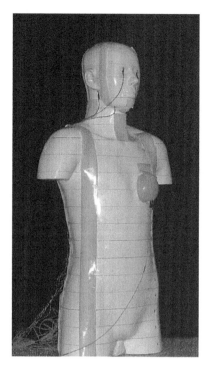

Figure 4.1 A typical anthropomorphic phantom used for empirical dosimetry estimations showing the cables that connect to the dosimeters that are embedded in different sections of the phantom. As shown here, anthropomorphic dose phantoms normally come in slabs that can be separated to allow for the insertion of the dosimeters in the different organs. (Image courtesy of Dr. Shuji Koyama, from Koyama S. et al., 2010. Radiation dose evaluation in tomosynthesis and C-arm cone-beam CT examinations with an anthropomorphic phantom. *Medical Physics* 37: 4298–4306. © American Association of Physicists in Medicine. With permission.)

radiophotoluminescence glass dosimeters (RPLs), and metal oxide semiconductor field effect transistors (MOSFETs), among others (Rogers and Cygler 2009). TLDs and RPLs are small crystalline detectors that absorb and retain energy following interaction with an x-ray. The absorbed energy, stored in the form of electrons elevated to an excited state, is released in the form of light when heated (TLDs) or exposed to ultraviolet light (RPLs). The amount of light released is a consequence of the number of excited electrons present, which is proportional to the incident x-ray energy during image acquisition. To be able to convert this light output measurement into an estimate of absorbed dose, a calibration curve is obtained for each dosimeter before it is used.

Since each point dosimeter only provides information on the local dose at the point where the dosimeter was located in the phantom during acquisition, to obtain an estimate of the mean absorbed dose to an entire organ, it is important to use various dosimeters placed throughout the phantom's representation of each organ. This is especially important for large organs, where the dose distribution can vary greatly between the areas of the organ closer and those further away from the radiation source. A number of studies, which will be discussed later, have used dosimeters of various types and anthropomorphic phantoms to evaluate dose in different tomosynthesis imaging applications (Bacher et al. 2009; Sarkar et al. 2009; Koyama et al. 2010; Mermuys et al. 2010; Cavagnetto et al. 2011; Machida et al. 2011). As can be seen from some of these studies, for a comprehensive characterization of the dose to the body, a high number of detectors positioned throughout the phantom may need to be used during a study, with all relevant organs and various locations in each organ sampled.

An alternative and also well-established method for estimating the dose involved in imaging studies is using computer-based Monte Carlo simulations. In Monte Carlo simulations, the appropriate geometry (including imaging system and patient description) and physics relevant to the acquisition are input into the computer and the acquisition of the images is simulated. An example Monte Carlo simulation setup is shown in Figure 4.2. During these simulations, a very large number of x-rays are simulated leaving the x-ray source and their fate as they travel through the simulation geometry and interact in the patient tissue and/or with the imaging system is simulated based on the relevant physics phenomena applicable to each condition. When given the correct inputs, and enough x-rays are simulated, the results obtained with Monte Carlo simulations should reflect the results that would be obtained for an empirical measurement.

Monte Carlo simulations for dosimetry studies have several advantages over empirical studies. In the first place, as opposed to empirical estimates in which the dose to an entire organ must

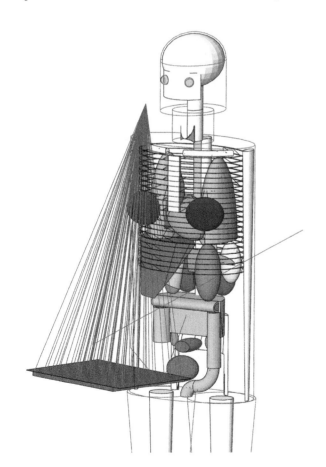

Figure 4.2 A typical Monte Carlo simulation geometry for a radiation dose study, including an anthropomorphic mathematical phantom based on the Cristy and Eckerman phantom (Cristy 1980). The organs in the human body are represented by simplified geometrical shapes. The ribs are hidden from view to enhance visibility of the organs. The imaging system is simulated as only a point x-ray source and a flat-panel detector. Some example x-rays are also shown.

be estimated from various point measurements throughout the organ, in Monte Carlo simulations, the absorbed dose to an organ can be obtained directly, since the computer program can report the total energy deposited within the volume of the organ and the mass of the organ is known. In addition, detailed information on dose variations with location within tissues and organs can be obtained. As an example, Thacker and Glick (2004) and Sechopoulos et al. (2010) reported on the dose distribution within the breast during mammography and dedicated breast computed tomography (CT) imaging acquisitions.

In addition, when performing a comprehensive image acquisition optimization study, it is possible that a number of acquisition parameters are varied, each with several possible values, resulting in possibly hundreds of different conditions being studied. These studies would be very challenging if not impossible to perform empirically; however, varying parameter values in computer simulations is trivial, so all these combinations of conditions can be studied easily. Another advantage of Monte Carlo simulations for these types of studies is the fact that they can be performed before an actual imaging system is built. Therefore, the impact that different design decisions might have on dosimetry (and image quality) can be studied using Monte Carlo simulations before these decisions are finalized.

These advantages, combined with the increase in computing power over the years, have resulted in increasing popularity for the use of Monte Carlo simulations for acquisition optimization and dosimetry characterization studies in medical imaging. This can be seen when one considers just the number of dosimetry studies based on Monte Carlo simulations in tomosynthesis imaging alone (Sechopoulos et al. 2007; Ma et al. 2008; Sechopoulos and D'Orsi 2008; Sabol 2009; Båth et al. 2010; Quaia et al. 2010; Svalkvist et al. 2010; Dance et al. 2011; Feng and Sechopoulos 2012).

Of course Monte Carlo simulations also have limitations that do not apply to empirical measurements. Most importantly, computer simulations require some degree of simplification of the problem. Typically, the imaging system and the imaged patient are not simulated in their entirety to the last detail, and several assumptions need to be made. Although this is normally not problematic, it is imperative that the impact that these simplifications have on the results be well understood. In addition, Monte Carlo simulations need to be validated before their results can be trusted, and this normally involves some degree of new empirical work and/or comparison to previously reported studies.

4.2.4 EQUIVALENT DOSE

As mentioned, the same amount of absorbed dose deposited by an x-ray compared to that deposited by an alpha particle is not related to the same increase in stochastic risk. This is because each radiation type has a specific *relative biological effectiveness* (*RBE*), which is reflected by assigning each radiation type a *radiation weighting factor*, denoted w_R. These factors, last revised in 2007 (International Commission on Radiological Protection 2007), are used in the following equation to compute *equivalent dose*, H:

$$H = \sum_R w_R D_R \qquad (4.1)$$

where w_R denotes the radiation weighting factor of radiation type R and D_R denotes the absorbed dose due to radiation type R. As for absorbed dose, the unit of equivalent dose is also joules per kilogram, but for equivalent dose, this is given the special name sievert (Sv).

Equivalent dose, however, just like effective dose discussed next, is a special type of metric because it is not purely based on physical phenomena. The radiation weighting factors are recommended based on radiobiological findings and ultimately arrived at by a commission, with the final values affected by judgments of practicality and usability (International Commission on Radiological Protection 2007). For example, all photon radiation is assigned a w_R of 1, irrespective of energy, although the RBE actually varies with energy. This simplification leads the International Commission on Radiological Protection (ICRP) to state that equivalent dose is "not intended for retrospective assessment of individual risks of stochastic effects from radiation exposures" (International Commission on Radiological Protection 2007), so care must be taken in not overassigning significance to this type of radiation metric.

4.2.5 EFFECTIVE DOSE

Different organs in the body have different levels of sensitivity to radiation. When determining increase in stochastic risk due to absorbed dose from ionizing radiation, this variation in radiosensitivity is taken into account by the *effective dose*, which combines the equivalent dose to different organs via a weighted sum using the *tissue weighting factors*, denoted w_T, as the weighting factor. Therefore, the effective dose, E, which also has units of sieverts, is computed according to the following equation:

$$E = \sum_T \left(w_T \sum_R w_R D_{T,R} \right) = \sum_T w_T H_T \qquad (4.2)$$

where w_T denotes the tissue weighting factor of organ T, w_R denotes the radiation weighting factor of radiation type R, $D_{T,R}$ denotes the absorbed dose in organ T due to radiation type R, and H_T denotes the equivalent dose in organ T. The latest versions of the tissue weighting factors were recommended by the ICRP in 2007 (International Commission on Radiological Protection 2007), which revised previous recommendations from 1977 and 1990 (International Commission on Radiological Protection 1977; International Commission on Radiological Protection 1991). The sum of all the tissue weighting factors is unity, so that the magnitude of effective dose is equivalent to that of equivalent dose (and absorbed dose if due to photon irradiation) if the dose is uniformly distributed throughout the body.

As is the case for the radiation weighting factors and equivalent dose, the tissue weighting factors and effective dose are based on radiobiology and epidemiology studies, but ultimately arrived at by consensus of a commission. Therefore, effective dose is also not a metric based on physical phenomena, and as such has many limitations, including the fact that the tissue weighting factors are averages for all ages and both sexes. As a consequence, care must be taken how effective dose is used, since it is not intended for personalized risk assessments. Aside from radiological protection calculations, an appropriate

use of effective dose estimations relevant to medical imaging could be for comparison of image acquisition protocols during optimization of new imaging technologies that irradiate multiple organs, for example, chest or abdominal tomosynthesis exams, or for the comparison of imaging techniques with different exposure geometries, for example, comparing CT or conventional projection imaging techniques to tomosynthesis.

Currently, there is a growing movement to record the effective dose resultant from an imaging procedure in the patient's medical history. Although this could provide useful information on dose trends at an institution or for a specific exam over periods of time or help identify patients that underwent acquisitions with problematic protocols, it is important to not attribute too much significance to these records. For example, effective dose is currently estimated in CT using the dose-length product (DLP) and study-specific conversion factors. But the DLP is based on the volumetric CT dose index (CTDI_{VOL}), which is based on a standard cylindrical phantom. Therefore, a record of a high effective dose for a CT study could actually not reflect a normal patient dose level to a large patient. In tomosynthesis dosimetry, especially in body applications, any future estimates of patient dose that could be recorded in the clinical history could very well also be based on standardized anthropomorphic phantoms, and therefore also would not reflect the variability in patient sizes, making patient-specific conclusions on stochastic risk based on these records suspect.

4.2.6 BREAST GLANDULAR DOSE

In breast imaging, additional dose metrics from those described above are commonly used. Since in mammography the rest of the organs of the body aside from the breast are outside the primary x-ray beam and are therefore minimally exposed (Sechopoulos et al. 2008), metrics that communicate the breast dose rather than the dose to the whole body are of interest. Of course, the absorbed dose, as described in Section 4.2.3, could be used. Furthermore, if one considers the tissue composition of the breast, it can be determined that absorbed dose to the whole breast is also not the most relevant dose metric in breast imaging. This is because the breast is grossly composed of three different tissues: glandular tissue, surrounded by adipose tissue, and both covered by a layer of skin. Moreover, since breast cancer mostly only develops in glandular tissue, Hammerstein et al. proposed that for the study of the risk involved in breast imaging with ionizing radiation, only the energy deposited in the glandular tissue of the breast is relevant (Hammerstein et al. 1979). This has given rise to the concept of the breast *glandular dose*, sometimes denoted D_g, which includes only the dose to the glandular tissue in the imaged breast, excluding the dose to the adipose and skin tissues. Since the glandular dose is a form of absorbed dose, it also has units of gray.

To communicate and compare the dosimetry of different imaging technologies, systems and/or protocols, some form of normalization of the dose values is required. Therefore, when communicating relative dose values, the breast glandular dose is given per unit air kerma at a reference point, usually being the air kerma incident upon the entrance (x-ray tube side) breast surface. This is denoted the *normalized glandular dose*, and sometimes given the symbol D_gN. In theory, the normalized glandular dose is unitless, but in practice it is quoted as having units of milligray per milligray (mGy/mGy) air kerma. Previously, before the use of air kerma became the recommended metric for x-ray intensity, many studies reported normalized glandular dose per unit reference exposure, using units of milligray per roentgen (mGy/R), but this is now discouraged.

The use of the glandular dose as the preferred metric for breast dosimetry introduces a major difficulty however. As can be seen from any pair of patient breast images (see Figure 4.3), the amount and distribution of glandular tissue in the human breast can vary widely from patient to patient following a somewhat random pattern. Therefore, how can the dose to a randomly varying tissue be estimated? To address this issue, the concept of the homogeneous breast and the *mean glandular dose* was introduced (Hammerstein et al. 1979). Instead of attempting to estimate the energy deposited in the glandular tissue when this tissue is pseudo-randomly distributed inside the breast, something clearly impossible to do, Hammerstein et al. proposed that an "average" breast could be assumed to consist of a homogeneous mixture of 50% adipose tissue and 50% glandular tissue, surrounded by a layer of skin and adipose tissue. This gives rise to the concept of the *mean glandular dose*, which specifies the dose to the glandular portion of the homogeneous mixture of adipose and glandular tissue that encompasses the whole breast, aside from the surrounding skin layer. It should be noted that although the mean glandular dose is normally always the one quoted, the term *mean* is very often omitted, and the metric is many times called only the glandular dose.

Of course, given the assumptions necessary to define and estimate mean glandular dose, it is a limited dose metric in a similar fashion to the effective dose for body dosimetry. Specifically, and as clearly pointed out by Hammerstein et al., the use of the mean glandular dose (and the normalized mean glandular dose) is only appropriate for comparison of

different breast imaging techniques, protocols, systems and/or technologies, and as such is not a measure of individual patient risk. For the latter, an estimate of the actual amount of energy deposited in each portion of glandular tissue as actually distributed in each specific patient would be needed. Given recent technology advances, to a limited extent, such estimates could be possible. Specifically, using dedicated breast CT, which provides tomographic high-resolution breast images that capture the individual's glandular tissue distribution, combined with advanced segmentation algorithms (Packard and Boone 2007; Nelson et al. 2008; Li et al. 2009; Yang et al. 2011), and voxel-based Monte Carlo simulations (Sechopoulos et al. 2010), estimates of energy deposited only in the glandular tissue in breasts of specific patients could be obtained. However, the applicability of these methods is limited by the substantial amount of work and computer processing time involved, combined with the inaccuracies introduced by noise in the acquired images, by the cone-beam tomographic reconstructions and by the automated segmentations. In addition, and most importantly, a reliable and accurate estimate of stochastic risk due to the resultant estimates of the energy deposition at each location would also be required. Clearly, these are very challenging, if not impossible requirements, and therefore the use of the mean glandular dose as a relative measure is currently preferred.

This does not mean that research should not be performed on improving how mean glandular dose is defined and estimated so as to better represent the "average" breast. For example, in a recent, very important study, only recently made possible due to the introduction of some of the technologies and methods just mentioned, Yaffe et al. showed that a mixture of approximately 14% glandular and 86% adipose is a better representation of the "average" composition of the breast than a 50%/50% mixture, and in fact 95% of women studied had a volumetric breast density (with skin included) lower than 45% glandular tissue (Yaffe et al. 2009). In another important study using dedicated breast CT, Huang et al. determined that the average breast skin thickness is 1.45 mm, considerably less than what had been previously assumed for breast dosimetry studies (Huang et al. 2008). Continued refinements of this type to the breast model for dosimetry in breast imaging should continue to be pursued.

4.3 ACQUISITION PARAMETERS AND DOSE

Tomosynthesis imaging involves many acquisition parameters, some that are usually part of the system design and therefore unchangeable from scan to scan (e.g., total acquisition angular range, number of projection views) and others that may be specified for each acquisition (e.g., tube voltage). Most, if not all, of these parameters have an impact on the resultant dose to the patient. Therefore, a comprehensive understanding of how each of these parameters affects dose is important to be able to minimize the radiation dose while maximizing image quality and diagnostic performance.

4.3.1 TUBE VOLTAGE AND X-RAY SPECTRUM SHAPE

The applied voltage that accelerates the electrons from the cathode to the anode of the x-ray tube is called the x-ray tube voltage,

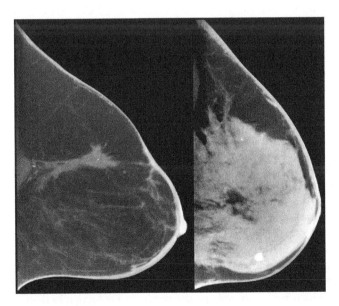

Figure 4.3 A slice from two different dedicated breast CT images showing the variability in amount and distribution of glandular breast tissue between patients. This variability makes patient-specific breast dosimetry extremely challenging and is the main reason for assuming the breast consists of a homogeneous mixture of adipose and glandular tissue.

System design

and determines the maximum x-ray energy in the produced x-ray spectrum. Owing to this fact, the x-ray tube voltage is also called the peak kilovoltage (kVp). Normally, the tube voltage is the simplest parameter that can be varied to change the shape of the x-ray spectrum. Other parameters that can affect the shape of the x-ray spectrum are the material used for the target in the x-ray tube, and the material and amount of added filtration present at the output of the x-ray tube. These two parameters involve hardware items that in many systems are not usually user-selectable. For example, all body tomosynthesis systems have a tungsten target. The target material in breast imaging systems can vary, however, with molybdenum, rhodium, and tungsten being the current choices. The possibility of varying filter material (with a set thickness for each choice) is more common in tomosynthesis systems.

A higher tube voltage results in an x-ray spectrum with higher mean and maximum x-ray energies, and with a higher fluence at all x-ray energies for the same tube current–exposure time product, as shown in Figure 4.4. Of course, generally, x-rays of higher energies are more penetrating than lower-energy x-rays, and the differences in attenuation among different tissues are smaller at higher energies. Therefore, usually a higher tube voltage, with a corresponding lowering of the tube current–exposure time product resulting in an equal signal level at the detector, results in a lower dose to the patient with an image of lower contrast. However, this does not always mean that if lower contrast is clinically acceptable, then a higher tube voltage should be used. Many additional considerations need to be taken into account, and therefore acquisition optimization based on the tube voltage parameter selection is not trivial and is usually application-specific.

4.3.1.1 Body tomosynthesis

The tube voltage in body tomosynthesis imaging applications is normally set to a constant value, while only varied for special acquisitions (e.g., dual energy (Gomi et al. 2011)). For example, in chest tomosynthesis, the x-ray tube voltage is commonly set at 120 kVp (Vikgren et al. 2008; Dobbins 2009; Quaia et al. 2010; Jung et al. 2012), while for urinary tract stone imaging, the x-ray

tube voltage is set to 80 kVp (Mermuys et al. 2010). However, in most cases, the selected tube voltage for tomosynthesis imaging is a holdover from that used for its planar equivalent. In chest tomosynthesis, a comprehensive study to optimize tube voltage has not been reported to date (Dobbins 2009), while a study has been reported on substantial dose savings and adequate image quality being obtained by performing chest tomosynthesis for a specific application with a lower tube voltage (100 kVp), an added copper filter (0.3 mm thick) and a lower tube current–exposure time product (Kim et al. 2010).

Clearly, optimization studies to investigate the optimal x-ray spectrum shape for the different tomosynthesis applications are needed. The problem with these studies is that often they do not provide a definite answer. As an example, two studies aimed at optimizing the tube voltage for flat-panel-based digital planar chest radiography have arrived at different conclusions. One patient study comparing posterior–anterior (PA) chest radiographs acquired at 90, 121, and 150 kVp concluded that the lowest tube voltage is optimal (Uffmann et al. 2005), while a phantom study comparing 100, 120, and 140 kVp concluded that 120 kVp was the optimal tube voltage (Metz et al. 2005). It is very possible that the optimal tube voltage for chest tomosynthesis differs from that for digital planar chest radiography, but the differing conclusions arrived at by optimization studies for a much more established, but similar, imaging technology emphasize the need to further study this area for chest tomosynthesis.

4.3.1.2 Breast tomosynthesis

With the advent of digital mammography, the use of a fully automated automatic exposure control (AEC) involving the acquisition of a low-dose scout image to determine the correct acquisition technique for each patient, including tube voltage, became commonplace. In digital mammography, the tube voltage used for imaging usually varies from 22 kVp up to 32 kVp (Hendrick et al. 2010), although this range varies among system manufacturers and some higher tube voltages are being used with the most modern systems. Currently, some commercial breast tomosynthesis systems use the compressed breast thickness to set the tube voltage to be used during tomosynthesis acquisition (Ren et al. 2010; T. Mertelmeier, 2011, personal communication).

For mammography and breast tomosynthesis, the normalized glandular dose (i.e., the mean glandular dose for an incident air kerma of 1 mGy) increases with increasing tube voltage (Dance 1990; Wu et al. 1991, 1994; Boone 1999; Sechopoulos et al. 2007; Sechopoulos and D'Orsi 2008). However, as mentioned, a higher-energy x-ray beam has a higher output fluence and penetrability for a constant tube current–exposure time product, resulting in a lower image noise level. For constant image noise levels, which are achieved by reducing the tube current–exposure time product, the mean glandular dose in absolute terms is actually inversely proportional to tube voltage. Of course, as mentioned previously, this increase in tube voltage also results in a reduction in tissue contrast.

To better understand the trade-offs between dose, noise, and image contrast, several studies have been performed on optimizing the x-ray spectrum to be used in breast tomosynthesis. Ren et al., in an experimental study using a maximum tube

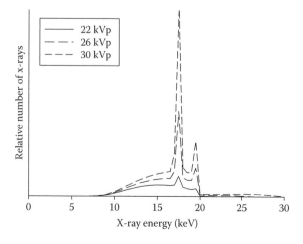

Figure 4.4 Typical x-ray spectra used in breast imaging showing the variation in fluence at each energy with varying tube voltage. A higher tube voltage results in a higher maximum x-ray energy, higher mean energy, and higher fluence at all energies.

voltage of 39 kVp, found that higher-energy spectra improved the contrast-to-noise ratio (CNR) in the projections of signals on a homogeneous background (Ren et al. 2005). To characterize signal detectability in a homogeneous background, Glick et al. used a serial cascaded model in projection space with varied x-ray spectra up to a tube voltage of 40 kVp, finding that the detectability metric is maximized with a tungsten target and a rhodium filter, but as opposed to the study by Ren et al., with lower mammographic energies (Glick and Gong 2006). Wu et al. performed an empirical study in which the optimal x-ray spectra were determined by acquisition of the 0° projections of a circular signal on homogeneous backgrounds and comparison of the resulting CNR in the projections normalized by the mean glandular dose, finding that this figure of merit was improved for the harder x-ray spectra only for the thicker breasts (Wu et al. 2006). In a simulation study that compared the signal-to-noise ratio (SNR) in the 0° and 25° projections of simulated microcalcifications in a homogeneous background with x-ray spectra of up to 40 kVp, Zhao et al. found it advantageous to increase the tube voltages slightly for all but the thickest breasts, for which an increase of 9 kVp was found to be optimal (Zhao et al. 2005).

As can be seen, just as in digital chest radiography, the optimization of tube voltage for breast tomosynthesis imaging is still an open question. However, using Monte Carlo simulations, the impact that different tube voltages and x-ray tube target material and added filtration has on breast glandular dose in tomosynthesis imaging has been well characterized (Sechopoulos et al. 2007; Ma et al. 2008; Sechopoulos and D'Orsi 2008; Dance et al. 2011), providing the data necessary to help address this optimization question.

Although reports on optimal tube voltage for breast tomosynthesis have yet to arrive at a single conclusion, for one of the commercial breast tomosynthesis systems (Selenia Dimensions, Hologic, Inc., Bedford, MA), it was found, using breast phantoms, that mammography and tomosynthesis acquisitions of the same breast thickness and density are performed with different tube voltages and filters (see Table 4.1) (Feng and Sechopoulos 2012).

These differences in the x-ray source settings result in considerably different x-ray spectra as shown in Figure 4.5. In

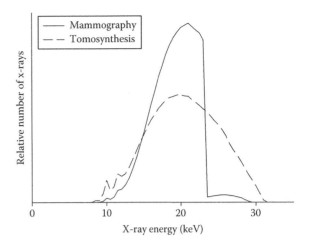

Figure 4.5 X-ray spectra used for mammography and tomosynthesis for breasts of the same compressed breast thickness (5 cm). As can be seen, the spectra were generated using a different tube voltage and the use of a different filter material introduces important shape variations. Each spectrum is normalized to its own area under the curve. (Adapted from Feng, S. S. J., I. Sechopoulos. 2012. *Radiology* 263: 35–42.)

addition, although the tube voltage for the breast tomosynthesis acquisitions is higher than that in mammography for all breast thicknesses, due to the use of different filters, the first half value layer (HVL) of the tomosynthesis spectra are actually lower for all but the three thickest breasts. The use of higher tube voltages but with a lower atomic number filter for tomosynthesis may be due to the need to minimize the tube current on time to reduce focal spot blur, since this is a continuous motion system (Ren et al. 2010).

In summary, although the dosimetric consequences of acquiring breast tomosynthesis images with different tube voltages and target/filter materials are well understood, combining this data with image quality assessment is not trivial. This has resulted in several groups performing these optimization studies with different methods, and some arriving at different conclusions on what is optimal. However, in a current commercial tomosynthesis system, it has been found that higher-energy spectra are used only for the thickest breasts when comparing tomosynthesis acquisitions to digital mammography imaging.

Table 4.1 Tube voltage, filter material, and measured first half value layer used by the Hologic Selenia Dimensions tomosynthesis systems, which have a tungsten anode for breast phantoms of different thickness

BREAST THICKNESS (CM)	MAMMOGRAPHY			TOMOSYNTHESIS		
	FILTER	TUBE VOLTAGE (kVp)	HVL (mm Al)	FILTER	TUBE VOLTAGE (kVp)	HVL (mm Al)
2	Rh	25	0.453	Al	26	0.441
3	Rh	26	0.504	Al	28	0.476
4	Rh	28	0.526	Al	29	0.505
5	Rh	29	0.551	Al	31	0.541
6	Rh	31	0.567	Al	33	0.572
7	Ag	30	0.586	Al	35	0.600
8	Ag	32	0.611	Al	38	0.660

Source: Feng, S. S. J., I. Sechopoulos. 2012. *Radiology* 263: 35–42.

4.3.2 TUBE CURRENT AND EXPOSURE TIME

Aside from the x-ray spectrum shape, the current between the cathode and the anode in the x-ray tube also determines the x-ray flux output by the x-ray source. Combined with the amount of time that the current is on (the exposure time), the resulting tube current–exposure time product is directly proportional to the fluence output by the x-ray source. Therefore, if all other parameters are kept constant, the tube current–exposure time product is directly and linearly proportional to the patient dose and to the signal at the detector during a tomosynthesis acquisition, and therefore inversely proportional to image noise (see Figure 4.6). As such, the appropriate setting for this parameter is crucial to obtain an image with the appropriate noise levels while maintaining an appropriate dose level.

Unfortunately, setting the appropriate tube current–exposure time product for acquisition is a laborious task in which many considerations need to be taken into account. Probably the major consideration is identification of the diagnostic task to be fulfilled. For example, the appropriate dose levels for lung nodule detection might be considerably lower than that for other general diagnostic tasks in chest tomosynthesis (Dobbins 2009). Additional important considerations are patient size, tissue composition (e.g., breast density in breast tomosynthesis), and detector efficiency, among others. Other more clinical factors are also relevant: for example, is this a screening test applied to the general (and mostly healthy) population, a diagnostic test applied to symptomatic patients, or a test to monitor treatment response in patients being treated with a serious disease? Of course the answer to this question, combined with the consequences of a false-positive or false-negative error

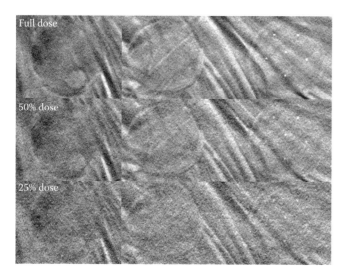

Figure 4.6 Regions of interest of tomosynthesis images of a breast phantom showing a mass, a fiber, and a microcalcification cluster, acquired at three different tube current–exposure time product settings. The full dose image was acquired using the system's automatic exposure control. The other two images were acquired at 50% and 25% of the automatic setting. The means ± standard deviations in a homogeneous portion of the reconstructed images were 480.5 ± 14.2 digital units (DU), 483.3 ± 17.6 DU, and 487.0 ± 22.0 DU, for the full, 50%, and 25% dose images. A clear loss in visibility of the lesions with an increase in image noise due to the decrease in tube current–exposure time product used can be seen.

due to an image that is too noisy, affects what the appropriate fluence level should be, which in turn determines the resultant noise level.

4.3.2.1 Distribution of tube current–exposure time product

Particular to tomosynthesis imaging is the issue of how to divide the available total tube current–exposure time product among the several projections to be acquired. Of course, in planar imaging, this is not an issue since only one projection is acquired. In CT, the tube current was historically constant throughout acquisition. This changed with the introduction of tube current modulation algorithms, which automatically vary the tube current in a very complex fashion in an attempt to maintain the signal level at the detector constant by compensating for the axial and angular variations in the attenuation of the anatomy included in the x-ray paths. In tomosynthesis, the default choice has been to divide the tube current–exposure time product equally among the projections. For chest and other types of body tomosynthesis, there have not been any reports of deviation from the use of uniform exposure distributions among the tomosynthesis projections. However, it has been suggested that for breast tomosynthesis there could be more optimal ways to divide the available total fluence. Specifically, Nishikawa et al. suggested that if the total tube current–exposure time product was set so that a single breast tomosynthesis acquisition resulted in a breast dose equivalent to a two-view mammographic examination, then it could be advantageous to use 50% of the available exposure to acquire the 0° projection (equivalent to a mammogram) and then to divide the remaining 50% over the other projections (Nishikawa et al. 2007). Furthermore, Nishikawa et al. suggested that for the detection of microcalcifications only the 0° projection be interpreted by a physician after analysis by a CAD system, and that only a lower-resolution version of the tomosynthesis reconstruction be shown to the physician to detect soft tissue masses.

Additional studies have been reported on the resultant image quality when nonuniform distributions of exposure are used. Specifically, Das et al. tested the Nishikawa et al. proposal using simulated tomosynthesis images based on a benchtop-dedicated breast CT image of a mastectomy specimen (Das et al. 2009). The simulated images were used to perform a localization receiver operating characteristics (LROC) study using medical physicists as observers. Das et al. reported that for microcalcification detection, the area under the LROC curve was higher when viewing the reconstruction from the projections using a uniform exposure distribution than when viewing only the 0° projection acquired with 50% of the total available exposure. Furthermore, it was found that there was no significant difference in mass detection between the reconstruction using the uniform exposure distribution and that using the variable exposure distribution. However, as opposed to Nishikawa et al.'s proposal, CAD was not used for the microcalcification detection task, and it is not clear how the use of medical physicists as expert observers impacts the results compared to using radiologist experts in breast imaging, those interpreting the images in the clinical realm. Another study by Hu and Zhao used an ideal numerical observer based on a cascaded linear system model to investigate the use of variable exposure distribution for detection of microcalcifications in breast imaging (Hu and Zhao 2011). They found that the conspicuity of small microcalcifications may be increased by using a higher tube current–exposure time product in the central projections (close to the 0° projections) than in the wider-angle projections. However, this study did not report on what effect, if any, this type of distribution might have on mass visibility.

Since Sechopoulos et al. (2007) showed that the normalized breast glandular dose can vary considerably with projection angle, these forms of nonuniform distribution of the available total tube current–exposure time product have interesting consequences for the breast glandular dose. Therefore, the resulting total breast glandular dose from nonconventional acquisitions involving variable tube current–exposure time products could not necessarily vary as expected at first glance. To correctly determine the impact that nonuniform distribution of the tube current–exposure time product has on dosimetry, the variation of dose with projection angle must be taken into account, as discussed in the following section.

4.3.3 ACQUISITION GEOMETRY

Optimization of tomosynthesis imaging acquisitions is also different than for planar and CT imaging due to the number and importance of parameters related to acquisition geometry that should be considered. In tomosynthesis imaging, three particular parameters of the acquisition geometry are of special interest due to their impact on image quality and dose and to the lack of equivalent parameters in established imaging technologies from which usual values can be migrated. These parameters are the number of projections to be acquired, the total angular range to cover during acquisition, and the distribution of the projections throughout this angular range. As expected, the impact of these parameters on tomosynthesis image quality has resulted in a large number of studies on this matter, both in body tomosynthesis (Deller et al. 2007; Machida et al. 2010; Svalkvist et al. 2010) and in breast tomosynthesis (Maidment et al. 2005, 2006; Zhou et al. 2007; Chawla et al. 2009; Sechopoulos and Ghetti 2009; Reiser and Nishikawa 2010; Gang et al. 2011; Lu et al. 2011). In addition to the studies performed to study the impact of these acquisition geometry parameters on image quality, other studies have been reported on how these parameters affect dose, discussed next.

4.3.3.1 Chest tomosynthesis

In body tomosynthesis, as opposed to breast tomosynthesis, varying projection angle can result in different organs being inside and outside the primary x-ray beam. Therefore, from a dosimetric perspective, understanding the impact that varying acquisition geometry has on dosimetry is important.

In an early study on chest tomosynthesis, Deller et al. studied the variation in SNR with the total dose used to acquire a complete chest tomosynthesis image and with the total number of projections acquired (Deller et al. 2007). The authors studied six different values for total number of projections per tomosynthesis acquisition, from five total projections to 60 total projections.[*] For each of the geometries studied, Deller et al. included three

settings for total dose. It was found, not surprisingly, that the main factor determining SNR was the total dose. However, it was also found that the number of projections used to deliver that total dose had a minor impact on SNR.

Sabol performed a Monte Carlo simulation-based chest tomosynthesis dosimetry study simulating a specific clinical body tomosynthesis system (Definium 8000, VolumeRAD, GE Healthcare, Chalfont St. Giles, UK) (Sabol 2009). For this system, for chest imaging, the center of rotation for the x-ray tube is 9.9 cm above the detector entrance surface with a total x-ray tube angular range of ±15°. Sabol used the PCXMC Monte Carlo simulation package that represents the patient with a mathematical anthropomorphic phantom (Cristy 1980). As expected, Sabol reported that the absorbed dose to different organs in the body varies in different manner with projection angle, depending on the position of the organ in the body and that there is a resulting variation in effective dose with projection angle (see Figure 4.7). Although this variation in effective dose with projection angle results in a reduction in the effective dose per tube current–exposure time product of approximately 10% (J. Sabol, 2011, personal communication), another factor must be taken into account when estimating the effective dose from chest tomosynthesis. During acquisition of a chest tomosynthesis image with this system, a scout image, equivalent to a PA chest radiograph, is first acquired, and a "dose ratio" parameter is specified. The total tube current–exposure time product to be used for the complete tomosynthesis acquisition is calculated by multiplying the input dose ratio by the tube current–exposure time product used for the scout acquisition as determined by the AEC. According to the manufacturer's suggestions, this dose ratio is typically 10×. During acquisition of the tomosynthesis projections, however, the actual tube current–exposure time product the imaging system uses in each projection is the nearest setting immediately below the value that would be necessary to obtain the desired total tube current–exposure time product. For example, if the PA radiograph resulted in a tube current–exposure time product of 2 mAs, and the 10× dose factor were specified for the tomosynthesis acquisition, then the desired tube current–exposure time product setting for each of the 60 tomosynthesis projections would be 2 mAs × 10/60 projections = 0.333 mAs per projection. However, the system will actually acquire each tomosynthesis projection with the closest available tube current–exposure time product that is below 0.333 mAs, which, for example, could be 0.3 mAs, resulting in a total actual tube current–exposure time product for the tomosynthesis acquisition of 18.0 mAs, instead of the requested 20.0 mAs. When taking this factor into account in the dosimetry estimations and the comparison of tomosynthesis dose to radiographic dose, Sabol found that using the dosimetry values for planar PA chest radiography and multiplying them using the input dose ratio would overestimate the dose from a PA chest tomosynthesis acquisition by about 25%. As mentioned, less than 10% of this overestimation is due to the variation in effective dose with projection angle due to geometrical variations in organ exposure and source-to-imager distance. The rest of the overestimation is due to the rounding down of the tube current–exposure time product per projection. When estimating dose from chest tomosynthesis with these types of imaging systems, care must be

---

[*] It should be noted that the current commercial body tomosynthesis systems include 60 and 74 projections per acquisition (Båth et al. 2010; Yamada et al. 2011).

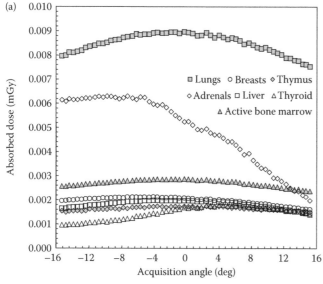

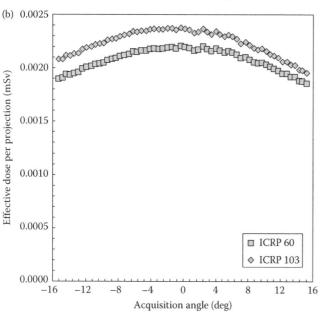

Figure 4.7 (a) Variation in absorbed dose with projection angle for various organs in the body during acquisition of a chest tomosynthesis image. (b) Variation in resultant effective dose with projection angle during acquisition of a chest tomosynthesis image. The effective dose calculations are performed with the tissue weighting factors recommended by the ICRP Report 60 in 1991 and by the ICRP Report 103 in 2007. (Images courtesy of Dr. John Sabol, from, Sabol, J. M. 2009. A Monte Carlo estimation of effective dose in chest tomosynthesis. *Medical Physics* 36: 5480–5487. © American Association of Physicists in Medicine. With permission.)

taken to distinguish between the requested imaging technique and that actually used by the system, with the corresponding impact on the air kerma–area product.

In another study, Svalkvist et al. performed a Monte Carlo simulation study of a generic chest tomosynthesis imaging system to determine how the conversion factors for effective dose from air kerma–area product vary between a complete chest tomosynthesis examination and the 0° projection (Svalkvist et al. 2010). In essence, the authors studied how effective dose varies between a standard PA chest radiograph and a chest tomosynthesis acquisition for different imaging conditions. The imaging

conditions included were different-sized patients (in both height and weight), tube voltages, tomosynthesis total angular range, and dose distribution for the tomosynthesis projections (constant air kerma, constant air kerma–area product, constant tube current–exposure time product (the most clinically realistic), and constant effective dose). For this study, also using the PCXMC Monte Carlo package, the authors simulated a generic tomosynthesis system with the center of rotation of the x-ray source placed at the detector entrance surface. Svalkvist et al. found that the effective dose for the complete chest tomosynthesis acquisition differs by less than 10% from that resulting of a planar chest radiograph while maintaining the equivalent total air kerma–area product. In addition, for the protocol that maintains a constant tube current–exposure time product among tomosynthesis projections, which, as mentioned, is the most clinically relevant currently and probably in the future, the variation in total effective dose between the two imaging modalities is less than 5%. Therefore, for an approximate calculation of the effective dose resulting from a chest tomosynthesis acquisition, Svalkvist et al. concluded that normalized dose data from chest radiography is sufficient, to about 5–10%. These findings approximately match those reported by Sabol.

Finally, Båth et al. performed another Monte Carlo simulation study again using PCXMC, but using patient data from clinical acquisitions and again simulating the GE system (Definium 8000, VolumeRAD, GE Healthcare, Chalfont St. Giles, UK) to compare in absolute terms the effective dose from chest tomosynthesis to that of chest radiography on the same patients (Båth et al. 2010). In this study, the same conversion factor for effective dose from air kerma–area product was found for chest tomosynthesis and for PA view chest radiography. Therefore, according to Båth et al., and in accordance with Sabol and Svalkvist et al. to within less than 10%, the total effective dose of a chest tomosynthesis study spanning ±15° is equivalent to that of a PA chest radiograph, for the same total air kerma–area product.

4.3.3.2 Breast tomosynthesis

Several studies have been reported on how breast dose varies with tomosynthesis projection angle. In the first such study, Sechopoulos et al. proposed simplifying how breast tomosynthesis dosimetry is analyzed by decoupling the dose from the 0° projection from that of the other projections (Sechopoulos et al. 2007). Therefore, the authors characterized how the normalized glandular dose for the 0° projection, equivalent to the mammographic dose, varied with breast thickness, size and composition, x-ray tube voltage, target material, and added filtration. For the dose for the nonzero degree projection angles, the authors introduced the concept of the relative glandular dose (RGD), which was defined as the ratio of the breast glandular dose due to an acquisition at a certain tomosynthesis projection angle divided by the breast glandular dose at the 0° projection angle acquired with the same technique (tube target material, filter, voltage, and current–exposure time product) for the same breast. This decoupling of the two metrics allowed for an easier understanding of how the breast dose varies with projection angle, how this variation changes with different imaging conditions (breast size, thickness and composition, x-ray spectrum, etc.) and allowed for a more practical form to communicate all the necessary values to estimate the total breast glandular dose for any breast tomosynthesis acquisition protocol.

To perform this study, Sechopoulos et al. used the Geant4 toolkit for Monte Carlo simulations (Agostinelli et al. 2003; Allison et al. 2006), to simulate a generic breast tomosynthesis imaging system with a static detector and with compressed breasts with varying parameters in both the cranio-caudal (CC) and medio-lateral oblique (MLO) views. The authors found that the RGD can deviate considerably from unity, and therefore that when estimating the dose for a specific tomosynthesis projection, the dose due to the mammographic equivalent acquisition cannot be used. Furthermore, it was found that how RGD varies with angle depends mostly on geometrical aspects, specifically, the view (CC or MLO), the compressed breast thickness, and the breast size. The variation with breast size and thickness was found to be stronger in the MLO view than in the CC view, due to the location of the majority of the breast tissue toward the superior side of the detector when the breast is correctly positioned for the MLO view. This asymmetry results in a large variation in the distance between the x-ray tube and the breast tissue with projection angle, causing the large variation in RGD in the MLO view.

Other nongeometrical parameters studied, namely, the x-ray spectrum and the breast composition, do not introduce substantial variations in how RGD varies with projection angle. Finally, the position of the center of rotation of the x-ray tube, which was varied from on the detector entrance surface up to 8 cm above it, does not affect the variation of the RGD substantially either. Figure 4.8 shows how RGD varies with projection angle and how this variation is affected by some of the studied imaging parameters. As can be seen, for the angular range and other imaging conditions studied, for the MLO view, the RGD can vary in magnitude from ~0.5 to ~1.4. This entails, for example, that for a breast compressed in the MLO view, with a chest wall-to-nipple distance of 7 cm and a compressed breast thickness of 5 cm, a tomosynthesis projection acquisition at 30° would result in approximately 50% less glandular dose than that due to the 0° projection with the same acquisition parameters.

However, owing to the shape of the curve that describes the RGD variation with projection angle, when a complete tomosynthesis acquisition is considered, assuming a uniform exposure distribution among the projections and a symmetric distribution about the 0° position, the average RGD's deviation from unity is small, therefore decreasing the difference between the glandular dose due to a tomosynthesis acquisition and that due to a mammographic acquisition with the same imaging conditions and equal total tube current–exposure time product. Specifically, for the MLO view, for all the conditions studied,

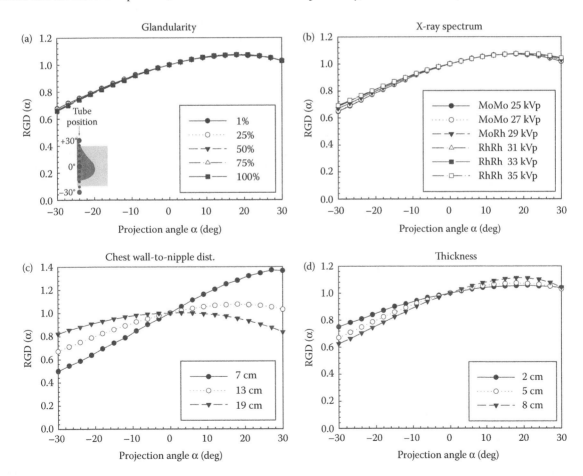

Figure 4.8 Examples of relative glandular dose (RGD) variation with projection angle for different imaging conditions: (a) breast glandular fraction; (b) incident x-ray spectrum; (c) breast size (described by the chest wall-to-nipple distance); and (d) compressed breast thickness. As can be seen, only the breast size and thickness introduce a variation in the RGD behavior with projection angle. (Image from Sechopoulos, I. et al., 2007. Computation of the glandular radiation dose in digital tomosynthesis of the breast. *Medical Physics* 34: 221–232. © American Association of Physicists in Medicine. With permission.)

the mean RGD for complete tomosynthesis acquisitions spanning ±30° varied from 0.91 to 1.01, while for the CC view, this range was 0.91–0.97. In another study, Sechopoulos and D'Orsi analyzed the dosimetric characteristics of performing breast tomosynthesis with x-ray tubes with tungsten targets (Sechopoulos and D'Orsi 2008). As expected, differences were found in the normalized glandular dose values for the 0° projections; however, the RGD was found to not vary considerably from those previously found for molybdenum and rhodium targets.

Therefore, for standard tomosynthesis acquisitions in which the projections are symmetrically and uniformly distributed about 0° with a constant tube current–exposure time product in each projection acquisition, using mammographic dose data results in a maximum error for tomosynthesis of up to ~10%. If more exact estimates are needed, then the impact of the RGD on the total glandular dose needs to be considered. In addition, as has been discussed previously, more advanced tomosynthesis acquisition protocols, with variable exposure distributions and nonuniform projection distribution throughout the total angular range have been proposed and continue to be studied. For such situations, care must be taken in how the dosimetric values of these protocols are estimated, and the impact of these advanced acquisition techniques on the RGD needs to be taken into account when estimating the resultant breast glandular dose. To facilitate these calculations, Sechopoulos et al. published fit equations and parameters to compute the RGD for any projection angle in the ±30° range for any x-ray spectrum and breast glandular fraction as a function of breast size (chest wall-to-nipple distance) and compressed breast thickness (Sechopoulos et al. 2007).

In a subsequent study, Ma et al. studied the dosimetric characteristics of breast tomosynthesis with a rotating detector, for x-ray tubes with tungsten targets, using the MCNPX Monte Carlo package (Ma et al. 2008). This study also included the compressed breast in the MLO and CC views, but did not decouple the normalized glandular dose for the 0° projection from that of the other projections using the RGD metric. It was found in this study that the introduction of rotating or tilting detectors to breast tomosynthesis does not impact the findings of the static detector studies substantially. This was later confirmed by Dance et al., and explained by the fact that in breast tomosynthesis most of the tissue is away from the detector edges, and therefore any x-ray beam variations that would be introduced by the rotating detector and the collimation that limits the beam to the detector edges would not affect how most of the breast tissue is irradiated (Dance et al. 2011). Another result reported by Ma et al. is that the normalized glandular dose varies linearly with breast glandularity (at least in the range studied: 30–70%), and that this linearity holds for non-0° projections. Taking advantage of this finding, comprehensive tables of normalized glandular dose for a range of projection angles (±40°), compressed breast thicknesses (3–7 cm), and tube voltages (20–50 kVp) are provided for the 50% glandular fraction breast with values for the slope for each combination of conditions so that the corresponding normalized glandular dose can be computed for any other glandular fraction.

In another study, Dance et al. extended the breast dosimetry protocol used in the United Kingdom, Europe, and by the International Atomic Energy Agency (IAEA) to include tomosynthesis imaging (Dance et al. 2011). In the original breast dosimetry protocol, the glandular dose to the breast (D_g) is estimated using the product of the incident air kerma at the top (x-ray tube side) surface of the compressed breast (K), a factor (g) that converts incident air kerma to glandular dose to a 50% glandular breast of a specified thickness due to an incident spectrum of a certain first HVL, a factor (c) that adjusts for different glandular fractions from 0.1% to 100% glandularity, and a factor (s) that adjusts for x-ray spectra generated with different target/filter material combinations:

$$D_g = Kgcs \qquad (4.3)$$

For tomosynthesis, Dance et al. introduce the factor t, which adjusts for the angle of the tomosynthesis projection for which the dose is being estimated, resulting in the new equation for the glandular dose to the breast due to a single tomosynthesis projection:

$$D_g = Kgcst \qquad (4.4)$$

In essence, the factor t is equivalent to the RGD metric proposed by Sechopoulos et al. In addition, Dance et al. propose the T factor, which is the mean of the individual t factors for each projection angle that are included in a complete tomosynthesis acquisition, resulting in the following equation for the total glandular dose to the breast due to a complete tomosynthesis image acquired with a total incident air kerma K_T:

$$D_{g,T} = K_T gcsT \qquad (4.5)$$

Given that these T factors are currently system-specific (no current commercial system at an advanced stage of development has capabilities to vary the acquisition geometry), Dance et al. published the T factors for three different commercial breast tomosynthesis systems as a function of compressed breast thickness only. For the two systems that use a "standard" tomosynthesis acquisition geometry, that is, a static or rotating full-field detector that acquires a certain number of full-field projection images while the x-ray tube rotates about the breast, the T factor range is from 0.960 to 0.997. Again, this shows that to within a small percentage (in this case 4%), mammographic dose estimates may be used for tomosynthesis dose estimations.

However, Dance et al. also published the equivalent T factors for the photon-counting scanning detector-based tomosynthesis prototype system developed by Sectra (Sectra Imtec AB, Linköping, Sweden, now Philips Healthcare, Best, The Netherlands). The geometry used by this system, described in Chapter 2, results in T factors that range from 0.759 to 0.983, showing that the breast dose from a tomosynthesis acquisition with this type of system is considerably different from that of a mammographic acquisition with a similar system, and therefore the T factor must be included in any breast dose estimate.

In summary, the data available on the impact that the tomosynthesis acquisition geometry has on breast dose show that under certain circumstances the variation of the dose with projection angle can be ignored, resulting in dose estimates accurate to within 5–10% of those predicted by mammography dosimetry. These circumstances include a uniform exposure distribution among the projections, a uniform and symmetric distribution of the projections among the total angular range, and the use of a full-field acquisition tomosynthesis system. In addition, this has only been investigated up to a maximum angular range of about ±30° to ±40°, so this same conclusion should not be made for wider angular ranges. However, RGD and/or t factors are essential and should be used if the dose due to a single projection is being estimated, a more accurate estimate with an error of <10% is required, a scanning detector tomosynthesis system similar to that of Sectra is being studied, an advanced acquisition protocol is used in which either the exposure is not distributed uniformly throughout the projections or the projections themselves are not distributed uniformly throughout the total angular range.

4.4 CLINICAL TOMOSYNTHESIS DOSIMETRY

So, how much radiation dose is actually involved during the acquisition of a clinical tomosynthesis image? This section will address this most important aspect that directly impacts the feasibility and applicability of tomosynthesis imaging for routine clinical use.

4.4.1 CHEST TOMOSYNTHESIS

In the first study on dose in absolute terms from chest tomosynthesis, mentioned in Section 4.3.3.1, Sabol analyzed 286 PA and 104 lateral clinical planar chest radiography acquisitions of "medium-sized" patients acquired with the standard protocol (120 kVp, no additional filtration) (Sabol 2009). Using a system equivalent to that used for the clinical acquisitions, the tube current–exposure time product relationship to air kerma was determined, and the average incident air kerma for the PA and lateral studies calculated. The author assumed that the chest tomosynthesis acquisitions for a patient would be performed according to the manufacturer's suggestions, that is, with a "dose ratio" of 10×. Using a Monte Carlo simulation, estimations of the average incident air kerma for the PA and lateral chest radiographs and for the chest tomosynthesis acquisitions including the rounding down of the tube current–exposure time product for the individual tomosynthesis projections, the effective dose to a medium-sized patient for the three images (PA, lateral, and tomosynthesis) was estimated. The effective dose due to a PA planar chest radiograph was found to be 0.018 mSv, due to a lateral planar chest radiograph it was found to be 0.050 mSv, and for chest tomosynthesis it was 0.134 mSv. It should be noted that according to the suggested use for the system investigated (Definium 8000, VolumeRAD, GE Healthcare, Chalfont St. Giles, UK), a chest tomosynthesis acquisition should be performed following a scout image acquisition, which is equivalent to a PA radiograph. Therefore, the total effective dose from a complete chest tomosynthesis examination, which also produces an image equivalent to the PA radiograph, is 0.152 mSv.

Again using the GE tomosynthesis system, Quaia et al. also reported on a prospective clinical study comparing the clinical performance of chest tomosynthesis to radiography using CT

as truth (Quaia et al. 2010). For this study, which included 228 patients, effective dose for tomosynthesis and radiography was estimated using the air kerma–area product and a Monte Carlo simulation, while for CT it was estimated using the DLP and established conversion factors. The authors found that the average effective dose for two-view radiography was 0.06 mSv, while for the complete tomosynthesis acquisition, including the scout image, it was 0.2 mSv. The CT effective dose was estimated at 2–4 mSv.

Båth et al. estimated the effective dose resulting from chest radiography and chest tomosynthesis acquisitions for 40 patients imaged with the commercial GE system using the recorded acquisition technique and Monte Carlo simulations (Båth et al. 2010). In this case, the effective dose due to a PA planar chest radiograph was found to be 0.014 mSv, due to a lateral planar chest radiograph it was found to be 0.039 mSv, and for chest tomosynthesis it was 0.122 mSv. For the cases where the tomosynthesis image was acquired after the radiography images, a separate scout image was acquired, which resulted in an average effective dose of 0.011 mSv, so a complete tomosynthesis examination with inclusion of the equivalent PA acquisition resulted in an average effective dose of 0.133 mSv. It can be seen that all three of these studies that analyzed the same commercial system using the standard imaging technique arrived at similar dosimetry results.

In two clinical studies using the GE system, Kim et al. report on the use of a chest tomosynthesis imaging protocol modified to result in a reduced patient dose compared to the standard protocol (Kim et al. 2010, 2012). For this, the authors modified the imaging technique by reducing the tube voltage to 100 kVp, added copper filtration, and reduced the dose ratio to 5× the scout image. These modifications were estimated to result in an effective dose of 0.05 mSv for the complete tomosynthesis examination, including the scout image, a reduction of about two-thirds over the effective dose reported for the standard technique. The tomosynthesis effective dose was estimated using radiographic conversion factors, so any variability in effective dose with projection angle was ignored. Upon comparison of performance for different clinical tasks, the tomosynthesis images were found superior to the radiographs, using CT as truth. No comparison on the performance between these reduced dose tomosynthesis images and tomosynthesis images with standard dose levels was reported.

Using a different imaging system than the one used in all the studies discussed above (Sonialvision Safire, Shimadzu Corp., Kyoto, Japan), Koyama et al. studied the dose from tomosynthesis imaging using an anthropomorphic phantom and photodiode dosimeters (Koyama et al. 2010). The authors found that the effective dose for chest tomosynthesis using the routine imaging technique suggested for this imaging system was 0.92 mSv.

As can be seen, with the current recommended imaging techniques, chest tomosynthesis results in a higher effective dose than a complete two-view radiographic examination (~0.15 mSv to 0.9 mSv versus ~0.06 mSv), with the difference between the two modalities depending on the imaging system used. However, the clinical performance of tomosynthesis seems to be a substantial improvement over radiography. Furthermore, one study reported on a technique that resulted in an effective dose similar, and even lower, to that of radiography with good results. Finally, if chest tomosynthesis can be used to replace chest CT for any clinical indication (Johnsson et al. 2010), then the dose savings would be very substantial, since the effective dose from CT is in the range of 4.0–18.0 mSv (Mettler et al. 2008).

4.4.2 BREAST TOMOSYNTHESIS

Although various studies on the normalized breast glandular dose due to breast tomosynthesis imaging, and how it compares to that of mammography, were reported even before tomosynthesis was being used in the clinic (Sechopoulos et al. 2007; Ma et al. 2008; Sechopoulos and D'Orsi 2008; Dance et al. 2011), these studies did not address the magnitude of glandular dose in tomosynthesis in absolute terms, and how these absolute values compare to mammography. As a baseline to compare mammography to tomosynthesis dose, it can be considered that using data from 5021 acquisitions during the DMIST trial, Hendrick et al. reported that the average of the breast mean glandular dose per mammographic view was 2.37 mGy for screen-film mammography and 1.86 mGy for digital mammography (Hendrick et al. 2010).

Many early clinical studies with breast tomosynthesis report dose values with little detail. In a study of 30 women comparing digital mammography to tomosynthesis using two views of one breast per patient, Good et al. specify that the tomosynthesis dose was comparable to a digital mammography examination, with an average mid-breast dose of ~2 mGy per view (Good et al. 2008). For this study, an early Hologic prototype was used that is different from the one currently in clinical use. However, no further details are given. In a later report, Gur et al. also gives little detail but states that the radiation dose for tomosynthesis was the same as that of a mammographic acquisition (Gur et al. 2009). It is not clear which system or systems were used for this study. In another early clinical tomosynthesis study, also using an early Hologic prototype, Poplack et al. specify that the tomosynthesis acquisition resulted in approximately 4 mSv, presumably in equivalent dose, to a breast of average thickness, resulting in double the dose per view in digital mammography (Poplack et al. 2007). In a large study with 513 participants by Teertstra et al., again using a Hologic prototype, it is reported that the tomosynthesis exposure was set so that the tomosynthesis dose was "comparable" to that due to digital mammography (Teertstra et al. 2010). The authors report that the dose between the two modalities was essentially equivalent when determined using a specific breast phantom. Although they state that for the patient images the mammography technique was set by the AEC and the tomosynthesis technique by a "technique chart," it is not immediately clear that the equivalence between mammography and tomosynthesis dose was consistent among the different imaged breasts.

With a tomosynthesis prototype from another manufacturer (based on the Mammomat Novation[DR], Siemens, Erlangen, Germany), Andersson et al. specify that the same target/filter material combination and tube voltage were used for both the digital mammography and the breast tomosynthesis acquisitions, but that the tube current–exposure time product for tomosynthesis was set at double that of digital mammography, which was set by the AEC (Andersson et al. 2008). Therefore, in this study, to within ~10% according to the previously discussed results on tomosynthesis dose variation with projection angle, the tomosynthesis dose was approximately double the mammography dose.

Finally, in another large study, with a prototype system based on a third manufacturer (based on the Senographe DS, GE Healthcare, Buc, France), Gennaro et al. report on a study that compared dual-view mammography to single-view tomosynthesis (Gennaro et al. 2010). For this study, the single-view tomosynthesis acquisition protocol was manually set so that the resultant breast dose would not be higher than that from a two-view mammographic examination. This was achieved using a thickness-dependent technique chart, resulting in essentially equivalent dose values between tomosynthesis and mammography, except for the thickest breasts (>70 mm thick), in which the glandular dose for tomosynthesis did not increase further, while for mammography they continued to do so.

It is apparent that depending on the system manufacturer, the development stage, and the foreseen protocol for the acquisition(s) of a tomosynthesis exam (one vs. two views), the breast glandular dose from tomosynthesis in these studies ranged from being equivalent to mammography to being twice that of mammography. However, although all these studies discuss tomosynthesis dosimetry either in absolute terms and/or in comparison to mammography, their utility in answering the question posed at the beginning of this section is limited. This is because all of these studies are patient trials, as opposed to retrospective reviews of imaging techniques used during tomosynthesis imaging performed as standard of care. In addition, it is not apparent that any of these studies were performed using the commercial versions of the systems that are being or will be used clinically. These limitations are a major consequence of the novelty of breast tomosynthesis, and the very limited time it has been in use as part of routine standard of care. It is certain that once the technology is more widespread in the clinic, more studies will report on the technique factors and the resulting breast dose obtained via a retrospective chart review. It is certain that, at least in the early stages, most of the clinical tomosynthesis acquisitions will be performed as an adjunct to digital mammography acquisitions,[*] so a clear comparison of the dosimetric characteristics of each will be able to be performed on the same patients.

An alternative method of obtaining a comparison of the dosimetry of mammography to that of tomosynthesis was reported by Feng and Sechopoulos (2012). For this study, the authors used homogeneous breast phantoms of varying thickness and glandular fraction equivalent to characterize the technique factors automatically chosen by a clinical breast tomosynthesis system (Selenia Dimensions, Hologic, Inc., Bedford, MA) to perform mammography and tomosynthesis acquisitions. After modeling the characteristics of each x-ray spectrum used by the system, the authors used a Monte Carlo simulation to estimate the normalized mean glandular dose resulting from the acquisition techniques automatically selected for each

---

[*] In fact, the first FDA-approved breast tomosynthesis system for clinical use in the United States, the Hologic Selenia Dimensions (Hologic, Inc., Bedford, MA), has been approved for use as an adjunct to mammography only.

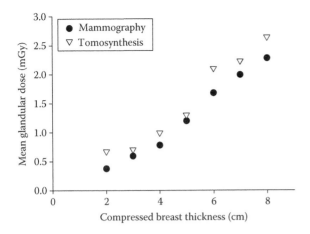

Figure 4.9 Mean glandular dose variation with compressed breast thickness for mammography and tomosynthesis for a breast with 50% glandular fraction. These values were estimated using a Monte Carlo simulation and breast phantoms to determine the imaging technique selected by the automatic exposure control of the Selenia Dimensions tomosynthesis imaging system (Hologic, Inc., Bedford, MA). (Adapted from Feng, S. S. J., I. Sechopoulos. 2012. *Radiology* 263: 35–42.)

breast phantom. The combination of these results allowed the authors to estimate the mean glandular dose resulting from both mammography and tomosynthesis acquisitions with this system. The authors found that for the average breast (defined as 5 cm thick with a composition of 50% glandular tissue) the mean glandular dose was 1.30 mGy for tomosynthesis and 1.20 mGy for mammography, yielding a difference of only 8%. These values varied considerably with breast size and composition however, resulting in a range of the ratio between mean glandular dose due to tomosynthesis to that due to mammography of 0.670–2.45. Figure 4.9 shows how the mean glandular dose varied with compressed breast thickness for a breast with 50% glandular fraction imaged with either mammography or tomosynthesis.

It is of interest that the glandular dose values for both tomosynthesis and mammography for the average breast with this system are lower than the average glandular dose from digital mammography (and of course screen-film) from the DMIST trial as reported by Hendrick et al. for all system manufacturers for breasts with comparable breast thickness (Hendrick et al. 2010). This may be due to advancements in digital detectors that have allowed the manufacturers to reduce the dose required to obtain an adequate image quality. It must be noted that this study, as mentioned, involved the use of breast phantoms to characterize the AEC response, which might result in some deviations from the behavior with patient breasts. As discussed above, once retrospective chart reviews of patient mammographic and tomosynthesis studies are available, these will provide the most reliable estimates of tomosynthesis dose in absolute terms and comparisons to mammography dose.

4.4.3 OTHER CLINICAL APPLICATIONS

Bacher et al. compared the dose characteristics and image quality of tomosynthesis, body multislice CT, and dedicated cone-beam CT (CBCT) for paranasal sinus imaging using the Rando anthropomorphic phantom and TLDs (Bacher et al. 2009). The authors found that for all imaging modalities the salivary glands, brain, and thyroid were the organs receiving the highest dose. In addition, the dedicated CBCT modality resulted in the lowest effective dose (30 μSv), compared to tomosynthesis (65 μSv) and low- and standard-dose CT (200 and 1400 μSv, respectively).

Regarding organ dose to the eye lens from sinus imaging, a concern for deterministic effects, specifically induction of cataracts, Machida et al. found that the absorbed dose to the eye lens from tomosynthesis imaging was 112 μGy versus 32,500 μGy from multislice CT (Machida et al. 2011). In another clinical application but also involving imaging the head, the use of tomosynthesis for pediatric facial bone imaging was investigated (King et al. 2011). It was found that while the image quality of tomosynthesis was improved over radiography and was good enough to replace CT, the absorbed dose to the eye lens from tomosynthesis varied from 0.3 mGy to 3.7 mGy, while for digital radiography it was between 0.1 mGy and 0.4 mGy, and for CT it was 26 mGy. Clearly, in contrast to CT acquisitions, the possibility of performing the tomosynthesis acquisition of the head in the PA view, and therefore having the eye lens shielded by the head from the direct exposure to the incident x-rays greatly reduces the dose to this radiosensitive tissue.

In the same paper mentioned previously, Koyama et al. also investigated the effective dose due to head, abdomen, and hip-joint tomosynthesis imaging using an anthropomorphic phantom and photodiode dosimeters (Koyama et al. 2010). For these clinical applications, they found effective dose values of 0.07, 1.12, and 0.82 mSv, respectively. By way of comparison, average effective dose values reported for radiographs of the head, abdomen, and hip are 0.1, 0.7, and 0.7 mSv, respectively (Mettler et al. 2008). For CT scans, head, abdomen, and pelvis CT result, on average, in effective dose values of 2, 8, and 6 mSv, respectively (Mettler et al. 2008). Clearly, the use of tomosynthesis for any of these applications results in similar effective dose compared to planar radiography and could introduce significant dose savings if it replaces CT for any clinical application. Interestingly, although Koyama et al.'s findings on effective dose for chest imaging are considerably higher than those reported by other investigators, the value reported for head imaging (0.07 mSv) is similar to that found by Bacher et al., as described above.

Tomosynthesis has also been investigated for the detection of urinary stones, again comparing its performance and dosimetric characteristics to digital radiography and using CT as the gold standard. In one report, Mermuys et al. found that the effective dose from tomosynthesis imaging for this clinical application was slightly above that for digital radiography (0.85 mSv vs. 0.5 mSv, respectively), while providing significant dose savings over CT (2.5 mSv and 12.6 mSv for low- and high-dose multislice CT, respectively) (Mermuys et al. 2010).

Imaging of the wrist and hands of 30 patients with rheumatoid arthritis with digital radiography and tomosynthesis was reported by Canella et al. (2011). The authors found that while clinical performance with tomosynthesis was significantly superior to radiography for detection of hand and wrist bone erosions, the average effective dose for tomosynthesis was 0.11655 μSv compared to 0.0441 μSv for radiography. Of course, these dose values are minimal so their increase with tomosynthesis should not limit the introduction of this technology for this application.

In summary, it seems that from the dosimetric point of view, tomosynthesis tends to result in a mild increase in patient dose

compared to its planar radiography/mammography equivalent. In some applications and according to some reports, however, there is no dose difference, while for some this increase might be twofold. However, this possible increase in dose seems to be accompanied by very substantial improvements in clinical performance. This is even more important considering the relative lack of experience with tomosynthesis compared to the well-established radiography/mammography modalities, from both the technology and the observer point of view. In addition, further technique optimization and validation studies are still needed, as shown by the study by Kim et al., in which it seems that the dose may have been reduced by two-thirds for a specific chest tomosynthesis clinical application while still providing adequate diagnostic image quality. Finally, tomosynthesis, even in its present unoptimized form, results in typically at least one order of magnitude in dose savings over CT. It is still not clear in which clinical applications, if any, tomosynthesis could replace CT, but given the large dose savings this possibility is well worth investigating.

ACKNOWLEDGMENT

I thank Dr. John Sabol for the very useful scientific discussions.

REFERENCES

Agostinelli, S., J. Allison, K. Amako et al. 2003. Geant4—A simulation toolkit. *Nuclear Instruments and Methods in Physics Research Section A: Accelerators, Spectrometers, Detectors and Associated Equipment* 506: 250–303.

Allison, J., K. Amako, J. Apostolakis et al. 2006. Geant4 developments and applications. *IEEE Transactions on Nuclear Science* 53: 270–278.

Andersson, I., D. Ikeda, S. Zackrisson et al. 2008. Breast tomosynthesis and digital mammography: A comparison of breast cancer visibility and BIRADS classification in a population of cancers with subtle mammographic findings. *European Radiology* 18: 2817–2825.

Bacher, K., K. Mermuys, J. Casselman, H. Thierens. 2009. Evaluation of effective patient dose in paranasal sinus imaging: Comparison of cone beam CT, digital tomosynthesis and multi slice CT. *World Congress on Medical Physics and Biomedical Engineering*, Munich, Germany, Springer, Berlin, Heidelberg. 25/3: 458–460.

Båth, M., A. Svalkvist, A. von Wrangel, H. Rismyhr-Olsson, Å. Cederblad. 2010. Effective dose to patients from chest examinations with tomosynthesis. *Radiation Protection Dosimetry* 139: 153–158.

Boone, J.M. 1999. Glandular breast dose for monoenergetic and high-energy X-ray beams: Monte Carlo assessment. *Radiology* 213: 23–37.

Canella, C., P. Philippe, V. Pansini et al. 2011. Use of tomosynthesis for erosion evaluation in rheumatoid arthritic hands and wrists. *Radiology* 258: 199–205.

Cavagnetto, F., R. Bampi, M. Calabrese, F. Chiesa, G. Taccini. 2011. The use of metal oxide semiconductor field effect transistor (MOSFET) in digital breast tomosynthesis. *European Congress of Radiology*, Vienna, Austria.

Chawla, A. S., J. Y. Lo, J. A. Baker, E. Samei. 2009. Optimized image acquisition for breast tomosynthesis in projection and reconstruction space. *Medical Physics* 36: 4859–4869.

Cristy, M. 1980. Mathematical phantoms representing children of various ages for use in estimates of internal dose. *ORNL/NUREG/TM-367*. Oak Ridge; Tennessee: Oak Ridge National Laboratory.

Dance, D. R. 1990. Monte Carlo calculation of conversion factors for the estimation of mean glandular breast dose. *Physics in Medicine and Biology* 35: 1211–1219.

Dance, D. R., K. C. Young, R. E. van Engen. 2011. Estimation of mean glandular dose for breast tomosynthesis: Factors for use with the UK, European and IAEA breast dosimetry protocols. *Physics in Medicine and Biology* 56: 453.

Das, M., H. C. Gifford, J. M. O'Connor, S. J. Glick. 2009. Evaluation of a variable dose acquisition technique for microcalcification and mass detection in digital breast tomosynthesis. *Medical Physics* 36: 1976–1984.

Deller, T., K. N. Jabri, J. M. Sabol et al. 2007. Effect of acquisition parameters on image quality in digital tomosynthesis. *Medical Imaging 2007: Physics of Medical Imaging*, San Diego, CA, USA, SPIE. 6510: 65101 L-11.

Dobbins, J. T., 3rd. 2009. Tomosynthesis imaging: At a translational crossroads. *Medical Physics* 36: 1956–1967.

Feng, S. S. J., I. Sechopoulos. 2012. Clinical digital breast tomosynthesis system: Dosimetric characterization. *Radiology* 263: 35–42.

Gang, G. J., J. Lee, J. W. Stayman et al. 2011. Analysis of Fourier-domain task-based detectability index in tomosynthesis and cone-beam CT in relation to human observer performance. *Medical Physics* 38: 1754–1768.

Gennaro, G., A. Toledano, C. di Maggio et al. 2010. Digital breast tomosynthesis versus digital mammography: A clinical performance study. *European Radiology* 20: 1545–1553.

Glick, S. J., X. Gong. 2006. Optimal spectra for indirect detector breast tomosynthesis. *Proceedings of SPIE* 6142: 61421 L-9.

Gomi, T., M. Nakajima, H. Fujiwara, T. Umeda. 2011. Comparison of chest dual-energy subtraction digital tomosynthesis imaging and dual-energy subtraction radiography to detect simulated pulmonary nodules with and without calcifications: A phantom study. *Academic Radiology* 18: 191–196.

Good, W. F., G. S. Abrams, V. J. Catullo et al. 2008. Digital breast tomosynthesis: A pilot observer study. *American Journal of Roentgenology* 190: 865–869.

Gur, D., G. S. Abrams, D. M. Chough et al. 2009. Digital breast tomosynthesis: Observer performance study. *American Journal of Roentgenology* 193: 586–591.

Hammerstein, G. R., D. W. Miller, D. R. White et al. 1979. Absorbed radiation dose in mammography. *Radiology* 130: 485–491.

Hendrick, R. E., E. D. Pisano, A. Averbukh et al. 2010. Comparison of acquisition parameters and breast dose in digital mammography and screen-film mammography in the american college of radiology imaging network digital mammographic imaging screening trial. *American Journal of Roentgenology* 194: 362–369.

Hu, Y.-H. and W. Zhao. 2011. The effect of angular dose distribution on the detection of microcalcifications in digital breast tomosynthesis. *Medical Physics* 38: 2455–2466.

Huang, S.-Y., J. M. Boone, K. Yang, A. L. C. Kwan, N. J. Packard. 2008. The effect of skin thickness determined using breast CT on mammographic dosimetry. *Medical Physics* 35: 1199–1206.

International Commission on Radiological Protection. 1977. *ICRP Publication 26: Recommendations of the International Commission on Radiological Protection*. Oxford; New York: Published for the International Commission on Radiological Protection by Pergamon Press.

International Commission on Radiological Protection. 1991. *ICRP Publication 60: 1990 Recommendations of the International Commission on Radiological Protection*. Oxford; New York: Published for the International Commission on Radiological Protection by Pergamon Press.

International Commission on Radiological Protection. 2007. *ICRP Publication 103: The 2007 Recommendations of the International Commission on Radiological Protection*. Oxford; New York:

Published for the International Commission on Radiological Protection by Pergamon Press.

Johnsson, Å. A., J. Vikgren, A. Svalkvist et al. 2010. Overview of two years of clinical experience of chest tomosynthesis at Sahlgrenska University Hospital. *Radiation Protection Dosimetry* 139: 124–129.

Jung, H. N., M. J. Chung, J. H. Koo, H. C. Kim, K. S. Lee. 2012. Digital tomosynthesis of the chest: Utility for detection of lung metastasis in patients with colorectal cancer. *Clinical Radiology* 67: 232–238.

Kim, E. Y., M. J. Chung, Y. H. Choe, K. S. Lee. 2012. Digital tomosynthesis for aortic arch calcification evaluation: Performance comparison with chest radiography with CT as the reference standard. *Acta Radiologica* 53: 17–22.

Kim, E. Y., M. J. Chung, H. Y. Lee et al. 2010. Pulmonary mycobacterial disease: Diagnostic performance of low-dose digital tomosynthesis as compared with chest radiography. *Radiology* 257: 269–277.

King, J. M., S. Hickling, I. A. Elbakri, M. Reed, J. Wrogemann. 2011. Dose and diagnostic image quality in digital tomosynthesis imaging of facial bones in pediatrics. *Medical Imaging 2011: Physics of Medical Imaging*, Lake Buena Vista, Florida, USA, SPIE. 7961: 79611B-7.

Koyama, S., T. Aoyama, N. Oda, C. Yamauchi-Kawaura. 2010. Radiation dose evaluation in tomosynthesis and C-arm cone-beam CT examinations with an anthropomorphic phantom. *Medical Physics* 37: 4298–4306.

Li, C. M., W. P. Segars, G. D. Tourassi et al. 2009. Methodology for generating a 3D computerized breast phantom from empirical data. *Medical Physics* 36: 3122–3131.

Lu, Y., H. Chan, J. Wei et al. 2011. Image quality of microcalcifications in digital breast tomosynthesis: Effects of projection-view distributions. *Medical Physics* 38: 5703.

Ma, A. K. W., D. G. Darambara, A. Stewart, S. Gunn, E. Bullard. 2008. Mean glandular dose estimation using MCNPX for a digital breast tomosynthesis system with tungsten/aluminum and tungsten/aluminum + silver x-ray anode-filter combinations. *Medical Physics* 35: 5278–5289.

Machida, H., T. Yuhara, T. Mori et al. 2010. Optimizing parameters for flat-panel detector digital tomosynthesis. *Radiographics* 30: 549–562.

Machida, H., T. Yuhara, M. Tamura et al. 2012. Radiation dose of digital tomosynthesis for sinonasal examination: Comparison with multi-detector CT. *European Journal of Radiology* 81: 1140–1148.

Maidment, A., M. Albert, S. Thunberg et al. 2005. Evaluation of a photon-counting breast tomosynthesis imaging system. *Proceedings of SPIE* 5745: 572–582.

Maidment, A. D. A., C. Ullberg, K. Lindman et al. 2006. Evaluation of a photon-counting breast tomosynthesis imaging system. *Proceedings of SPIE* 6142: 61420B-11.

Mermuys, K., F. De Geeter, K. Bacher et al. 2010. Digital tomosynthesis in the detection of urolithiasis: Diagnostic performance and dosimetry compared with digital radiography with MDCT as the reference standard. *American Journal of Roentgenology* 195: 161–167.

Mettler, F. A., Jr., W. Huda, T. T. Yoshizumi, M. Mahesh. 2008. Effective doses in radiology and diagnostic nuclear medicine: A catalog. *Radiology* 248: 254–263.

Metz, S., P. Damoser, R. Hollweck et al. 2005. Chest radiography with a digital flat-panel detector: Experimental receiver operating characteristic analysis. *Radiology* 234: 776–784.

Nelson, T. R., L. I. Cervino, J. M. Boone, K. K. Lindfors. 2008. Classification of breast computed tomography data. *Medical Physics* 35: 1078–1086.

Nishikawa, R. M., I. Reiser, P. Seifi, C. J. Vyborny. 2007. A new approach to digital breast tomosynthesis for breast cancer screening. *Proceedings of SPIE* 6510: 65103C-8.

Packard, N., J. M. Boone. 2007. Glandular segmentation of cone beam breast CT volume images. *Proceedings of SPIE* 6510: 651038-8.

Poplack, S. P., T. D. Tosteson, C. A. Kogel, H. M. Nagy. 2007. Digital breast tomosynthesis: Initial experience in 98 women with abnormal digital screening mammography. *American Journal of Roentgenology* 189: 616–623.

Quaia, E., E. Baratella, V. Cioffi et al. 2010. The value of digital tomosynthesis in the diagnosis of suspected pulmonary lesions on chest radiography: Analysis of diagnostic accuracy and confidence. *Academic Radiology* 17: 1267–1274.

Reiser, I., R. M. Nishikawa. 2010. Task-based assessment of breast tomosynthesis: Effect of acquisition parameters and quantum noise. *Medical Physics* 37: 1591–1600.

Ren, B., C. Ruth, J. Stein et al. 2005. Design and performance of the prototype full field breast tomosynthesis system with selenium based flat panel detector. *Proceedings of SPIE* 5745: 550–561.

Ren, B., C. Ruth, T. Wu et al. 2010. A new generation FFDM/tomosynthesis fusion system with selenium detector. *Proceedings of SPIE* 7622: 76220B-11.

Rogers, D. W. O., J. E. Cygler. 2009. *Clinical Dosimetry Measurements in Radiotherapy: Proceedings of the American Association of Physicists in Medicine Summer School, Colorado College, Colorado Springs, Colorado, June 21–25, 2009*. Madison, Wisconsin: Medical Physics Pub.

Sabol, J. M. 2009. A Monte Carlo estimation of effective dose in chest tomosynthesis. *Medical Physics* 36: 5480–5487.

Sarkar, V., C. Shi, P. Rassiah-Szegedi et al. 2009. The effect of a limited number of projections and reconstruction algorithms on the image quality of megavoltage digital tomosynthesis. *Journal of Applied Clinical Medical Physics* 10: 155–172.

Sechopoulos, I., C. J. D'Orsi. 2008. Glandular radiation dose in tomosynthesis of the breast using tungsten targets. *Journal of Applied Clinical Medical Physics* 9: 161–171.

Sechopoulos, I., S. S. J. Feng, C. J. D'Orsi. 2010. Dosimetric characterization of a dedicated breast computed tomography clinical prototype. *Medical Physics* 37: 4110–4120.

Sechopoulos, I., C. Ghetti. 2009. Optimization of the acquisition geometry in digital tomosynthesis of the breast. *Medical Physics* 36: 1199–1207.

Sechopoulos, I., S. Suryanarayanan, S. Vedantham, C. D'Orsi, A. Karellas. 2007. Computation of the glandular radiation dose in digital tomosynthesis of the breast. *Medical Physics* 34: 221–232.

Sechopoulos, I., S. Suryanarayanan, S. Vedantham, C. J. D'Orsi, A. Karellas. 2008. Radiation dose to organs and tissues from mammography: Monte carlo and phantom study. *Radiology* 246: 434–443.

Svalkvist, A., L. G. Månsson, M. Båth. 2010. Monte Carlo simulations of the dosimetry of chest tomosynthesis. *Radiation Protection Dosimetry* 139: 144–152.

Teertstra, H., C. Loo, M. van den Bosch et al. 2010. Breast tomosynthesis in clinical practice: initial results. *European Radiology* 20: 16–24.

Thacker, S. C., S. J. Glick. 2004. Normalized glandular dose (DgN) coefficients for flat-panel CT breast imaging. *Physics in Medicine and Biology* 49: 5433–5444.

Uffmann, M., U. Neitzel, M. Prokop et al. 2005. Flat-panel–detector chest radiography: Effect of tube voltage on image quality. *Radiology* 235: 642–650.

Vikgren, J., S. Zachrisson, A. Svalkvist et al. 2008. Comparison of chest tomosynthesis and chest radiography for detection of pulmonary nodules: human observer study of clinical cases. *Radiology* 249: 1034–1041.

Wu, T., B. Liu, R. Moore, D. Kopans. 2006. Optimal acquisition techniques for digital breast tomosynthesis screening. *Proceedings of SPIE* 6142: 61425E-11.

Wu, X., G. T. Barnes, D. M. Tucker. 1991. Spectral dependence of glandular tissue dose in screen-film mammography. *Radiology* 179: 143–148.

System design

Wu, X., E. L. Gingold, G. T. Barnes, D. M. Tucker. 1994. Normalized average glandular dose in molybdenum target-rhodium filter and rhodium target-rhodium filter mammography. *Radiology* 193: 83–89.

Yaffe, M. J., J. M. Boone, N. Packard et al. 2009. The myth of the 50–50 breast. *Medical Physics* 36: 5437–5443.

Yamada, Y., M. Jinzaki, I. Hasegawa et al. 2011. Fast scanning tomosynthesis for the detection of pulmonary nodules: diagnostic performance compared with chest radiography, using multidetector-row computed tomography as the reference. *Investigative Radiology* 46: 471–477.

Yang, X., I. Sechopoulos, B. Fei. 2011. Automatic tissue classification for high-resolution breast CT images based on bilateral filtering *Proceedings of SPIE* 7962: 79623H.

Zhao, W., R. Deych, E. Dolazza. 2005. Optimization of operational conditions for direct digital mammography detectors for digital tomosynthesis. *Proceedings of SPIE* 5745: 1272–1281.

Zhou, J., B. Zhao, W. Zhao. 2007. A computer simulation platform for the optimization of a breast tomosynthesis system. *Medical Physics* 34: 1098–1109.

5

Tomosynthesis with circular orbits

Grant M. Stevens and Norbert J. Pelc

Contents

5.1 INTRODUCTION

5.1.1 BACKGROUND

As described in other chapters, tomosynthesis can be applied to a variety of clinical applications, and images can be obtained using a variety of system geometries. Because of the many options in image acquisition, in fact, Chapter 2 is entirely devoted to tomosynthesis system geometries. In general, as the relative motion of the x-ray source becomes more complicated, there is a greater opportunity to effectively blur the structures overlying the plane of interest. This chapter investigates the use of circular motion, and how it can lead to effective blurring of off-plane structures. In addition, design of filters to be used with (filtered) backprojection allows circular motion systems to produce images that appear to come from gantries with more sophisticated geometries.

5.1.2 IMAGE QUALITY TRADE-OFFS

At the highest level, the image quality in a tomosynthesis image can be considered to depend on image contrast, noise, and artifacts. For example, consider the overlying structures in Figure 5.1. If the goal of imaging is to visualize the sphere in the center, this will depend on

- How much contrast this sphere has relative to the background.
- How much noise is present in the image.
- How many artifacts tend to obscure the sphere or surrounding tissue.

All these factors depend on the acquisition and image reconstruction because these determine the system response. The amount of contrast for the unshaded sphere depends on how well the overlying spheres are blurred. This is affected by the blur response of the relative source motion in combination with filter modification. The amount of noise present in the image depends on the overall dose of the acquisition, and whether or not it is amplified by filtering. In general, structured noise (or out-of-plane objects that can still be visualized) is more of a determining factor than x-ray noise. This structured noise can also be thought of as image artifacts. In order to understand how the system geometry directly impacts these, a short background on the types of gantry motions is useful.

Figure 5.1 Simplified example of overlying structures.

5.1.3 TOMOSYNTHESIS MOTION

The previous chapters have detailed the historical development of tomosynthesis. While the development of digital detectors has made clinical tomosynthesis more feasible, these detectors still have to be mounted on gantries that move the x-ray source and/or detector relative to the patient. There are a number of types of motion of the source relative to the patient that can be achieved, including

- Linear (or pseudolinear, such as an arc)
- Two-dimensional (2D) linear (such as two orthogonal lines or a rectangular array of source points)
- Circular
- More complex (such as hypocycloidal)

While the simplest mechanical embodiment of tomosynthesis is a linear motion of the source/detector relative to the patient, the impact of this motion on visualization of out-of-plane objects is highly dependent on their orientation. Improper selection of the direction of motion in chest imaging, for instance, can lead to ribs that are still clearly defined in the tomosynthesis images. In the case of overlying structures with edges not entirely aligned in the preferred direction, it is difficult for linear motion to effectively blur the structures. As will be discussed below, the ability of a tomosynthesis system to effectively blur overlying structures improves as the motion becomes more complicated. The downside of this is that, as the complexity of the motion profile increases, gantries become harder to build and maintain. As will be shown below, however, reconstruction methods can also be employed to improve the compromise between mechanical limitations and anatomical visualization.

On a clinical tomosynthesis system, the patient is held stationary while the source and/or detector move relative to the patient. This is done because, while only the relative motion is critical for tomosynthesis, keeping the patient stationary is necessary for clinical imaging. As will be apparent in the sections below, experimentation done by the authors to test theoretical development was performed on a system in which the source and the detector remain fixed while a phantom is moved. While this approach prevents direct clinical investigations with such a system, it does allow greater flexibility in investigating the impact of the gantry geometry on tomosynthesis images. Moreover, this type of system can be used to understand the optimal setup prior to the construction of a gantry that can mechanically manipulate the source and/or detector as needed in a clinical environment.

5.1.3.1 Linear motion

It is useful to quickly revisit the motions here in order to understand the motivation for moving to a circular motion and the corresponding filtering work. In particular, the blur response of the system can be examined to evaluate expected performance.

For linear (or arc) motion, the source moves in a one-dimensional (1D) profile relative to the phantom/patient. Intuitively, this is a relatively easy motion to achieve mechanically. In fact, clinical tomosynthesis systems with this type of motion are currently available [GE, Hologic]. The impact of this motion is shown in Figure 5.2. For an infinite number of views, out-of-plane points are blurred into lines. For a finite number of views, the out-of-plane structures are blurred into a discretely sampled line, with the points becoming further apart as the distance away from the focal plane is increased. High-contrast objects off the focal plane can lead to linear streak artifacts, which can obscure anatomy in the tomosynthesis images [Macovski83].

As shown in Figure 5.2, only a small portion of Fourier's space is sampled when using linear motion. As described in detail below, filter modification cannot recover data in regions of the Fourier space that are not sampled. Instead, it can only be used to modify the weighting in frequencies that are sampled. Therefore, the best that one could hope to do is to recover a line-type blur function. This retains the shortcomings of a 1D blur response, which does not effectively remove sharp structures that are aligned parallel to the source motion. One could consider extending this to a 2D linear array, such as with orthogonal linear sweeps or a rectangular source array. For a multisource array [XinRay], this may be possible. For a conventional x-ray tube, however, this type of motion profile could be difficult to achieve in a reasonable imaging time due to the required path changes.

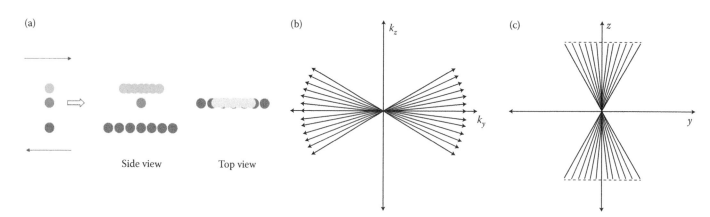

(a)   Side view   Top view
(b) k_z k_y
(c) z y

Figure 5.2 Linear motion tomosynthesis. (a) Spatial domain, (b) Fourier's domain sampling, and (c) impulse response.

System design

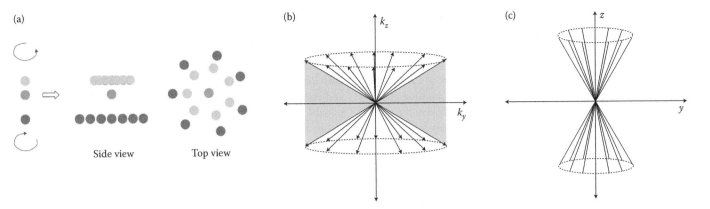

Figure 5.3 Circular motion tomosynthesis. (a) Spatial domain, (b) Fourier's domain sampling, and (c) impulse response.

5.1.3.2 Circular motion

For circular motion, the source moves in a circular profile relative to the phantom/patient. This type of acquisition, while still a relatively simple motion profile to achieve, brings the added benefit of a 2D blur response (see Figure 5.3). This can allow better visualization of in-plane versus out-of-plane structures than in the case of 1D blur responses. For an infinite number of views, out-of-plane points are blurred into circles. It can be shown that the radius of each circle is proportional to height of the object plane and the tangent of the tomographic angle [Stevens_th]. For a finite number of views, the out-of-plane structures are blurred into a discretely sampled circle, with the points becoming further apart along the circumference as the distance away from the focal plane is increased.

The sampling of Fourier space by circular tomosynthesis is not uniform. In computed tomography (CT), which also has nonuniform sampling, correction for this uneven sampling density improves the reconstructed images. In tomosynthesis, filtering can also be used to manipulate the effect of sampling density. This can improve the reconstruction of off-plane structures without corrupting in-plane structures. In addition, as shown in Figure 5.3, a larger portion of Fourier space is sampled using circular motion than with linear acquisition. Because of this, there are more frequencies sampled that are available for modification in a filtering step. This is discussed in detail below.

Another perspective is to interpret circular tomosynthesis as a limited-angle volumetric CT acquisition. Within this framework, one of the positive attributes of circular tomosynthesis is its flexibility. By adjusting the tomographic angle of the gantry, a trade-off between image quality and available space around the patient can be made. This can be seen by examining the three-dimensional (3D) impulse response (see Figure 5.3). As the tomographic angle decreases, the opening of the cone in the impulse response becomes smaller. The visual appearance of off-plane structures in the reconstructed image is modified by reducing the blurring radius. For an infinite number of views, larger tomographic angles provide better blurring of off-plane objects because the area of the blurring is larger (over which the contrast of the structure can be spread). For a limited number of views, there is a trade-off between the tomographic angle and the number of views required to adequately reconstruct an object, since the discrete nature of the impulse response becomes evident

sooner for a larger cone angle. This is demonstrated later for circular tomosynthesis.

An extended 3D object is affected by circular tomosynthesis in the same way as the point object described above. The extended object can be analyzed by treating it as a group of a number of small voxels, each of which acts like a point object. The portion of the object lying on the imaging plane is faithfully reconstructed while the portion off the plane is blurred as above. Alternatively, one may view this by convolving the 3D impulse response with the 3D object.

Because not all frequencies are sampled during acquisition (i.e., the cone in Figure 5.3), a perfect reconstruction is not possible in circular tomosynthesis. As the tomographic angle increases, the reconstructions improve since the region of Fourier space that is not sampled shrinks. In the limit of the tomographic angle going to $\pi/2$, all of Fourier space is covered (which is expected since this is the geometry for volumetric CT). In the limit of the tomographic angle going to 0, only the horizontal plane in Fourier's space is sampled. This yields no resolution in the z direction (as expected since this is the projection radiography geometry).

5.1.3.3 Complex motions

As discussed above, the type of blurring in tomosynthesis is dependent on the motion profile during acquisition. This leads to points being blurred into lines for linear motion and rings for circular motion. More complex motions, such as hypocycloidal, result in blur functions, which are more appealing (i.e., less disturbing) to the observer [Carter63]. More complex motions, however, require more costly equipment and can lead to longer exposures [Stanton88]. Because of this, it is desirable to have a relatively simple motion that can be used for acquisition, and to handle the data appropriately to generate images that show similar blur responses to more complex motions. An approach to this is described below.

5.1.4 TOMOSYNTHESIS IMAGE FORMATION

Chapters 7 and 8 of this book are focused on image reconstruction, including overviews and detailed descriptions. This chapter on circular motion tomosynthesis focuses on the use of filtered backprojection (Chapter 7). While this is not required for the investigation of circular motion, it does provide a natural framework for investigating how reconstruction can be modified to maximize the visualization improvement provided by tomosynthesis. Unlike computed tomography, the goal of tomosynthesis is not to faithfully reconstruct an object. Instead, the goal is to faithfully reconstruct a plane in the object, while blurring objects that lie above and below the plane of interest in a manner that is most easily (visually) ignored by a radiologist. This blurring function is dependent on both the system motion and the image reconstruction. The goal of this chapter is the investigation of this for circular motion systems.

5.2 DATA ACQUISITION

As discussed above, there is a great deal of flexibility in tomosynthesis, in both acquisition of the projection data and formation of the images. At the same time, this means that there are a number of choices that must be made during the process of generating a tomosynthesis image. In terms of acquiring the data, this includes selection of the tomographic angle and the number of projections. These parameters are interdependent, and also depend on the anatomical structures present and the image reconstruction. In addition, while not a choice made by the user, the actual geometry of the gantry during a scan must be accounted for in order to make high-quality images. The impact of the system geometry is discussed below; the selection of tomographic angle and number of projections are discussed in the following sections (with clinically relevant examples where the dependencies can be examined).

5.2.1 PHANTOM SYSTEM NOTE

As noted earlier, the x-ray tube and detector move about a stationary patient in a clinical system. For investigational work in circular tomosynthesis, however, a more flexible development environment is desirable. Therefore, the authors developed a phantom-based system in which the phantom rotates while the tube and detector remain stationary (see Figure 5.4). This is equivalent (up to a rotation of the detector) to rotating the

object while the tube and detector are fixed in space at an angle relative to the object's axis of rotation. The phantom system used was built using a prototype digital x-ray detector (GE Medical Systems), which was a 20 × 20 cm indirect conversion detector with 200 micron pixel pitch [Stevens_th]. Mounting the detector and x-ray tube/collimator assembly on a gantry with adjustments possible along multiple axes allowed a variety of system geometry setups for development work.

For the type of phantom system described in this chapter, or for a clinical tomosynthesis system, each of the projections in a tomosynthesis dataset looks like a conventional radiograph (albeit at an angle relative to the conventional imaging plane). As in CT, the information in the additional spatial dimension comes from the combination of all the projections. For the phantom system, we define the conventional imaging plane as a plane that is orthogonal to the axis of rotation of the positioning stage. It should be noted that the same formalism that is used with the phantom-based system can be adapted to a clinical system.

5.2.2 SYSTEM ALIGNMENT

One of the concerns with a tomosynthesis system consisting of heavy moving objects is how well the system can be aligned (and how well it remains in alignment). The required precision of the alignment that must be achieved and maintained is ultimately dependent on how sensitive the images are to system alignment. As a benchmark, if the impact of misalignments can be seen in the tomosynthesis images, the misalignments must be accounted for. This could be done by either more precisely aligning the system or developing a correction scheme to handle the misalignments.

It is realistic to expect that a tomosynthesis system will have mechanical misalignments prior to setup, during acquisition (such as due to sag during motion) or developing over time (such as due to mechanical wear). Therefore, it is useful to examine the effects of misalignments in order to gain an understanding of which ones are critical, and what artifacts are introduced by their presence. This can be done analytically, using computer simulations, or using data acquired with actual physical misalignments. Examples of how system misalignments can impact the performance of the circular tomosynthesis system developed by the authors included tomographic angle inaccuracy, detector rotation, detector shift, and axis of rotation misalignment [Stevens01a].

For example, a system that is perfectly aligned except that the tomographic angle is not known exactly results in images with structures that no longer focus at the expected plane. The images are now blurred for in-plane objects (see Figure 5.5). This produces an image that is shifted and stretched. The stretching scales linearly with distance from the center of the imaging volume, becoming more apparent for structures away from the patient center. Alternatively, a system with a detector rotated about its center (and about the central ray, but still normal to the central ray) results in images that are blurred and rotated. The plane in which the structure has the least amount of blurring is still the plane where the object lies, as is demonstrated in Figure 5.6. Finally, a system with a shifted detector (a shift in the plane orthogonal to the central ray) results in significant blurring and a shift of the plane with the least blurring (see Figure 5.7). A discussion here is not meant to fully characterize the system's sensitivity, but rather to depict the type of effects seen on reconstructed images. This can

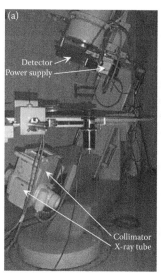

Figure 5.4 Phantom-based circular tomosynthesis system. (a) Digital detector mounted on an LU gantry (GE Healthcare) and (b) close-up of the custom rotational phantom positioning stage.

(a)

(b)

Figure 5.5 Example of impact of tomographic angle error. For the geometry of the system described in this chapter, simulations were performed to generate forward and backward projections of small spherical objects arranged in a grid (1 cm spacing). The objects were located at $z = 2$ cm, and planes were reconstructed using unfiltered backprojection (100 views) at $z = 2.00$, 1.91, 1.82, and 1.73 cm. (a) Reconstruction with the true tomographic angle (40°) and (b) reconstruction with a 4° error in the tomographic angle. (Reproduced from G. M. Stevens et al., *Medical Physics* **28**(7), 1475, 2001. With permission.)

(a)

(b)

(c)

Figure 5.6 Example of impact of detector rotation. For the geometry of the system described in this chapter, simulations were performed to generate forward and backward projections of small spherical objects arranged in a grid (1 cm spacing). The objects were located at $z = 2$ cm, and planes were reconstructed using unfiltered backprojection (100 views) at $z = 1.9$, 2.0, 2.1, and 2.2 cm. (a) Reconstruction with the correct system geometry, (b) reconstruction with a 3° error in the detector angle, and (c) difference of the two image sets. (Reproduced from G. M. Stevens et al., *Medical Physics* **28**(7), 1475, 2001. With permission.)

be the basis for specifications on system tolerances, or as motivation and targets for correction methods.

Based on these few examples, system misalignments can shift, stretch, and blur images. In addition, structures may appear better focused on a plane other than the one in which the object is

located. The particular impact depends on the specific geometrical details of the system, as well as the magnitude of the error. When there are errors in multiple parameters, which would be expected in a real system, their combined effect may be greater due to their interaction. This can be seen, for example, in Figure 5.10. This motivates the need to measure the system alignments, either for use in mechanical alignment or in a correction algorithm.

The demonstration of the sensitivity to geometrical misalignments presented above applies to an *in vitro* imaging system (as described above). While this could change somewhat for a clinical system in which the patient is stationary and the x-ray tube and detector move, the same type of approach can be taken. Certain misalignments will have smaller or larger effects between the system types, but the approach and a similar correction scheme to the one described below could also be performed on a clinical gantry.

5.2.2.1 System geometry determination

The above sensitivity examples highlighted the fact that circular tomosynthesis image quality is highly sensitive to the alignment of the image acquisition system. In order to either physically align or algorithmically correct the misalignments, the system geometry needs to be measured precisely. In the event of reproducible misalignments, either can be accomplished using scans of a calibration phantom prior to the tomosynthesis scan of interest.

The general approach for determining the system geometry is to use the system itself via multiple projections of small objects [Rougee93, Esthappan98, Hoffmann96, Gullberg90]. These images can then be used to find a set of geometrical parameters that are most consistent with the measured projections. This can be accomplished either by acquiring separate data of a calibration phantom (which requires the assumption that the gantry alignment is stable) or by using fiducial markers during the actual (phantom/clinical) scan. While investigators have used both techniques, the former approach is described here. The latter approach is identical, except that the projections of the markers are included in the object projections, rather than in an additional scan of a calibration phantom.

In order to fit the system geometry, an iterative process can be used to best match the geometry model with the projection data of the calibration phantom or fiducial markers. This can be accomplished using a calibration phantom consisting of small ball bearings (see Figure 5.8). The bearings are desirable because they allow easy identification of the projection center. While others [Rougee93, Hoffmann96, Webber97, Fahrig96] have used similar approaches in which at least the relative positions were

(a)

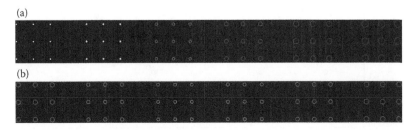

(b)

Figure 5.7 Example of impact of detector shift. For the geometry of the system described in this chapter, simulations were performed to generate forward and backward projections of small spherical objects arranged in a grid (1 cm spacing). The objects were located at $z = 2$ cm, and planes were reconstructed using unfiltered backprojection (100 views) at $z = 2.00$, 2.11, 2.22, 2.33, 2.44, and 2.55 cm (top left to lower right). (a) Reconstruction with the correct system geometry and (b) reconstruction with a 1-mm error in the detector shift (in both directions in-plane). (Reproduced from G. M. Stevens et al., *Medical Physics* **28**(7), 1477, 2001. With permission.)

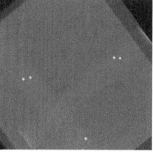

Figure 5.8 Two projection images of a calibration phantom. The phantom includes six small bearings separated by a layer of foam (relative location is not controlled). The positioning stage was rotated by 36° between the left and right images.

assumed to be known, the positions of the bearings need not be known in this method. A small number are needed, with a typical arrangement used by the authors being six bearings separated by plastic or foam over a 10 × 15 × 15 cm volume. The phantom is imaged using a small number of views, for example, 10 views over 360°, and the center of mass of the projection is determined for each bearing in each view.

A projection simulator, which generates noiseless projections of simulated point-objects, is seeded with an initial estimate of the geometry and an initial estimate of the location of each of the calibration objects. For the work performed by the authors, the projection simulator included the following geometrical parameters: (1) tomographic angle, (2) distance from the x-ray tube to the positioning stage, (3) distance from the tube to the detector, (4) incremental rotation angle of the positioning stage between projections, (5) relative locations of the objects in the calibration phantom, (6) (two) translations of the detector perpendicular to the central ray, and (7) (two) rotations of the detector relative to the central ray. A comparison of the measured and simulated projection centroids is performed, resulting in a computation of the root mean squared difference (RMS) between the predicted and measured centers. Small increments of the geometrical parameters (within an allowed range) are performed, and a new RMS statistic is calculated. These updates to the parameters are

then accepted if the fit with the measured projections is improved, and rejected otherwise. This iterative process is continued until an end criterion (such as RMS limit) is reached. An example of the output from this process is shown in Figure 5.9.

It should be noted that this approach may not give the true geometry of the system. Instead, it yields parameters that give projections that are consistent with the measured data. If the agreement is good (e.g., subpixel), however, the system can be expected to provide high spatial resolution at (or near) the location of the spheres using the fitted geometry. By using objects/markers throughout the volume of interest, the fit geometry can be used to generate high-quality backprojected images throughout the volume. This approach is acceptable because the goal ultimately is to reconstruct high-quality images, rather than necessarily to determine the system's true geometry.

5.2.3 GEOMETRICAL CORRECTION

Once the system geometry is known, it can be corrected (either mechanically or mathematically). One approach is to use a software technique to correct for gantry misalignments [Fahrig00, Concepcion92, Rizo94, Bronnikov99, Webber97, Webber00]. The goal of such an approach is to maintain both the imaging flexibility of the gantry and the low cost of the system. To do this, the system geometry is first fit (as described above). The geometric fit to the data can then be used to mathematically remap (resample) the data onto a perfectly aligned detector using the measured and fit geometries. While the remapping of the data prior to backprojection adds an interpolation step (which blurs the data), the remapping could be included in the backprojection algorithm. This would require only a single interpolation step in the entire process.

The phantom-based system developed by the authors was used to test the geometry correction technique. Qualitative comparison of images of a tomographic phantom with and without alignment (Figure 5.10) shows that the approach corrects for the misalignments. In addition, quantitative measurement of a bar pattern phantom shows that the frequency response of the reconstructed tomosynthesis image is consistent with expectations. The degradation of the square wave modulation

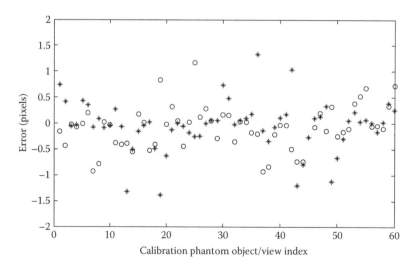

Figure 5.9 Residual errors (circles are errors in x, and asterisks are errors in y) after running the geometric fit routine on 10 views of a calibration phantom with six small bearings. The RMS statistic for this case is 0.4 pixels.

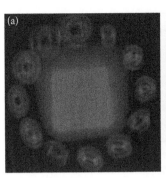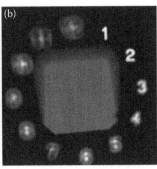

Figure 5.10 Reconstructions of a tomographic phantom (each numeral is separated in height from its neighbor by 1 mm) on the plane containing the numeral 1, for (a) data from a (misaligned) system that has not been corrected for geometrical misalignments and (b) data from a misaligned system that has been remapped using the correction scheme outlined in this chapter. (Reproduced from G. M. Stevens et al., *Medical Physics* **28**(7), 1479, 2001. With permission.)

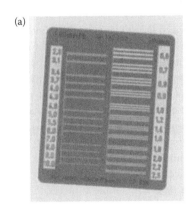

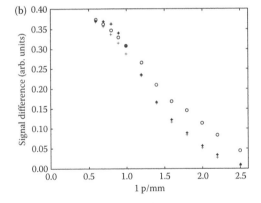

Figure 5.11 Bar pattern phantom (a) radiograph and (b) comparison of signal modulation (radiograph signal differences shown in circles, tomosynthesis in asterisks and radiation with interpolation theory in pluses) of a column of pixels in a tomosynthesis image compared with interpolation theory. (Reproduced from G. M. Stevens et al., *Medical Physics* **28**(7), 1479, 2001. With permission.)

relative to that of the radiograph is approximately as expected from the interpolation steps (see Figure 5.11).

5.3 IMAGE RECONSTRUCTION

5.3.1 FILTER DESIGN

As discussed in Chapter 2, a tomosynthesis image of a 3D object consists of a plane that is in focus, while the rest of the object is

blurred. The type of blurring of off-plane objects in tomosynthesis images is dependent on both the type of motion used during data acquisition and the image reconstruction. As mentioned earlier, this chapter will focus on the impact of filtered backprojection; alternative reconstruction techniques are described in other chapters in this book.

Many approaches have been taken to try to suppress the information from objects lying outside the plane of interest, including

- Selecting an optimal imaging geometry [Ruttiman89]
- Filtering the data [Badea98, Lauritsch98, Ghosh Roy85]
- Filling in the circle of x-ray source positions (i.e., using a more complex system motion) [Ruttiman89]
- Applying postprocessing filters [Ghosh Roy85]
- Using a linear systems approach to design filters [Lauritsch98]

With filter design, the 3D impulse response of circular tomosynthesis can be modified.* In particular, this can be used to achieve results similar to that of systems with more complex motions. For example, one could consider blurring an off-focal plane point into a disk (rather than a ring). This is similar to (but even more uniform than) the behavior of hypocycloidal motion. The filtering approach is desirable because the uniform blurring is created by manipulation of the data after collection using a relatively simple geometry, rather than by specialized mechanical motion of the x-ray tube and detector during acquisition of the projection data.

This type of approach is possible because the impulse response of the system depends on the frequencies that were sampled, and their weighting. By modifying the weighting of the frequencies that are sampled (based on the system geometry), the impulse response is changed. For example, a circular tomosynthesis dataset can be used to reconstruct a radiograph from a source position that was not acquired (if this source position is within the circular path). In particular, this applies to a radiograph with a source position located at the center of the circular path. This is possible because the region in frequency space sampled by the radiograph is a subset of the region sampled by circular tomosynthesis. Because the radiograph (at a tomographic angle of 0) samples only the horizontal plane in Fourier's space, all other frequencies in the filtered data must be set to zero, and the frequencies within the horizontal plane must be properly weighted. The resulting image would look like a radiographic image except for differences in the noise level, as discussed below.

Similarly, data acquired with circular tomosynthesis with a given tomographic angle can be filtered to produce an impulse mimicking a system with a smaller tomographic angle. This would allow an out-of-plane point to be blurred into a ring with a smaller radius. This approach can then be generalized to blurring a point to a uniform disk [Stevens01f]. Since the goal is to create a filter that will blur off-plane objects as uniformly as possible while faithfully depicting on-plane objects, the reduction in high-frequency content of a disk (as compared with

---

* While the same framework can also be used with simpler motions, the small region of frequency space that is sampled limits the potential usefulness of the filter modification.

a ring) is potentially useful. In addition, as is discussed below, no spatial information for the on-plane object is lost by the filtering.

The discussion here has focused on the generation of a filter to blur an off-focal plane point into a disk. The framework for generating the disk function, however, is more general. In particular, one could customize the filter design to avoid generating objects that mimic pathology. For instance, if a radiologist is looking for circular masses, a disk blur filter may not be an optimal choice. Again, because of the relatively large volume in 3D frequency space that is sampled by circular tomosynthesis, there is a relatively large degree of freedom in filter design/selection.

5.3.2 FILTER PERFORMANCE

In order to test the performance of the disk filter described above, both simulations and *in vitro* scans were performed (using the phantom-based system). Simulations allow the investigation of basic blurring properties, while *in vitro* scans can be used to demonstrate the potential clinical utility. Also, datasets with very large numbers of projections show capabilities, while datasets with smaller numbers of projections show performance in more clinically realistic scenarios.

An example of the change in the blur response of circular tomosynthesis due to the disk filter is shown in Figure 5.12. For unfiltered backprojection, the sphere is faithfully reconstructed on the plane in which it lies, and blurs into a ring with increasing radius and decreasing amplitude. The amplitude of the ring is inversely proportional to the diameter of the ring, and therefore falls off as the distance from the focal plane increases. As discussed previously, filtering the data prior to backprojection changes the manner in which the object is blurred. Using the disk filter, the sphere is still faithfully reconstructed on the plane in which it lies. Now, however, the sphere blurs into a uniform disk away from the focal plane. The amplitude of the disk is inversely proportional to the area of the disk, and falls off more quickly as the distance from the focal plane increases. This results in lower contrast for each pixel in the (off focal-plane) blurred image with the disk filter. In addition, there is less high-frequency information in the blurred points, as can be noted in Figure 5.13.

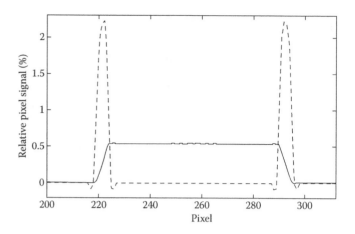

Figure 5.13 Cross-sectional plot of an off-plane reconstruction (1000 projections) of a small spherical object for unfiltered (solid) and disk-filtered (dashed) backprojection. (Reproduced from G. M. Stevens et al., *Medical Physics* **28**(3), 377, 2001. With permission.)

These are expected to lead to less obtrusive out-of-plane features in images, as can be evaluated in the clinical examples below.

While such simulations show the blur response in the limit of a very large number of projections, this is unlikely the case in a clinical setup. In order to investigate how the blur response changes with the number of views, the same simulation can be repeated using a smaller number of projections. Examples of the same simulated sphere reconstructed with 100 and 20 projections are shown in Figures 5.14 and 5.15.

With 100 projections, the ring blur transforms to discrete points, which are especially visible at increasing distances from the object. The disk blur is still uniform in the central region, but becomes discrete on the boundary. The region of smooth blurring is further reduced for both approaches in the case of 20 projections. Interestingly, in the view-starved cases, the disk profile is flat in the central region of the disk, while the profile degrades to discrete dots (with negative side lobes) in a ring-like pattern at the edge of the disk. This is an extension of what happens in unfiltered backprojection. For planes near the sphere, unfiltered reconstructions with 100 views are equivalent to those

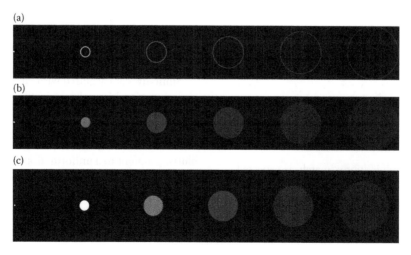

Figure 5.12 (a) Unfiltered and (b) disk-filtered backprojections of a small spherical object with 1000 views. The object is located at $z = 0$, and the reconstruction plane heights from left to right are $z = 0, 1, 2, 3, 4, 5$ cm. The disk-filtered images are shown with intensities increased 10-fold in (c) in order to allow visualization of the off-plane intensity falloff.

(a)

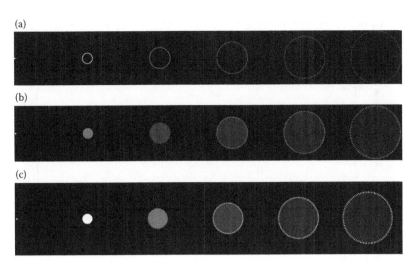

(b)

(c)

Figure 5.14 (a) Unfiltered and (b) disk-filtered backprojections of a small spherical object with 100 views. The object is located at $z = 0$, and the reconstruction plane heights from left to right are $z = 0, 1, 2, 3, 4, 5$ cm. The disk-filtered images are shown with intensities increased 10-fold in (c) in order to allow visualization of the off-plane intensity falloff.

with 1000 views. Eventually, as the distance from the object increases, the discrete nature of the backprojection is evident and the rings become discrete dots in the ring pattern. Similarly, the reconstructions with filtered backprojection with 100 and 1000 views are the same near the object. Farther from the object, the discrete dots are evident on the circumference of the disk. A plot of the pixel intensities along the circumference of the disks is shown in Figure 5.16 for unfiltered and disk-filtered backprojection with 1000, 100, and 20 views for a reconstruction 8 cm away from the object. The pixel intensities are normalized to the intensity of the on-plane image. Away from the circumference of the disk (i.e., in the interior of the disk), the relative intensity of the disk blurring function is the same as in the 1000 view case.

While such simple simulations can provide insight into the blur response and potential trade-offs, they do not adequately demonstrate the impact on clinically relevant structures. As an example of more complex structures, circular tomosynthesis images of a desiccated dog lung [Henry] are shown in Figure 5.17. The

image shows the image plane containing the first bifurcation of the bronchial tube, for both unfiltered and disk-filtered backprojection. When comparing the same image plane, one reconstructed using 100 views and one using 20 views, the plane reconstructed from 20 views shows more discrete artifacts arising from off-plane objects for both reconstructions. This is especially noticable for high spatial frequency structures (e.g., smaller airways and the edges of the bronchioles). In both cases, application of the disk filter improves the blurring of the off-plane objects. The blurring of surrounding objects is more uniform in the filtered lung reconstructions, since the filter smoothes the artifacts as the reconstruction plane moves away from the object. This appears to help define which objects are located in the plane of interest, and improves the visualization of the objects in the reconstructed plane.

In addition to the image contrast and artifact level, filtering impacts the image noise level. This can be evaluated using an analysis of the relative signal-to-noise ratio (SNR) [Stevens01f]. The dependence of the relative SNR (on the focal plane of the

(a)

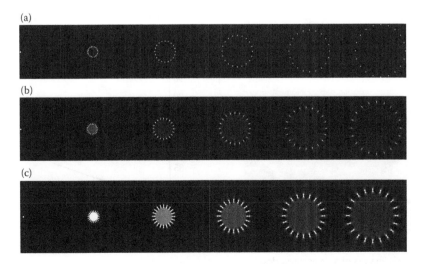

(b)

(c)

Figure 5.15 (a) Unfiltered and (b) disk-filtered backprojections of a small spherical object with 20 views. The object is located at $z = 0$, and the reconstruction plane heights from left to right are $z = 0, 1, 2, 3, 4, 5$ cm. The disk-filtered images are shown with intensities increased 10-fold in (c) in order to allow visualization of the off-plane intensity falloff.

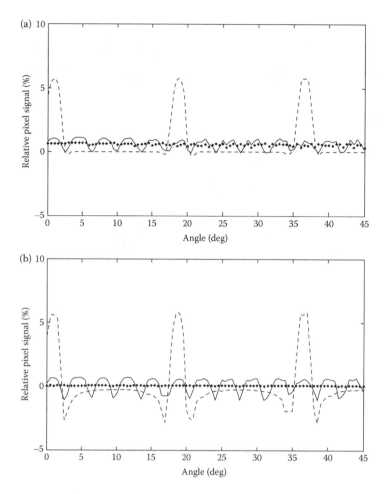

Figure 5.16 Pixel intensities along a portion of the circumference of off-plane reconstructions of a small sphere resulting from (a) unfiltered and (b) disk-filtered backprojection for 1000 views (dotted), 100 views (solid), and 20 views (dashed).

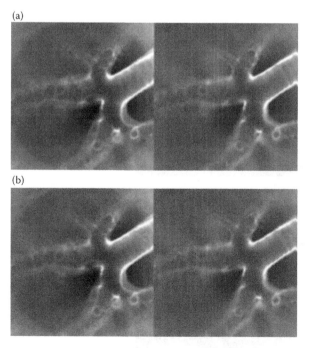

Figure 5.17 Circular tomosynthesis images of a desiccated dog lung phantom using unfiltered (left) and disk-filtered (right) backprojection, with 100 projections (a) and 20 projections (b). (Reproduced from G. M. Stevens et al., *Medical Physics* **28**(3), 378–379, 2001. With permission.)

reconstructed image) on the tomographic angle is shown in Figure 5.18 (and was confirmed by simulations). For tomographic angles in the range of 10–120°, the SNR is reduced by ~10–15% for the disk-filtered backprojections relative to the unfiltered reconstructions. Because the increase in noise due to the filtering is small, and the blurring of off-plane objects is improved, the benefits may outweigh this small noise penalty.

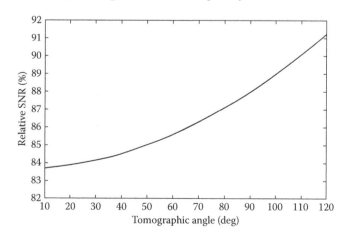

Figure 5.18 Signal-to-noise ratio for disk-filtered backprojection relative to unfiltered backprojection. (Reproduced from G. M. Stevens et al., *Medical Physics* **28**(3), 375, 2001. With permission.)

5.4 CIRCULAR TOMOSYNTHESIS EXAMPLES

There are several chapters in this book dedicated to clinical applications of tomosynthesis. In order to give more examples of how clinically relevant structures appear in circular tomosynthesis images, and to further demonstrate the performance of the disk filter, several phantom and *in vitro* examples are provided here. These include cervical spine imaging (using an anthropomorphic phantom) and breast imaging (using a phantom and *in vitro* tissue).[*]

5.4.1 CERVICAL SPINE IMAGING

As an example of how circular tomosynthesis performs with relevant anatomical structures, the authors imaged the cervical spine region of an anthropomorphic phantom [Anth_phantom]. This anthropomorphic phantom is made of bone- and tissue-equivalent plastic in the shape of a human skull and cervical spine. Tomosynthesis images were reconstructed using both unfiltered backprojection and filtered backprojection using the disk filter. Both the radiographic and tomosynthesis images were also postprocessed using unsharp masking because of the high contrast differences present in the phantom (resulting in large dynamic range requirements for image display). As in the rest of this chapter, the goal of the evaluation here is to show the relevance of the use of filtered backprojection for image reconstruction.

For comparison purposes, lateral and A–P view radiographs using the same system are shown in Figure 5.19. While the lateral view clearly depicts the cervical spine, overlying structures obscure the upper spine in the A–P view. The use of circular tomosynthesis, as shown in Figure 5.20, allows the visualization of these vertebrae. While unfiltered backprojection (with the corresponding ring blur) improves the image, the blurring of the overlying structures is made more uniform with the use of the disk filter. A large number of views are needed to adequately blur the out-of-plane objects because of the size and complex nature of the overlying objects. If the number of views is not sufficient, artifacts are present in the tomosynthesis images. This can be seen in Figure 5.20 for the case of a 50° tomographic angle dataset with 50 projections, and increasingly for the case of 20 projections.

Because the number of views that may be acquired can be limited, an approach to reduce artifacts with a smaller number of views is the reduction of the tomographic angle. The use of a 20° tomographic angle, for instance, allows the use of 50 views (see Figure 5.21). For this tomographic angle, 20 views still appears to be insufficient. The trade-off with the smaller tomographic angle is a larger slice thickness, which reduces the ability to effectively blur out-of-plane objects. Such a compromise, as evident by the loss of some detail in Figure 5.21 relative to Figure 5.20 (and more apparent in Figure 5.22), may be acceptable if the protocol requires a reduction in the number of views.

Imaging the cervical spine highlights the trade-off in tomosynthesis imaging between slice thickness in the reconstructed image and the number of views required (to limit artifacts). One approach is to first determine the slice thickness

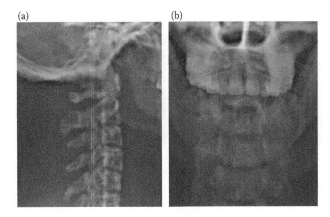

Figure 5.19 (a) Lateral and (b) A–P view radiographs of an anthropomorphic phantom. (Reproduced from G. M. Stevens et al., *Radiology* **228**, 573, 2003. With permission.)

needed to image the region of interest. This determines the minimum tomographic angle, which (along with the overall thickness of the object) then dictates the number of views needed. The artifacts present in the reconstructed images suggest that at least 100 views are needed for a 50° tomographic angle dataset of the head phantom. This tomographic angle is sufficient to adequately remove the overlying structures in order to visualize the upper cervical spine. If a larger slice thickness is acceptable, then a 20° tomographic angle can be used with 50 views. It is uncertain

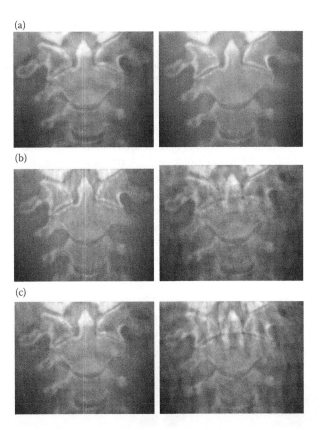

Figure 5.20 Circular tomosynthesis images using unfiltered (left) and disk-filtered (right) backprojection, with 100 views (a), 50 views (b), and 20 views (c) of an anthropomorphic phantom. The images were acquired with a 50° tomographic angle and were postprocessed using unsharp masking. (Reproduced from G. M. Stevens et al., *Radiology* **228**, 574, 2003. With permission.)

---

[*] As discussed below, the phantom system was not optimized for mammographic applications.

(a)

(b)

(c)

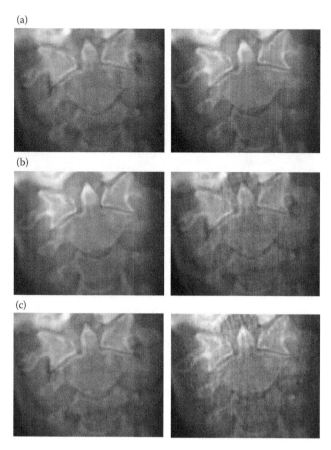

Figure 5.21 Circular tomosynthesis images using unfiltered (left) and disk-filtered (right) backprojection, with 100 views (a), 50 views (b), and 20 views (c) of an anthropomorphic phantom. The images were acquired with a 20° tomographic angle and were postprocessed using unsharp masking. (Reproduced from G. M., Stevens et al. *Radiology* **228**, 574, 2003. With permission.)

whether or not this scenario would be sufficient, however, since out-of-plane objects begin to become apparent in images of C-1 and C-2 for a 20° tomographic angle. A 10° tomographic angle yields images corresponding to very thick slices.

Comparing the tomosynthesis images of the cervical spine with the radiographs, the visualization of C-1 and C-2 is far superior to a standard A–P radiograph. Tomosynthesis provides flexibility in the choice of imaging parameters with tomosynthesis (based on the desired final images and system/patient limitations). The use of the

(a)　　　　(b)

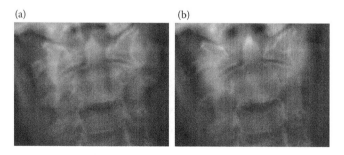

Figure 5.22 Circular tomosynthesis images using 100 views of the anthropomorphic phantom using unfiltered (a) and disk-filtered (b) backprojection. The images were acquired with a 10° tomographic angle and were postprocessed using unsharp masking. (Reproduced from G. M. Stevens et al., *Radiology* **228**, 574, 2003. With permission.)

disk filter to improve the visualization of these vertebrae by more uniformly blurring the overlying jaw structures appears promising.

5.4.2 BREAST IMAGING

The authors also investigated the use of circular tomosynthesis for breast imaging. As discussed below, the tomosynthesis system on which the imaging was performed was not intended for mammographic applications. Therefore, one should not interpret the results as absolute capabilities of circular tomosynthesis. The relative performance of circular tomosynthesis to projection radiographs, however, is still valid. Given these limitations, the ACR Mammography QC phantom [Mammo_phantom] was examined both with and without additional overlying structures. This phantom allows the investigation of typical tasks in mammography: 3D location of calcifications and visualization of low-contrast masses and fibers. The phantom layout is shown in Figure 5.23, as well as radiographs. In order to evaluate the benefits of tomosynthesis in a more clinically relevant scenario, a fresh mastectomy sample was imaged.

For mammography, the ability of tomosynthesis to provide (limited) depth information can be used to better determine the 3D location of calcifications. For this phantom, however, all of the objects are located in approximately the same plane. Therefore, images were acquired with the phantom tilted[*] during acquisition by approximately 20°. With the tilted phantom, planes reconstructed planes parallel to the sample positioning stage axis should focus on a calcification, while neighboring calcifications are blurred (because they fall on different planes in the tilted phantom). The relative blurring of the off-plane calcifications is indicative of the ability to resolve the 3D location of a calcification in breast tissue.

As shown in Figure 5.24, a radiograph of the tilted ACR phantom contains no depth information. Circular tomosynthesis images of the tilted phantom, however, clearly depict the 3D distribution of the calcifications (see Figure 5.25). For each image plane, a subset of the calcifications is in focus, while the off-plane calcifications are blurred into rings. Also shown in Figure 5.25, a smaller tomographic angle (which increases the slice thickness) makes a larger volume of the object effectively in focus for each plane. Calcifications separated in height by a smaller amount than this slice thickness appear to lie in approximately the same plane. This means that a smaller tomographic angle decreases the ability to resolve the 3D location of the calcifications.

The use of the disk filter is only a slight improvement over the unfiltered backprojection images in this simplified case (see Figure 5.25). Because of the thin phantom and close proximity of the calcifications, unfiltered backprojection produces rings that have relatively small radii. The small radii means a small area over which the disk filter can place part of the blurred intensity. Therefore, judgment should be reserved on the relative performance of the disk filter to this application.

Like other applications, breast imaging is confounded by the presence of overlying structures. The ability to visualize suspicious structures in a breast is limited not only by SNR but

---

[*] Alternatively, a different plane or surface could be used for the image reconstruction. This is another example of the flexibility of tomosynthesis, in which the focal planes can be selected retrospectively.

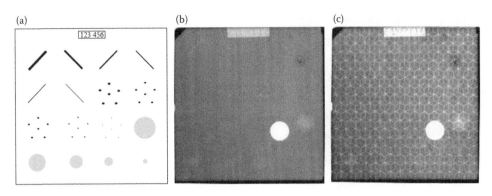

Figure 5.23 ACR Mammographic Quality Control phantom. (a) Layout, (b) radiograph, and (c) radiograph with a layer of beeswax to mimic overlying breast structures. The bright white disk in (b) and (c) is generally used to determine film contrast.

Figure 5.24 Radiograph of the largest calcification group in the tilted ACR Mammographic QC phantom. (Reproduced from G. M. Stevens et al., *Radiology* **228**, 571, 2003. With permission.)

also by the superposition of other objects with similar or greater contrast. In order to investigate the ability of tomosynthesis to effectively blur overlying structures using the ACR phantom, images were acquired with the phantom covered with a layer of structured beeswax. The beeswax, which was a sheet with a

3D honeycomb-like shape (see Figure 5.26), provides a similar impediment to visualization as overlying structures present in breast tissue.

The addition of the structured beeswax to the phantom makes the visualization of the calcifications, fibers, and masses more difficult (see Figure 5.23). Circular tomosynthesis improves the ability to see the structures of interest, as shown in Figure 5.27. For a 50° tomographic angle, the use of 100 projections appears adequate, while the wax layer is more visible in the case of 20 projections. In this case, the overlying structures were approximately 1 cm from the plane of interest. As this distance increases, the number of views required to adequately blur the overlying objects increases. While the use of 20 views is marginal here, one would predict that a larger number would be required for a 4-cm compressed breast. As in other examples, as the tomographic angle decreases, the larger slice thickness limits the ability of tomosynthesis to blur out-of-plane objects.

In addition to the ACR phantom, a fresh mastectomy sample was imaged by the authors prior to its pathology workup. Both

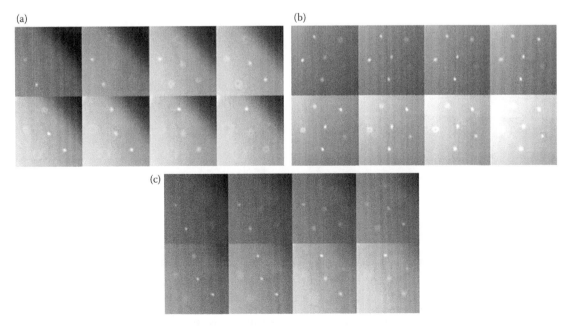

Figure 5.25 Circular tomosynthesis images with (a) 50° tomographic angle and unfiltered backprojection, (b) 20° tomographic angle and unfiltered backprojection, and (c) 50° tomographic angle and filtered backprojection with the disk filter. Images were generated using 100 projections, and each view in the series is separated from its neighbor by 1 mm. (Reproduced from G. M. Stevens et al., *Radiology* **228**, 572, 2003. With permission.)

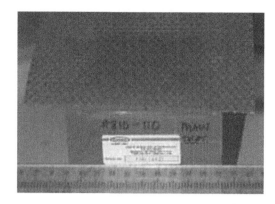

Figure 5.26 The ACR Mammographic QC phantom covered with a thin layer of structured beeswax to emulate overlying structures.

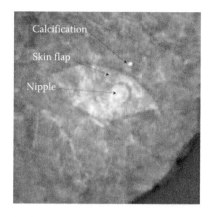

Figure 5.28 Radiograph of a fresh mastectomy sample, wrapped in a towel. The image was postprocessed with unsharp masking. (Reproduced from G. M. Stevens et al., *Radiology* **228**, 573, 2003. With permission.)

the radiographic and tomosynthesis images were filtered using unsharp masking to handle the display range requirements. As shown in Figure 5.28, overlying structures make the visualization of the tissue difficult. Circular tomosynthesis images show improved visualization of the tissue sample by the blurring of the overlying structures. The use of the disk filter more effectively blurs the skin flap, nipple, and calcification. (Figure 5.29 shows a plane reconstructed just below the skin flap.) For a 50° tomographic angle, images created using a dataset with 20 projections show some artifacts (e.g., discrete copies of the calcification), and more artifacts for the case of only 10

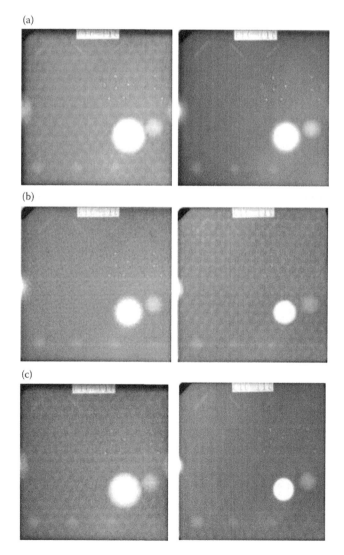

Figure 5.27 Circular tomosynthesis images with (a) 50° tomographic angle and 100 projections, (b) 50° tomographic angle and 20 projections, and (c) 20° tomographic angle and 100 projections. Images on the left were generated using unfiltered backprojection, and images on the right used filtered backprojections with the disk filter.

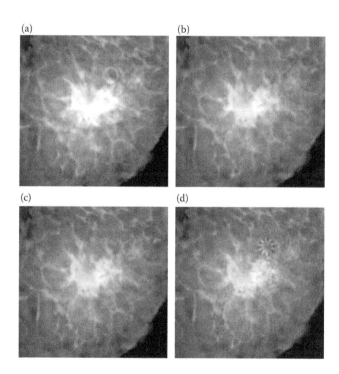

Figure 5.29 Circular tomosynthesis images with (a) 100 projections and unfiltered backprojection, (b) 100 projections and disk-filtered backprojection, (c) 20 projections and disk-filtered backprojection, and (d) 10 projections and disk-filtered backprojection. The images were acquired with a 50° tomographic angle and were postprocessed using unsharp masking. (Reproduced from G. M. Stevens et al., *Radiology* **228**, 573, 2003. With permission.)

projections. With these artifacts, the region below the skin flap can be visualized in the 20 projection image. Images from the case of 10 projections, however, show even the structure from the towel in which the sample was wrapped. This suggests that, for this specimen at a 50° tomographic angle, 20 views is close to the minimum number required to provide good image quality. This may need to be increased for *in vivo* imaging, where the amount of breast tissue would likely be larger.

For the breast phantom and tissue samples shown here, a cautionary note is required. The phantom system used to generate the images was built using a general-purpose (i.e., not mammographic specific) x-ray source, generator, and detector. Because of this, both the x-ray spectrum and spatial resolution are not consistent with the state-of-the-art mammographic machines. In order to be fair for the circular tomosynthesis evaluation, reconstructed images were compared with radiographs obtained on the same systems. Comparison of phantom images on mammographic-specific systems showed that the degradation of images by overlying structures was due to the radiographic imaging process, rather than the suboptimal imaging conditions of the prototype system.

5.5 CONCLUSIONS

This chapter has demonstrated the ability of tomosynthesis with circular orbits to more effectively blur overlying structures than with a simpler motion profile. In addition, careful selection of the filter using in filtered backprojection can improve the effective "removal" of out-of-plane objects. This places circular tomosynthesis as a good intermediate option between simpler motions with less optimal blurring and more complex motions with the associated costs. This can be understood from basic imaging principles, and was examined with simulations and phantom-based experimentation. Application of these techniques (and evaluation of key imaging parameters) to several relevant clinical scenarios appears promising.

REFERENCES

[Anth_phantom]   Radiology Support Devices, Long Beach, California.

[Badea98]   C. Badea, Z. Kolitsi, and N. Pallikarakis, A wavelet-based method for removal of out-of-plane structures in digital tomosynthesis, *Computerized Medical Imaging and Graphics* **22**, 309–315, 1998.

[Bronnikov99]   A. V. Bronnikov, Virtual alignment of x-ray cone-beam tomography system using two calibration aperture measurements, *Optical Engineering* **38**, 381–386, 1999.

[Carter63]   S. J. Carter, J. J. Martin, J. H. Middlemiss, and F. G. M. Ross, Polytome tomography, *Clinical Radiology* **14**, 405–413, 1963.

[Concepcion92]   J. A. Concepcion, J. D. Carpinelli, G. Kuo-Petravic, and S. Reisman, CT fan beam reconstruction with a nonstationary axis of rotation, *IEEE Transactions on Medical Imaging* **11**, 111–116, 1992.

[Esthappan98]   J. Esthappan, H. Harauchi, and K. R. Hoffman, Evaluation of imaging geometries calculated from biplane images, *Medical Physics* **25**, 965–975, 1998.

[Fahrig96]   R. Fahrig, A. Fox, and D. Holdsworth, Characterization of a C-arm mounted XRII for 3D image reconstruction during interventional neuroradiology, *Proceedings of the SPIE Medical Imaging Conference, Image Processing* **2708**, 30–38, 1996.

[Fahrig00]   R. Fahrig and D. Holdsworth, Three-dimensional computed tomographic reconstruction using a C-arm mounted XRII: Image-based correction of gantry motion nonidealities, *Medical Physics* **27**, 30–38, 2000.

[GE]   GE Healthcare VolumeRAD (http://www.gehealthcare.com/euen/radiography/products/applications/volumerad.html).

[Ghosh Roy85]   D. N. Ghosh Roy, R. A. Kruger, B. Yih, and P. Del Rio, Selective plane removal in limited angle tomographic imaging, *Medical Physics* **12**, 65–70, 1985.

[Gullberg90]   G. T. Gullberg, B. M. W. Tsui, C. R. Crawford, J. G. Ballard, and J. T. Hagius, Estimation of geometrical parameters and collimator evaluation for cone beam tomography, *Medical Physics* **17**, 264–272, 1990.

[Henry]   Dog lung phantom provided by Dr. Robert W. Henry, Department of animal science, University of Tennessee, Knoxville, TN.

[Hoffmann96]   K. R. Hoffmann, J. Esthappan, S. Li, and C. Pelizzari, A simple technique for calibrating imaging geometries, *Proceedings of the SPIE Medical Imaging Conference, Image Processing* **2708**, 371–376, 1996.

[Hologic]   Hologic Selenia Dimensions (http://www.hologic.com/en/breast-imaging/selenia-dimensions-tomosynthesis/).

[Lauritsch98]   G. Lauritsch and W. H. Härer, A theoretical framework for filtered backprojection in tomosynthesis, *Proceedings of the SPIE Medical Imaging Conference, Image Processing* **3338**, 1127–1137, 1998.

[Macovski83]   A. Macovski, *Medical Imaging Systems* (Prentice Hall, New Jersey, 1983).

[Mammo_phantom]   American college of radiology (ACR) mammography quality control phantom.

[Rizo94]   P. Rizo, P. Grangeat, and R. Guillemaud, Geometric calibration method for multiple-head cone-beam spect system, *IEEE Transactions on Nuclear Science* **41**, 2748–2760, 1994.

[Rougee93]   A. Rougée, C. Picard, C. Ponchut, and Y. Trousset, Geometrical calibration of x-ray imaging chains for three-dimensional reconstructions, *Computerized Medical Imaging and Graphics* **17**, 295–300, 1993.

[Ruttiman89]   U. E. Ruttiman, X. Qi, and R. L. Webber, An optimal synthetic aperture for circular tomosynthesis, *Medical Physics* **16**, 398–405, 1989.

[Stanton88]   L. Stanton, Conventional tomography, In *Radiology: Diagnosis-Imaging-Intervention*, J. M. Taveras and J. T. Ferrucci, eds., **3**, Ch. 12 (J.B. Lippincott Company, Philadelphia, PA, 1988).

[Stevens01a]   G. M. Stevens, R. Saunders, and N. J. Pelc, Alignment of a volumetric tomography system, *Medical Physics* **28**(7), 1472–1481, 2001.

[Stevens01f]   G. M. Stevens, R. Fahrig, and N. J. Pelc, Filtered backprojection for modifying the impulse response of circular tomosynthesis, *Medical Physics* **28**(3), 372–380, 2001.

[Stevens_th]   G. M. Stevens, *Volumetric Tomographic Imaging* (PhD thesis, Stanford University, 2000).

[Stevens03]   G. M. Stevens, R. L. Birdwell, C. F. Beaulieu, D. M Ikeda, and N. J. Pelc, Circular tomosynthesis: Potential in imaging of breast and upper cervical spine--preliminary phantom and in vitro study, *Radiology* **228**(2), 569–575, 2003.

[Webber97]   R. L. Webber, Self-Calibrated Tomosynthetic, Radiographic-Imaging system, method, and device, U.S. Patent # 5,668,844, 1997.

[Webber00]   R. L. Webber, Method and system for creating task-dependent three-dimensional images, U.S. Patent # 6,081,577, 2000.

[XinRay]   XinRay Systems (http://www.xinraysystems.com/index2.php?option=com_content&do_pdf=1&id=20).

System design

6

Tomosynthesis system modeling

Ingrid Reiser, Beverly Lau, and
Robert M. Nishikawa

Contents

6.1 INTRODUCTION

System modeling is an important aspect of designing and optimizing an x-ray imaging device. While x-ray imaging in general is well understood, the challenge in modeling tomosynthesis systems arises from the system geometry, which can lead to oblique x-ray trajectories with respect to the object and the detector surfaces. Depending on the system geometry (see Chapter 2), x-ray fluence can vary substantially across the object. Furthermore, x-rays entering the detector at oblique angles increase detector blur, which potentially degrades the image (Mainprize et al., 2006).

There are several different approaches to system modeling. One is a particle-based approach, where the history of individual photons, and potentially all subsequent particles that are generated, is tracked until all energy has been deposited. This approach is known as *Monte-Carlo method*, described by Bauer as follows: "Briefly, the Monte Carlo method consists in formulating a game of chance or a stochastic process which produces a random variable whose expected value is the solution of a certain

problem" (Bauer, 1958). Monte-Carlo simulations are used extensively in physics and related fields, including medical physics (Rogers, 2006). The advantage of using Monte-Carlo simulations in x-ray imaging is the natural incorporation of secondary particles and interactions. The drawback of this method is its high computational cost because of the large number of particles that need to be tracked.

Another approach to system modeling is to use ray tracing to obtain the primary x-ray image of an object, and then to derive approximate models for the secondary processes that occur during imaging. These processes tend to degrade the primary x-ray image and are sometimes called "physical factors." Often, their modeling involves convolving the entire x-ray field-of-view (FOV) with a point-response function (PRF). The PRF is the system response to a point object, which can be obtained from measurements, Monte-Carlo methods, or first principles. The convolution approach has been applied to modeling focal spot blur, scatter, and detector blur (Boone and Cooper, 2000; Zhou et al., 2007). However, while this approach can be more practical in terms of computation, there is potential for oversimplification.

Convolution produces a constant spatial correlation across the x-ray field. Therefore, convolution models tend to fail at object boundaries (Boone and Cooper, 2000). Using this method for modeling should be done with care to ensure that the assumptions are met.

Cascaded linear systems analysis (CSA) has become a standard method for analyzing and modeling x-ray imaging systems. Developed by Rabbani, Van Metter, and Shaw (Rabbani et al., 1987), and further developed independently by Cunningham (Cunningham et al., 1994) and by Siewerdsen (Siewerdsen et al., 1997), CSA follows the transfer of energy through the x-ray detector and through postacquisition processing (such as image reconstruction) to the final displayed image. The transfer of energy within the detector is modeled in the spatial frequency domain. Each energy conversion can be described as one of two steps: amplification and stochastic scattering. For example, the interaction of x-ray quanta in the phosphor is an amplification process, with an amplification factor less than 1. The x-ray energy is converted into optical photons (amplification process) and the light quanta scatter within the screen before escaping. The light scattering gives rise to stochastic scattering. By cascading the different amplification and scattering steps, one can obtain the modulation transfer function (MTF) and the noise power spectrum (NPS) of the detector. This allows the noise equivalent quanta (NEQ) and the detective quantum efficiency (DQE) to be calculated. An excellent treatise on this topic has been given by Cunningham (Beutel et al., 2000, Chapter 2). One can then add sampling and linear reconstruction or image processing to the model to obtain the characteristics of the whole system (Tward and Siewerdsen, 2008; Zhao and Zhao, 2008b).

CSA is a powerful tool for understanding limitations of an x-ray system and it can be used for system optimization. It has been used to evaluate mammography, general radiography, computed tomography (CT), contrast-enhanced, and dual energy systems (Siewerdsen et al., 1998; Zhao et al., 2003; Richard and Siewerdsen, 2008; Tward and Siewerdsen, 2008; Fredenberg et al., 2012). In tomosynthesis, several researchers have used CSA to study the effects of acquisition parameters (Glick and Gong, 2006; Tward and Siewerdsen, 2008; Zhao and Zhao, 2008b; Zhao et al., 2009; Mainprize and Yaffe, 2010; Hu and Zhao, 2011). Zhao and her colleagues have used CSA to study the effect of scanning angle on the 3D MTF, NPS, and DQE, and they have examined the effects of reconstruction filters on aliasing, particularly in the planes perpendicular to the detector (Hu and Zhao, 2011).

CSA makes several assumptions. It assumes the system is linear or linearizable, and shift invariant, and the statistics are stationary. It also ignores the depth of interaction of the x-ray in the detector. Depending on the detector characteristics, the depth at which the x-ray interacts in the phosphor screen can affect the MTF, the Swank noise factor, and the shape of the x-ray quantum NPS (Lubinsky et al., 1987; Nishikawa et al., 1989; Nishikawa and Yaffe, 1990). For high-resolution imaging, these factors are important and can limit the accuracy of CSA when the depth of x-ray interactions is not considered. Depth-dependent x-ray interactions produce geometric blur for oblique incidence, which can become important in

tomosynthesis (see also Section 6.5.1), and which is generally nonstationary because x-ray entrance angles can vary strongly across the projection FOV.

Finally, CSA cannot account for advanced iterative image reconstruction algorithms, which generally are neither linear nor stationary, but which are well suited for solving sparse sampling problems such as tomosynthesis and tend to outperform linear reconstruction algorithms (Wu et al., 2004).

Given these limitations, the remainder of this chapter focuses on the system modeling that has been specifically developed to account for oblique x-ray paths that occur in tomosynthesis systems. Therefore, the first two approaches to system modeling will be described in detail below, followed by a discussion on object model, and concluding with a discussion of system model validation.

6.2 GENERAL X-RAY IMAGING EQUATION

X-ray fluence (Φ) is defined as the number of x-ray photons traveling through an area dA with a normal vector parallel to the x-ray beam direction $\hat{\Phi}$.

Now we wish to determine the number of photons N received by a flat detector bin of area A located at a distance d from the x-ray source. The normal vector of the detector surface, $\hat{n}_d$, is assumed to point along the x-ray beam direction, and d is assumed to be much greater than the detector bin width Δx, so that all x-rays are incident on the detector surface at an approximately normal angle. Under these conditions, the number of photons detected in the detector bin is given by $N = \Phi_0 A$ (Figure 6.1a).

In general, however, the detector bin may not intersect the x-ray beam at a normal angle (Figure 6.1b). In this situation, fewer photons will strike the detector bin as it tilts away from the beam, and x-rays will enter the detector surface under an oblique angle γ. Combining this effect with the inverse square law, the number of x-ray photons N detected in a detector bin located at $\vec{r}_d$ becomes

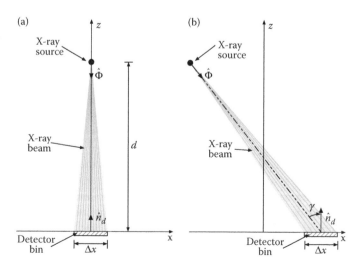

Figure 6.1 X-ray fluence for (a) normal incidence or (b) oblique incidence. The ray spacing is constant. It can be seen that fewer rays strike the detector bin in (b).

$$N(\vec{r}_d, \vec{r}_s) = \Phi_0 A \frac{d^2}{|\vec{r}_s - \vec{r}_d|^2} \cos(\gamma(\vec{r}_s, \vec{r}_d)), \qquad (6.1)$$

where $\vec{r}_s$ are the coordinates of the x-ray source (Figure 6.1b). Note that γ depends on the location of the detector bin relative to the x-ray source.

When an object is placed between the detector and the x-ray source, the transmitted x-ray fluence is

$$\Phi(\vec{r}_d, \vec{r}_s) = \int_0^{E_{max}} \phi(E) \exp\left[-\int_C \mu(\vec{r}, E)\,ds\right] dE, \qquad (6.2)$$

where $\mu(\vec{r}, E)$ is the spatial distribution of linear x-ray attenuation coefficients that describe the object, and C is the integration path. C is a straight line connecting the x-ray source at $\vec{r}_s$ with the detector bin at $\vec{r}_d$. E_{max} is the maximum energy of the x-rays, and $\phi(E)$ is the x-ray photon fluence in energy bin dE, $\phi(E) = d\Phi(E)/dE$.

Combining Equation 6.2 with Equation 6.1 results in the fundamental x-ray imaging equation for a general imaging geometry:

$$N(\vec{r}_d, \vec{r}_s) = \int_0^{E_{max}} \phi(E) A \frac{d^2}{|\vec{r}_s - \vec{r}_d|^2} \cos(\gamma(\vec{r}_s, \vec{r}_d))$$
$$\times \exp\left[-\int_C \mu(\vec{r}, E)\,ds\right] dE. \qquad (6.3)$$

This equation allows the computation of the radiographic image of an object imaged by a point source. With this basic x-ray imaging equation established, the next step is to turn to the object that is being imaged.

Within this system modeling paradigm, the object is described through the spatial distribution of its constituents. An object could be a phantom of well-specified shape and material(s), or a body part composed of different tissue types defined on a numerical grid. The chemical composition of the tissues or materials determines the x-ray attenuation properties of the object. Mass attenuation coefficients (μ/ρ) for all elements and a wide array of materials and tissues can be found in Hubbell and Seltzer (2012) and ICR (1989, 1992). For compounds and mixtures, the mass attenuation coefficient is obtained by combining mass attenuation coefficients by weight, that is

$$(\mu/\rho)_{mix} = \sum_i^N w_i(\mu_i/\rho_i), \qquad (6.4)$$

where N is the number of elemental constituents of a material, and w_i is the constituent's proportion by weight. Linear attenuation coefficients of breast tissues, including tumor tissue, can be found in Johns and Yaffe (1987).

X-ray mass attenuation is strongly energy dependent. Thus, knowledge of the x-ray spectrum is important. Boone et al. have developed a spectral model using interpolating polynomials, and

have published coefficients to generate tungsten anode spectra for energies up to 140 keV, as well as spectra for Mo, Rh, and W anodes up to 40 keV (Boone and Seibert, 1997; Boone et al., 1997).

The radiographic image of an object, imaged using an x-ray beam with a given energy spectrum, can be computed by the use of Equation 6.3. However, the resulting image does not include the many *physical factors* that can degrade an actual x-ray image, which include quantum noise, finite beam and detector element sizes (partial volume), focal spot blur, detector blur and noise, and secondary x-rays that do not contribute to image contrast (x-ray scatter).

6.3 FOCAL SPOT BLUR AND MOTION

In most tomosynthesis systems, x-ray projections from multiple angles are acquired by moving the x-ray source across a defined trajectory such as an arc (see Chapter 2 for a detailed discussion on tomosynthesis geometries). In general, the scan time needs to be short in order to minimize patient motion.

There are two broad classes of x-ray tube motion: continuous or step-and-shoot. Continuous motion gives rise to motion blur, depending on the x-ray tube travel speed and detector pixel size. On the other hand, step-and-shoot can result in x-ray source vibrations that lead to inaccuracies in the system geometry, which causes calibration errors in the reconstructed image. Generally, a scan with continuous tube motion can be performed faster than a scan in step-and-shoot mode. In continuous motion scans, the x-ray beam is usually pulsed to minimize motion blur.

To date, the effect of the acquisition mode has been modeled mainly as a modification of the system MTF. At the time of this writing, there was no experimental work on this topic.

Zhao et al. have modeled focal spot blur as an additional aperture term in the MTF, with the aperture width given by the path length a_{FS} covered by the focal spot during the exposure (Zhao and Zhao, 2008b)

$$\mathrm{MTF}_{FS}(f, a_{FS}) = \mathrm{sinc}(a_{FS} f). \qquad (6.5)$$

Within this model, the total MTF_{tot} consists of detector MTF and motion blur and is given by

$$\mathrm{MTF}_{tot} = \mathrm{MTF}_{FS} \cdot \mathrm{MTF}_{det}. \qquad (6.6)$$

Since a_{FS} varies with tube velocity and object height above the detector, focal spot blur depends on the overall system geometry.

Shaheen et al. have compared the effect of continuous or step-and-shoot motion on microcalcification and mass conspicuity in tomosynthesis through simulation (Shaheen et al., 2011). The continuous motion of the x-ray tube was modeled by convolving projections with a PRF that included motion blur along the scan direction; that is, the PRF was wider in the scan direction than in all other directions. Geometric blur was included in the PRF for both scan modes. For step-and-shoot, a 8–9% increase of microcalcification peak contrast was found compared to a continuous scan. For masses, the difference in contrast was small and not significant.

Acciavatti et al. investigated trade-offs between continuous or step-and-shoot tube motion, including patient motion

(Acciavatti and Maidment, 2012). They developed analytic expressions for projections of a sinusoidal signal, accounting for both tube and detector motion. These projections were reconstructed using filtered back-projection to obtain a reconstructed sinusoid, whose frequency was varied to generate an MTF-like curve characterizing the resolution in the reconstructed image. For continuous tube motion, they determined that intermediate scan times, of the order of 3 s, maximized MTF. For longer scan times, the MTF was degraded due to patient motion, and gantry motion blur degraded the MTF at shorter scan times. In their study, patient motion was modeled as having a constant velocity of the order of 30 μm/s, which was estimated by tracking microcalcifications in a tomosynthesis scan.

6.4 SCATTER MODELS FOR TOMOSYNTHESIS

Scatter can degrade image quality. In the aerial image transmitted through an object, scattered x rays impinge on the detector at locations that do not correspond to the unattenuated x-rays. Reconstructed data from these projection view images will not accurately reflect the attenuation coefficients of the materials in the object (Siewerdsen et al., 2006; Wu et al., 2009a). In order to yield appropriate attenuation values in the reconstructed voxels, scatter in the projection view images needs to be reduced.

Modeling scatter, then, has two purposes: to understand the effects of scatter on image quality and to remove it from clinical images to improve lesion conspicuity and reduce reconstruction artifacts.

The remainder of this section will first discuss general properties of x-ray scatter, and then proceed to describe methods to model scatter in tomosynthesis.

6.4.1 SCATTER SIMULATION WITH MONTE-CARLO

In the diagnostic energy range (i.e., 20–150 keV), x-ray attenuation in tissue is mainly due to photoelectric absorption, and elastic and inelastic scattering. The photoelectric absorption cross section (σ_{PE}) varies strongly with atomic number, thus producing the x-ray tissue contrast that is observed in diagnostic images. Inelastic, or Compton's scattering, depends weakly on the atomic number and therefore does not produce much x-ray contrast. Figure 6.2 shows the variation of the three dominant interaction cross sections with x-ray energy for carbon, as representative of tissue. Photoelectric absorption and Compton's scatter cross sections are equal at about 20 keV. At higher x-ray energies, Compton's scatter becomes the dominant effect.

In an x-ray image of a uniform object, scatter produces an approximately uniform offset that reduces radiographic contrast in screen-film systems. This is ameliorated by the use of an antiscatter grid, which accepts primary and forward-scattered radiation, but rejects large-angle scatter such as Compton's scattered x-ray photons. Tomosynthesis systems generally do not use a scatter grid, because the x-ray geometry depends on the projection angle, in most systems. Therefore, considerably more scatter is present in the tomosynthesis projection views than what would be expected in a mammogram. Furthermore, tomosynthesis images tend to be acquired with higher-energy

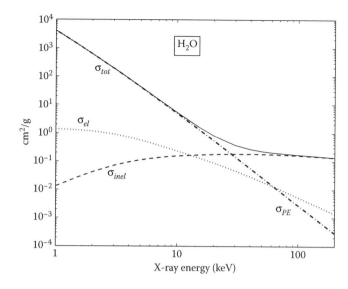

Figure 6.2 Cross sections of the dominant interactions of diagnostic x-rays with water: elastic and inelastic scatter (σ_{el} and σ_{inel}, respectively), and photoelectric (σ_{PE}). The total cross section is indicated by σ_{tot}. (From Berger, M. 1998. XCOM: Photon Cross Sections Database.)

x-ray beams compared to mammography, incurring more scatter. Lastly, in tomosynthesis, x-ray path lengths through the object can vary across the object surface, even with compression, which produces a highly nonuniform x-ray scatter distribution.

A comprehensive investigation of factors that influence scatter in mammography was conducted by Boone et al. (2000b). Using a Monte-Carlo simulation, the scatter-to-primary ratio (SPR) was determined for uniform slabs with semicircular shape. The linear x-ray attenuation coefficient of the slab was varied to represent breast tissue of glandular fractions ranging between 0% (purely adipose) and 100% (entirely glandular).

A strong dependence of SPR on object thickness and x-ray FOV was found. SPR varied linearly with object thickness, and it varied depending on the location within the object (and FOV). For a smaller FOV, increasing the air gap reduced the SPR, but this effect was weaker for larger FOV.

A weak dependence of SPR on x-ray beam energy was found. Also, glandular fraction had little effect on SPR. On the other hand, water consistently produced higher SPR than breast tissue at any glandularity.

Sechopoulos et al. (2007a) investigated x-ray scatter in tomosynthesis using Monte-Carlo simulations. They determined scatter point-spread functions (sPSF) by detecting x-rays scattered from a pencil beam incident on a scattering medium at oblique angles (Figures 6.3 and 6.4). As the obliquity increases, the sPSF becomes more anisotropic (Figure 6.3). Figure 6.4 shows line profiles through the center of the 2D sPSF. The sPSF was normalized so that

$$\sum_{x,y} \text{sPSF}(x, y) = \text{SPR},$$

(6.7)

where SPR is the scatter-to-primary ratio, defined as the ratio of integrated scattered energy to integrated primary x-ray energy, within a region of interest (Boone and Cooper, 2000).

$$S(x, y) = \iint_{x' y'} \text{FOV}(x', y') \text{sPSF}(x - x', y - y') \, dx' dy', \quad (6.8)$$

where $\text{FOV}(x, y)$ is 1 in regions that are illuminated by the primary x-ray beam and 0 otherwise. If the sPSF is normalized as in Equation 6.7, $S(x, y)$ is the scatter field relative to the primary x-ray field.

This model has been extended for oblique incidence by two groups. Wu et al. used the sPSF corresponding to the length of the oblique path through the scattering medium, t_α (Wu et al., 2009b). The incidence angle α for each tomosynthesis projection was determined at the center of mass of the irradiated material.

$$S(x, y) = \iint_{x' y'} \text{FOV}(x', y') \text{sPSF}(t_\alpha, x - x', y - y') \, dx' dy'. \quad (6.9)$$

Thickness-dependent $\text{sPSF}(t_\alpha)$ for normal x-ray incidence were obtained from Monte-Carlo simulations (Boone and Cooper, 2000). This approach produces reasonable agreement with angle-dependent Monte-Carlo simulation of scatter distributions calculated by Sechopoulos et al. However, the authors note that this method produces errors in the scatter estimate at the material edges.

Park et al. simulated angle-dependent sPSF using Monte-Carlo calculation (Park et al., 2012). The scatter field was estimated from

$$S(x, y) = \text{SPR}(\alpha, x, y) \iint_{x' y'} P(x', y') \text{sPSF}_n$$
$$\times (\alpha, x - x', y - y') \, dx' dy', \quad (6.10)$$

where sPSF_n is a normalized sPSF

$$\text{sPSF}_n = \text{sPSF}(\alpha, x, y) / \iint_{x y} \text{sPSF}(\alpha, x, y) \, dx dy. \quad (6.11)$$

SPR was estimated as the ratio of scattered to primary radiation through a uniform object using Monte-Carlo simulation; therefore, SPR varied across the detector plane (x, y) and was dependent on projection angle α. A single sPSF_n was used per projection view. Figure 6.5 shows a comparison of scattered radiation from an anthropomorphic breast phantom (Bakic et al., 2011) through Monte-Carlo simulation, for different tomosynthesis projection angles. The first column shows the distribution of primary x-rays. The second column shows the SPR, which was determined using a uniform phantom that had the same shape and average glandular fraction as the anthropomorphic phantom. The third column shows the distribution of scattered radiation obtained from Monte-Carlo simulation. The fourth column shows the distribution of scattered radiation obtained by Equation 6.10. Overall, there is qualitative agreement between both scatter distributions. However, the convolution method overestimates the magnitude of scattered radiation, and produces an overall smoother scatter field.

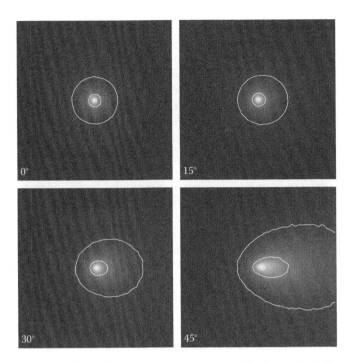

Figure 6.3 Scatter point-spread functions (sPSF) for four angles of incidence, through a 5-cm-thick slab of breast-equivalent scattering material. The air gap was 1 cm, and the detector was stationary and parallel to the scattering material for all angles of incidence. Image size is 10 × 10 cm². Contour lines are shown for 0.1, 0.5, and 0.9 relative maximum intensity.

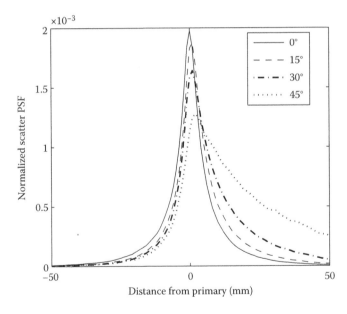

Figure 6.4 Line profiles through the scatter point-spread functions (sPSF) shown in Figure 6.3. The sPSF were normalized by the corresponding integrated primary x-ray beam, so that $\sum_{x,y} \text{sPSF}(x, y) = \text{SPR}$.

6.4.2 A CONVOLUTION MODEL FOR SCATTER IN TOMOSYNTHESIS

Boone et al. have proposed a convolution model for scatter (Boone and Cooper, 2000). Within this model, the scattered radiation $S(x, y)$ is approximated by a convolution of the sPSF with the x-ray FOV

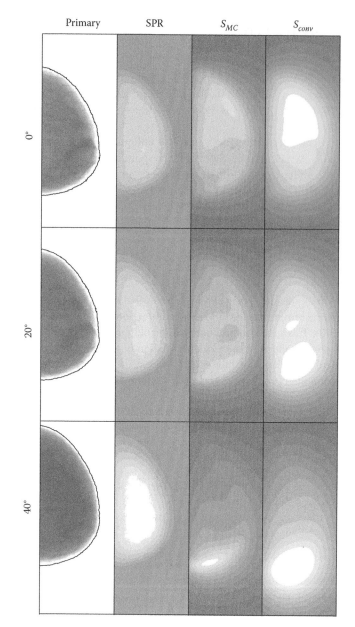

Figure 6.5 Comparison of scattered radiation from an anthropomorphic breast phantom computed through Monte-Carlo simulation (S_{MC}, third column) and via the convolution method (S_{conv}, fourth column). Also shown are the primary x-rays transmitted through the phantom (first column) and scatter-to-primary ratio (SPR) from Monte-Carlo simulation of x-rays traveling through a uniform phantom. All images are shown for 0°, 20°, and 40° x-ray source angles. Display windows: first column: [−0.7, 4]; second column: [−1, 0.7]; third and fourth column: [−0.3, 0.7]. (Image data courtesy of Subok Park, The Food and Drug Administration. See also Park, S. et al. 2012. *Proc. SPIE*, 8313, 83134S.)

6.5 DETECTOR MODELS FOR TOMOSYNTHESIS

This section is concerned with detector modeling in order to be able to simulate tomosynthesis projection views. X-ray detectors for tomosynthesis are discussed in detail in Chapter 3. Methods for simulating detector blur and noise based on MTF and NPS measurements (Saunders and Samei, 2003) and CSA (Siewerdsen et al., 1997; Zhao and Rowlands, 1997) have been developed, but

these models assume a stationary detector, and do not account for oblique x-ray incidence.

To date, full-field digital mammography (FFDM) and tomosynthesis use identical digital x-ray detector technology, thus, tomosynthesis might be considered as "enhanced full-field digital mammography." However, there are two major differences that affect detector choice and operation in tomosynthesis:

1. In tomosynthesis, the x-ray exposure, which is similar to that incurred in one FFDM image, is split between all projection views. As a result, the exposure to the detector is reduced by at least an order of magnitude (depending on the number of views of the tomosynthesis system). Care must be taken to reduce secondary noise sources in the detector, such as detector electronic noise, which could overwhelm the image and create "secondary quantum sinks." Quantum efficiency of a detector could be improved by increasing the thickness of the x-ray detection medium (see Table 6.1); however, increasing thickness also increases detector blur (Section 6.5.1).

2. In a typical FFDM system with a source-to-detector distance of 66 cm and a detector length of 30 cm, the maximum angle between x-rays and the detector normal about 12°. In a tomosynthesis system, this angle can be as large as 30°, with the actual maximum value depending on the system geometry. Furthermore, particularly for large-angle projections, the x-ray entrance angle varies strongly across the detector surface.

6.5.1 GEOMETRIC BLUR

The detector PRF is the output of a detector to an x-ray pencil beam incident on its surface. As x-ray photons enter the detector, they are absorbed along the beam trajectory, and secondary processes are initiated that will lead to the detection of the x-ray photons. Thus, x-ray interactions occur at all depths of the detector, with the number of interactions at a given depth depending on the x-ray path length through the *x-ray converter* (i.e., the detector material).

When x-rays enter the detector under an oblique angle, the width of the PRF increases due to "geometric blur." The principle of geometric blur is shown in Figure 6.6 and has been described previously (Que and Rowlands, 1995). The impact of this effect in terms of image degradation depends, for a given angle of incidence, on the thickness of the detector relative to the detector pixel size. The widening of the PRF due to oblique incidence has been demonstrated experimentally for CsI x-ray conversion screens (Badano et al., 2006; Mainprize et al., 2006).

6.5.2 AN INTERACTION-DEPTH DEPENDENT X-RAY CONVERTER MODEL

A model for blur from phosphor screens was originally developed by G. Lubberts, who found that the depth of x-ray interaction affected the distribution of optical quanta at the screen output, which is now known as the *Lubberts effect* (Lubberts, 1968). In an x-ray detection medium whose output is independent of the depth of x-ray interactions, the relative magnitude of the NPS is equal to the squared MTF, $NPS(f)/NPS(f = 0) = MTF^2(f)$. In a phosphor screen, due to the Lubberts effect, the optical photon distribution is depth dependent and therefore the NPS is not proportional to MTF^2.

Table 6.1 **Quantum efficiencies of different detector materials**

DETECTOR TYPE	CONVERTER MATERIAL	Z	QE ($E = 20$ KeV)		QE ($E = 30$ KeV)	
			$t = 100$ μm	$t = 200$ μm	$t = 100$ μm	$t = 200$ μm
Direct	a-Se	34	0.89	0.99	0.51	0.76
Indirect	CsI	53/55	0.7	0.91	0.34	0.56
Photon counting	Si	14	0.1	0.19	0.03	0.06

Source: Aslund, M. et al. 2007. *Med. Phys.*, 34(6):1918–1925.

Note: E denotes x-ray energy and t is material thickness. Si photon counting detectors are often operated in a slot-scan rather than a flat-panel configuration, which allows for greater material thickness and thereby increased quantum efficiency.

In his original model, Lubberts assumed a transparent phosphor screen. His model was later extended for turbid screens (Nishikawa and Yaffe, 1990; Van Metter and Rabbani, 1990). The extended model predictions for MTF, NTF, and DQE are in excellent agreement with experimental data (Nishikawa and Yaffe, 1990). Recently, we have adapted this model to account for oblique incidence by including both depth and in-plane spatial dependence (Reiser et al., 2009). A theoretical analysis of oblique incidence for this detector model can be found in Acciavatti and Maidment (2011).

Within this model, the x-ray converter is divided into n_l thin layers of thickness Δz (Figure 6.7). Assuming monoenergetic x-ray photons, the average number of x-ray photons absorbed in each layer at depth z is given by

$$\langle \mathbf{N}_{x,z}(\vec{r}_d, \vec{r}_s, z) \rangle = N(\vec{r}_d, \vec{r}_s) \mu_{xc} \frac{\Delta z}{\cos(\gamma(\vec{r}_d, \vec{r}_s))} \\ \times \exp\left[-\mu_{xc} \frac{z}{\cos(\gamma(\vec{r}_d, \vec{r}_s))}\right], \quad (6.12)$$

where μ_{xc} is the mass attenuation coefficient of the x-ray converter material, and $N(\vec{r}_d, \vec{r}_s)$ is the distribution of x-rays transmitted through the object, from Equation 6.3. The angle γ is defined as the angle between the x-ray beam direction $\vec{r}_d - \vec{r}_s$ and the detector bin normal $\hat{n}_d$. Equation 6.12 assumes a monochromatic x-ray beam. Note that because x-ray absorption is a stochastic

process, $\mathbf{N}_{x,z}$ is a Poisson-distributed random variable. In this section, bold-face will indicate a random variable.

The distribution of optical photons $\mathbf{N}_{opt,z}$ at the x-ray converter output is obtained by convolving the x-ray distribution at depth z with the converter's PRF, which, in general, is depth dependent, because the spread of optical photons, as well as the proportion of optical photons that escape the converter, depends on the distance traveled through the material. Then, $\mathbf{N}_{opt,z}$ can be written as

$$\mathbf{N}_{opt,z}(\vec{r}_d, \vec{r}_s, z) = \epsilon \frac{E_x}{E_{opt}} \xi(z) \left[\mathbf{N}_{x,z}(\vec{r}_d, \vec{r}_s, z) \otimes \mathrm{PRF}(z)\right], \quad (6.13)$$

where ϵ is the conversion efficiency of the x-ray converter, E_x and E_{opt} are the energies of the x-ray and the optical photons, $\xi(z)$ is the fraction of optical photons collected at the converter output, and $\otimes$ indicates convolution.

Finally, the contributions from all layers are added to produce the optical photon output from the x-ray converter

$$\mathbf{N}_{opt}(\vec{r}_d, \vec{r}_s) = \sum_{i=1}^{n_l} \mathbf{N}_{opt,z}(\vec{r}_d, \vec{r}_s, (i - 0.5)\Delta z). \quad (6.14)$$

Thus, this model allows to simulate an x-ray projection image of an object, including the detection of x-rays in a conversion material. This model accounts for both signal and noise transport through the converter, and accounts for oblique incidence naturally.

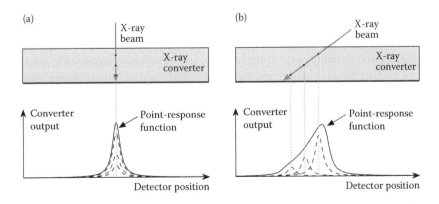

Figure 6.6 Geometric blur: Effect of x-ray entrance angle on converter point-response function for (a) normal incidence and (b) oblique incidence (30°). The solid line shows the PRF from x-ray interactions along the entire x-ray path through the conversion medium. The dashed lines show individual PRF from interactions at specific depths within the medium. The x-ray converter was assumed to be transparent to optical photons, and the shape of the optical photon distribution was independent of depth. The change in magnitude is due to the decrease in primary beam intensity. The linear x-ray attenuation coefficient of the converter was 216 cm⁻¹ (a-Se, 20 keV) and converter thickness was 100 μm. The PRF (solid line) was scaled by (a) 1/50 and (b) 1/20.

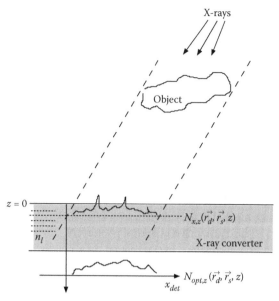

$N_{x,z}$: Distribution of x-rays absorbed in a thin layer at depth z
$N_{opt,z}$: Distribution of optical photons, originating from a thin layer at depth z, at converter output
$\vec{r}_d = (x_{det}, y_{det})$: Detector bin location
$\vec{r}_s$: Source location
z: Depth of layer on screen

Figure 6.7 Schematic of the x-ray converter model. (From Reiser, I. S., Nishikawa, R. M., and Lau, B. A. 2009. *Proc. SPIE*, 7258, 72585Z–72585Z–10.)

Implementation: The converter screen of total thickness T is divided into a finite number of layers n_l of thickness Δz. In addition, the converter is discretized along the parallel direction, resulting in n_u, n_v bins along directions $\hat{u}$ and $\hat{v}$. Once the discretization is determined, the next steps are (1) to determine the distribution of interacting x-rays in each layer, $N_{x,z}$, and (2) to obtain the distribution of optical photon created from x-ray interactions in a converter layer at depth z.

The number of interacting x-rays in each layer ($N_{x,z}$) can be obtained by projecting an object into a given detector layer, treating the converter material as part of the object, or, by propagating the x-ray distribution, which is incident on the detector surface, through the converter material. Quantum noise can then be added by sampling the number of interacting x-rays from a Poisson distribution with mean $N_{x,z}$.

The distribution of optical photons produced by x-ray interactions at a given depth in the converter depends strongly on the converter mechanism (direct, indirect x-ray conversion), as well as converter material and thickness. For a Gd_2O_2S converter, Nishikawa et al. have used an analytical model derived by Swank (1973) for the depth-dependent spread of optical photons, $\xi(z)$ MTF (z). By modeling the spread of optical photons in the converter screen, one can use this model to predict the changes in MTF, NTF, and DQE due to changes in screen thickness and the optical properties of the screen (Nishikawa and Yaffe, 1990). While such x-ray conversion screens are used primarily in mammographic screen-film systems, it is used here for the purpose of illustrating the detector model. Figures 6.8 and 6.9 show the effect of oblique incidence on the PRF in an 85-μm-thick Gd_2O_2S screen. Note that, for this conversion screen, the PRF becomes more anisotropic with increasing incidence angle, and also the peak of the PRF shifts away from the 0° position. In contrast, in Figure 6.6, the peak of the PRF for oblique incidence is close to the peak intensity for normal incidence. This difference in behavior is due to different converter properties: the Gd_2O_2S screen considered here is turbid, thus attenuating optical photons generated in upper layers more than those generated in layers near the converter exit. In the illustration in Figure 6.6, a transparent converter was assumed, without any loss of optical photons.

This detector model was used to simulate the image of an edge, as well as a set of 100 flat-field images at a dose of 11.8 mR. The x-ray energy was 20 keV and detector pixel size was 20 μm. Figure 6.10 shows the MTF, NPS, and DQE that were computed from 0.4 × 0.4 mm regions of interest (ROIs). For comparison, experimental data for zero degree incidence is shown as symbols.

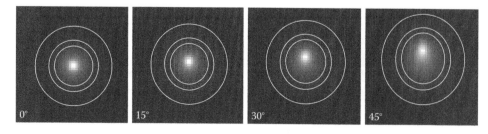

Figure 6.8 Angle-dependent point-response functions for an 85-μm-thick Gd_2O_2S screen. Simulated images of a 40-μm-diameter pinhole are shown. The pinhole was kept perpendicular to the x-ray beam, which was incident on a Gd_2O_2S conversion screen at 0° (normal incidence), 15°, 30°, and 45°. The image size is 0.46 × 0.46 mm with a pixel size of 0.9 μm. Contour lines are shown for 0.01, 0.05, 0.1 relative maximum intensity.

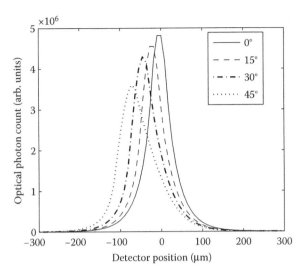

more x-ray photons are absorbed close to the surface of the detector. The optical photons thus need to travel a longer path to escape the converter, and are more likely to be absorbed in the phosphor material or scattered, thus producing wider PRFs. This causes a drop in DQE at low frequencies, and the increased width of the PRF makes the DQE narrower. The effect of oblique incidence on DQE shape is similar to the effect of increased converter thickness in that resolution is reduced. However, optical photon attenuation causes a drop in DQE that is not observed in thicker screen.

The depth-dependent optical photon distributions for structured CsI, an indirect converter material, has been determined through Monte-Carlo simulation (Freed et al., 2009; Badano et al., 2011) and Section 6.6.2. For direct a-Se conversion detectors, depth-dependent PRFs have been proposed (Badano et al., 2011).

Figure 6.9 Line profiles through the point-response functions shown in Figure 6.8.

Good agreement is found for MTF between experimental and simulated data. The DQE from simulated data overestimates the measured DQE, but the shape is preserved. Oblique x-ray incidence causes the MTF to become narrower and produces a slight increase in NPS. When x-ray incidence becomes oblique,

6.5.3 A FAST METHOD FOR COMPUTING ANGLE-DEPENDENT POINT-RESPONSE FUNCTIONS

Freed et al. have developed an analytic expression for the angle-dependent PRF of CsI conversion screens (Freed et al., 2010). Similar to the previous model, the distribution of optical quanta escaping the conversion screen (i.e., the PRF) is composed of three terms: N_1, the number of x-rays absorbed at depth z; n_2, the number of optical photons that escape the screen per x-ray

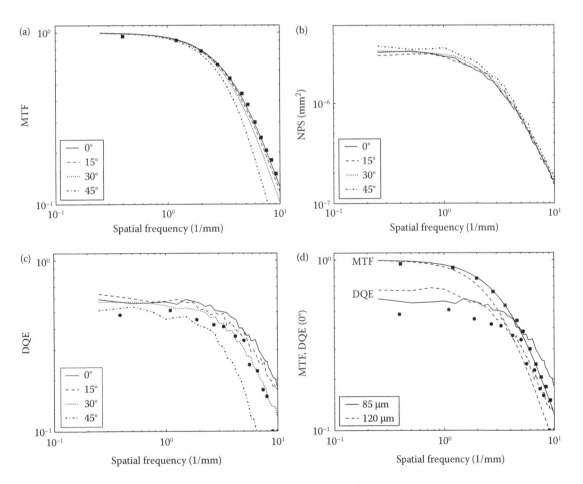

Figure 6.10 (a) MTF, (b) NPS, (c) DQE for a simulated 85-μm-thick Gd$_2$O$_2$S screen, and for different angles of x-ray incidence, and (d) DQE and MTF for x-rays normally incident on an 85- and 120-μm Gd$_2$O$_2$S screen, and MTF and DQE for normal x-ray incidence. The symbols (square: MTF; circle: DQE) are experimental data for normal incidence. (Adapted from Nishikawa, R. M. and Yaffe, M. J. 1990. *Med. Phys.*, 17:894–904.)

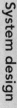

interaction; and L, the spatial distribution of optical photons produced at depth z, which escape the conversion screen, described by a Lorentzian. Each of these terms was approximated in the following way:

$$N_1(E, z, \theta, \phi) = N_0\, \mu_{pe}(E) \exp[-\mu_{xc}(E) t(z, \theta, \phi)] \quad (6.15)$$

$$\approx N_0\, \mu_{pe}(E)[a_0(E)z + a_1(E)] \quad (6.16)$$

$$n_2(E, z) = \epsilon(E)\xi(z) \quad (6.17)$$

$$\approx \epsilon(E)[b_0(z) + b_1] \quad (6.18)$$

$$L(x, y, z) = \frac{1}{1 + (2/\Gamma(z))^2 (x'^2 + y'^2)} \quad (6.19)$$

$$\approx \left[1 + \left(\frac{2}{g_0 z + g_1} \right)^2 [(x - z\tan(\theta))^2 + (y - z\tan(\theta))^2] \right]^{-1}, \quad (6.20)$$

so as to produce an expression for the screen PRF that could be analytically integrated

$$\mathrm{PRF}(x, y, \theta, \phi, E) = \int_0^{z_{max}} N_1(E, z, \theta, \phi) \cdot n_2(E, z) \cdot L(x, y, z)\, dz. \quad (6.21)$$

The analytic solution of this integral in terms of x, y, θ, ϕ, and E can be found in Freed et al. (2010), along with values for fit parameters a_i, b_i, and g_i ($i = 0,1$). Since the resulting expression is analytic (albeit with 12 terms), it allows for fast computation of PRFs. Figure 6.11 shows a comparison of the analytic PRFs with experimental PRFs and PRFs from Monte-Carlo simulation. Recently, this model was used to predict PRFs for a-Se direct conversion detectors, using empirical parameters (Badano et al., 2011).

6.5.4 ELECTRONIC NOISE

Electronic detector noise is a concern in tomosynthesis because it is incurred from each projection. Furthermore, the exposure to the detector at each view can be substantially lower in tomosynthesis, compared to full-field digital mammography, depending on the number of projection views acquired. Electronic noise at low exposure to the detector for a-Se direct conversion detectors has been studied extensively by Zhao et al. They found that the contribution of electronic noise to the total NPS to be below 10% even at a 0.4 mR exposure to the detector, indicating that this detector is quantum-noise limited in the tomosynthesis regime (Zhao and Zhao, 2008a). Glick et al. investigated the effect of electronic noise on detectability of a 5-mm-diameter mass in glandular tissue, using a serial cascade detector model (Glick and Gong, 2006). For breast thicknesses from 4 to 8 cm, electronic noise reduced the signal-to-noise ratio of smaller masses but this effect decreased as kVp increased.

For an earlier clinical prototype of a CsI indirect conversion detector, the electronic noise contribution accounted for

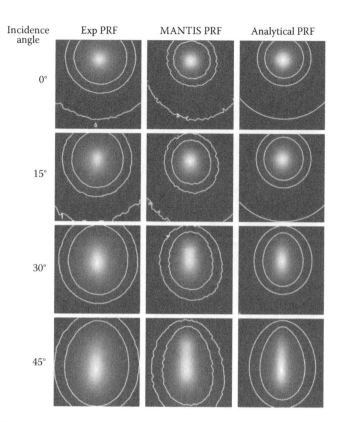

Figure 6.11 Comparison of angle-dependent point-response functions of a CsI scintillator that were determined experimentally, through Monte-Carlo simulation, and by the use of the analytic solution to Equation 6.22. (Reprinted with permission from Freed, M., Park, S., and Badano, A. 2010. *Med. Phys.*, 37(6):2593.)

about 20% of NPS at a detector entrance exposure of 1.3 mR (Vedantham et al., 2000). A more recent model of this detector has been shown to reduce electronic noise by a factor of 10 (Ghetti et al., 2008).

For the purpose of simulating electronic noise, experiments have confirmed that it has no frequency dependence and for some detectors, electronic noise variance is reported in the literature (Ghetti et al., 2008; Zhao and Zhao, 2008a), and Vedantham et al. describe their procedure for measuring electronic noise (Vedantham et al., 2000).

6.6 MONTE-CARLO SIMULATION

Monte-Carlo methods are a class of algorithms that involve repeated random sampling from defined probability distributions. These methods are useful when no deterministic solution exists for the problem at hand, or when the process to be simulated is random. A recent review (Rogers, 2006) finds that the main application of Monte-Carlo simulation in medical physics is for the purpose of radiation treatment planning.

In diagnostic imaging, Monte-Carlo simulations are used to model x-ray scatter and dose. Different Monte-Carlo code packages are being used. Examples include GEANT4, a Monte-Carlo package developed by researchers at CERN, used by Sechopoulos to determine scatter and radiation dose in tomosynthesis (Sechopoulos et al., 2007a, b). Boone et al. have developed independent Monte-Carlo code called SIERRA (simple investigative environment for radiological research applications)

(Boone et al., 2000a). They have published extensively on the properties of scattered radiation in mammography, as well as normalized glandular dose (DgN) and the CT dose index (CTDI) (Boone and Seibert, 1988; Boone, 1999, 2009; Boone et al., 2000b). Furthermore, Boone's group has used Monte-Carlo simulations to validate their convolution model for x-ray scatter in mammography (Boone and Cooper, 2000). PENELOPE (penetration and energy loss of positrons and electrons) is a Monte-Carlo package that simulates coupled transport of x rays, electrons, and positrons (Sempau et al., 1997, 2003). PENELOPE can further be used in conjunction with MANTIS (Monte-Carlo x-ray electron optical imaging simulation) to simulate optical photon transport (Badano and Sempau, 2006).

PENELOPE/MANTIS are discussed here in more detail because of a number of extensions that have been developed to facilitate x-ray imaging simulations, and that are freely available for download.

6.6.1 PENELOPE

The PENELOPE package includes algorithms to simulate Rayleigh's scattering, the photoelectric effect, Compton's scattering, and pair production for the x rays, and it also accounts for elastic and inelastic scattering along with Bremsstrahlung for the electrons and positron annihilation. Natively, the imaging geometry is defined through quadratic surfaces.

Several extensions make PENELOPE easier to use for imaging simulations. J. Sempau has developed penEasy, which provides an interface to PENELOPE and allows for a voxel-based definition of the imaging geometry, as well as additional tally options.

Further, researchers at the Food and Drug Administration (FDA) have developed penEasy_Imaging, which allows for a rectangular FOV, and provides an output tally as a radiographic image. Further, penEasy_Imaging includes an analytic ray-tracing model to generate noise-free primary image. penEasy_Imaging is available for download at http://code.google.com/p/peneasy-imaging. Two additional extensions are penMesh, which allow for object description with meshed surfaces (available at http://code.google.com/p/penmesh), as well as MC-GPM, a CUDA implementation of penEasy (available at http://code.google.com/p/mcgpu).

6.6.2 MANTIS

PENELOPE only simulates x-ray photons and electrons and thus cannot model the production of and subsequently track optical photons that are often involved in the x-ray detection process. MANTIS is a software package that is used in conjunction with PENELOPE to simulate the transport of optical photons (Badano and Sempau, 2006). This is particularly useful for modeling indirect detectors, which are common in medical imaging. The MANTIS code is available at http://code.google.com/p/mantismc.

Badano et al. have used MANTIS to simulate the needle-like structure that is used in CsI-based indirect conversion detectors. The columnar structure reduces scatter of optical photons by acting as a light-guiding structure, which allows to increase phosphor thickness without sacrificing spatial resolution. MANTIS was used to investigate anisotropic detector properties

in tomosynthesis (Badano and Sempau, 2006; Badano et al., 2007; Freed et al., 2009). A comparison of experimental and MANTIS-simulated PRF of a CsI scintillator was shown in Figure 6.11.

To speed up computation time, the use of graphics processing units (GPU) has been explored. A new code was developed, hybridMANTIS, which performs computations on both the CPU and a GPU. For this new configuration, a recent study found an improvement in computation time by almost a factor of 30, compared to computation on a single-thread CPU (Badal and Badano, 2009). In a recent development, CsI column location is computed "on the fly," and column roughness can be specified (Sharma et al., 2012). The code for hybridMANTIS is available at http://code.google.com/p/hybridmantis.

6.7 OBJECT MODELS

The physics models described above can be used to simulate the x-ray projection image of a virtual object. Ideally, this simulated image is identical to the x-ray projection of an actual object recorded by an actual x-ray imaging system. In the imaging context, virtual objects are often called *software phantoms* and contain information about the spatial distribution of x-ray interaction cross sections and material densities. A software phantom can be defined numerically or analytically, which can be described as voxel-based and surface-based, respectively. These two types of phantoms are discussed in the following sections.

6.7.1 VOXEL-BASED PHANTOM REPRESENTATION

Arguably the most straightforward way to mathematically describe a phantom is to assign linear x-ray attenuation coefficients to a regular grid, which can then be stored in computer memory as an array. Such a phantom is called a *voxelized*, *discrete*, or *numerical* phantom. While such a phantom is conceptually simple, it can require a large amount of computer storage, depending on the number of elements in the array, which depends on the physical size of the object to be represented and the resolution of the phantom.

The appropriate phantom resolution depends on object detail size and detector pixel size. If the true object contains details that are significantly smaller than the detector pixel size (assuming no magnification), we empirically found that the phantom resolution should be slightly greater than the detector resolution in order to avoid artifacts in the projection image, as seen in Figures 6.12 and 6.13.

Projecting a voxelized phantom onto a discrete image plane constitutes a discrete-to-discrete operation. As stated above, care needs to be taken to avoid artifacts. Projection involves computation of the line integral along a ray from the x-ray source to the detector bin. In *ray tracing*, the line integral is computed from a discrete set of points along the ray. Ray tracing can be performed along intersecting lines, as is the case with Siddon's method, which is a fast implementation of ray tracing (Siddon, 1985). In our work, ray tracing was performed via nearest neighbor or linear interpolation (Sidky et al., 2009), and Man and Basu have described a distance-driven method (Man and Basu, 2004).

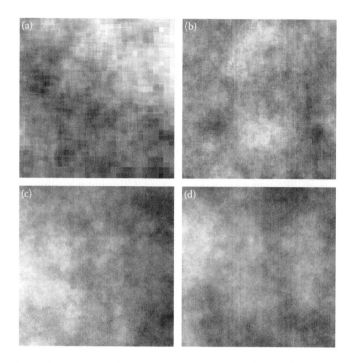

Figure 6.12 Artifacts from discrete-to-discrete projection for different ratios of phantom voxel side length (l_{phant}) to detector pixel side length (l_{det}): l_{phant}/l_{det} = (a) 4, (b) 2, (c) 1, and (d) 0.8. (Reprinted from Reiser, I. and Nishikawa, R. M. 2010. *Med. Phys.*, 37(4):1591–1600. With permission.)

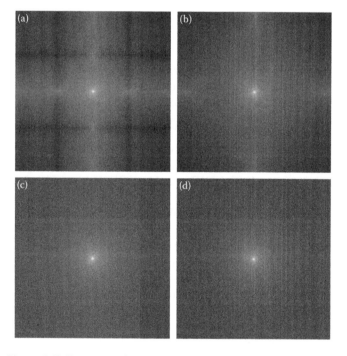

Figure 6.13 Frequency-domain manifestation of discrete-to-discrete projection artifacts that were shown in Figure 6.12. l_{phant}/l_{det} = (a) 4, (b) 2, (c) 1, and (d) 0.8. (Reprinted from Reiser, I. and Nishikawa, R. M. 2010. *Med. Phys.*, 37(4):1591–1600. With permission.)

6.7.2 SURFACE-BASED PHANTOM REPRESENTATION

If the phantom is composed of materials with geometric shapes such as planes, spheres, cylinders or cones, or intersections thereof, the phantom can be defined through a set of equations. For instance, a cube can be represented by six intersecting

planes. Such a phantom is called an *analytic phantom*. The storage requirements for such a phantom are minimal, and x-ray path integrals can be computed analytically by the use of a continuous-to-discrete projector, which eliminates discretization artifacts from the use of a discrete object. Analytic phantoms are useful when generating simple test phantoms, which often consist of simple geometric objects, such as the contrast-detail phantom shown in Figure 6.16. More complicated surfaces are often represented by meshed surfaces (Badal, 2008).

6.7.3 ANTHROPOMORPHIC PHANTOMS

Phantoms that approximate human anatomy are called *anthropomorphic phantoms*. Virtual anthropomorphic phantoms can be produced in two ways: (1) conversion of a physical object into a virtual one, and (2) generation of purely virtual objects using mathematical techniques.

Converting physical objects into virtual objects can be done using inanimate objects, live patients, or tissue samples. In any case, the object can be sampled for its textural properties through an imaging scan that can yield 3D data. This is typically done through a CT scan. Physical objects that have been converted into numerical phantoms include breast mastectomy specimens (O'Connor et al., 2010) and patient breasts from clinical CT data (Li et al., 2009).

Generation of purely virtual objects has been done for whole-body structures and for individual organs, in particular the breast. A whole-body anthropomorphic phantom is the NCAT phantom, which is constructed using NURBS surfaces (Segars, 2001). A historical review of the NCAT phantom is described in a review article (Segars and Tsui, 2009).

The need for realistic digital anthropomorphic breast phantoms has emerged in the past decade because of recent advances in 3D breast x-ray imaging techniques for breast cancer screening and diagnosis (such as MRI, CT, and tomosynthesis). With the increased interest in these new modalities, it has become essential to use anthropomorphic computer phantoms to

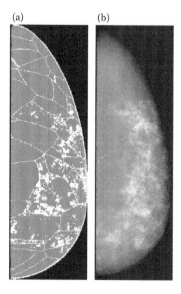

Figure 6.14 Statistically defined anthropomorphic breast phantom. (a) Slice through the phantom and (b) simulated mammogram.

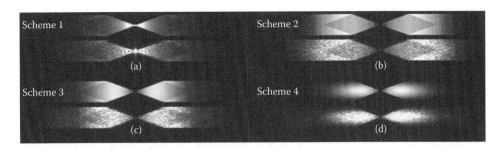

Figure 6.15 Comparison between modeled (top) and experimental (bottom) in-depth NPS for four reconstruction filtered schemes. (Reprinted from Zhao, B. et al. 2009. *Med. Phys.*, 36(1):240. With permission.)

optimize these systems in clinically relevant tasks. Many groups are working on creating three-dimensional breast phantoms based entirely on computer-generated models (Bakic et al., 2002a,b, 2003, 2011; Bliznakova et al., 2003; Zhang et al., 2008; Ma et al., 2009; Chen et al., 2011; Lau et al., 2012). In particular, Bakic's breast phantom, which simulates large-scale breast structure such as overall shape and Cooper's ligaments, is solely based on mathematical models (Bakic et al., 2002a, 2003; Pokrajac et al., 2012), and there is a working prototype for an analogous physical phantom (Carton et al., 2011). Small-scale breast anatomy, such as ducts and glands, has been modeled using power-law noise (Bliznakova et al., 2003; Gong et al., 2006; Chen et al., 2011). This small-scale tissue model has recently been added to compartments separated by Cooper's ligaments in Bakic's breast phantom, an example of which is shown in Figure 6.14 (Lau et al., 2012).

While phantoms based on patient data and mastectomy specimens have a higher chance of appearing more realistic, those phantoms generally require one patient per unique phantom. Moreover, phantoms made by imaging objects are limited by the pixel size of the imaging device and are subject to physical effects of the system, such as blurring and noise. Entirely computer-generated phantoms require no radiation exposure to patients, while also allowing a large number of unique samples to be generated in a reasonable amount of time with no imaging artifacts and at the desired resolution. Furthermore, entirely computer-generated models allow for rapid creation of many phantoms with a variety of tissue types. Having variety in tissue composition is important for representing the variety in the patient population.

6.8 VALIDATING A SIMULATION MODEL

A system model needs to be validated before it can be used to draw conclusions about the physical processes being modeled. There are two approaches for validating models: (1) to validate that general properties about the imaging method are accurate, and (2) to validate that the task yields accurate results. To validate general properties, the power spectrum can be measured. To validate based on the task, a simulation study can be performed and the results can be compared with experimental data. An example of each approach is given below.

Zhao et al. have validated their cascaded linear system model for breast tomosynthesis by comparing model predictions for NPS and MTF with actual measurements from tomosynthesis

images for different scan angles and reconstruction filters (Zhao and Zhao, 2008b; Zhao et al., 2009). Figure 6.15 shows a comparison of in-depth NPS for four reconstruction filter schemes, which demonstrates good overall agreement of the model predictions with the experimental data. Note that the NPS is a measure of stationary first- and second-order image statistics; thus, Figure 6.15 confirms that those can be predicted by cascaded systems analysis. However, since NPS ignores any nonstationary or higher-order image statistics, it is not clear whether such components are present in the image noise, and if so, they may not be predicted by the cascaded linear system model.

An alternative approach to system model validation involves the comparison of simulated phantom images to actual x-ray images of the same phantom. Typically, simple phantoms are chosen, which can be easily modeled for computer simulation.

Such a validation was performed using a contrast-detail phantom. This phantom consists of cylinders that increase in size along rows and decrease in height along columns. The actual phantom was imaged on a stereotactic breast-biopsy unit (Krupinski et al., 1995) that served as a surrogate tomosynthesis system by manually changing x-ray source angles. Next, simulated tomosynthesis projections were generated using a software phantom that matched the physical contrast-detail phantom (Reiser et al., 2007). Reconstructed phantom images from actual projections and simulated projections are shown in Figure 6.16. The appearance of both images is very similar.

(a) EM reconstruction of measured projection data   (b) EM reconstruction of simulated projection data

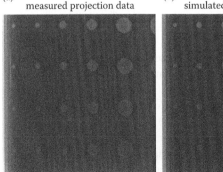

Figure 6.16 Slices through a tomosynthesis image of a contrast-detail phantom reconstructed from (a) actual data and (b) simulated projections. See also Figures 1 and 2 in the original paper. (Adapted from Reiser, I. et al. 2007. *Proc. SPIE*, 6510, 65103D–65103D–8.)

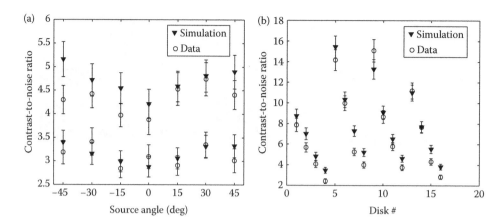

Figure 6.17 Comparison of contrast-to-noise ratio of disks, measured in actual and simulated x-ray images. (a) Tomosynthesis projections and (b) reconstructed tomosynthesis slice. See also Figures 4 and 5 in the original paper. (Adapted from Reiser, I. et al. 2007. *Proc. SPIE*, 6510, 65103D–65103D–8.)

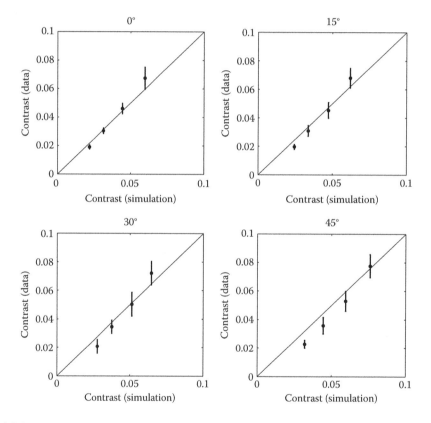

Figure 6.18 Comparison of disk contrast in actual and simulated projections. See also Figure 7 in the original paper. (Adapted from Reiser, I. et al. 2007. *Proc. SPIE*, 6510, 65103D–65103D–8.)

A more quantitative comparison of actual and simulated data was performed by measuring contrast-to-noise ratio (CNR) in projections and a reconstructed slice (Figure 6.17). Overall, CNR from simulated x-ray data was similar to CNR from the actual x-ray images. The accuracy of the simulation might be improved by using a more detailed detector model.

This approach to model validation has limitations. The overall goal of the system optimization study has to be kept in mind. In this study, the goal was to optimize tomosynthesis for computer-aided diagnosis. Therefore, it was verified that the simulation was able to predict feature values, such as signal contrast, with sufficient precision (Figure 6.18).

6.9 CONCLUSION

A number of models for physical factors that occur in tomosynthesis imaging have been presented. When simulation is used to address a specific question, the appropriate system model, in particular, the necessary amount of detail, needs to be considered carefully.

If possible, a systems model should be validated. However, this can be difficult or impossible to do if a configuration is being modeled that does not yet exist. It is then the responsibility of the investigator to carefully disclose all assumptions and approximations of their model when presenting their findings.

REFERENCES

1989. ICRU Report 44: Tissue Substitutes in Radiation Dosimetry and Measurement. Technical report, International Commission on Radiation Units and Measurements, Bethesda, MD.

1992. ICRU Report 46: Photon, Electron, Proton and Neutron Interaction Data for Body Tissues. Technical report, International Commission on Radiation Units and Measurements, Bethesda, MD.

Acciavatti, R. J. and Maidment, A. D. A. 2011. Optimization of phosphor-based detector design for oblique x-ray incidence in digital breast tomosynthesis. *Med. Phys.*, 38(11):6188–6202.

Acciavatti, R. J. and Maidment, A. D. A. 2012. Optimization of continuous tube motion and step-and-shoot motion in digital breast tomosynthesis systems with patient motion. *Proc. SPIE*, 8313, 831306–831312.

Aslund, M., Cederstroom, B., Lundqvist, M., and Danielsson, M. 2007. Physical characterization of a scanning photon counting digital mammography system based on Si-strip detectors. *Med. Phys.*, 34(6):1918–1925.

Badal, A. 2008. Development of Advanced Geometric Models and Acceleration Techniques for Monte Carlo Simulation in Medical Physics. Dissertation, Universitat Politecnica de Catalunya.

Badal, A. and Badano, A. 2009. Accelerating Monte Carlo simulations of photon transport in a voxelized geometry using a massively parallel graphics processing unit. *Med. Phys.*, 36(11):4878–4880.

Badano, A., Freed, M., and Fang, Y. 2011. Oblique incidence effects in direct x-ray detectors: A first-order approximation using a physics-based analytical model. *Med. Phys.*, 38(4):2095.

Badano, A., Kyprianou, I. S., Jennings, R. J., and Sempau, J. 2007. Anisotropic imaging performance in breast tomosynthesis. *Med. Phys.*, 34(11):4076.

Badano, A., Kyprianou, I. S., and Sempau, J. 2006. Anisotropic imaging performance in indirect x-ray imaging detectors. *Med. Phys.*, 33(8):2698.

Badano, A. and Sempau, J. 2006. MANTIS: Combined x-ray, electron and optical Monte Carlo simulations of indirect radiation imaging systems. *Phys. Med. Biol.*, 51(6):1545–1561.

Bakic, P., Albert, M., Brzakovic, D., and Maidment, A. 2002a. Mammogram synthesis using a 3D simulation. II. Evaluation of synthetic mammogram texture. *Med. Phys.*, 29(9):2140.

Bakic, P. R., Albert, M., Brzakovic, D., and Maidment, A. D. A. 2002b. Mammogram synthesis using a 3D simulation. I. Breast tissue model and image acquisition simulation. *Med. Phys.*, 29(9):2131.

Bakic, P. R., Albert, M., Brzakovic, D., and Maidment, A. D. A. 2003. Mammogram synthesis using a three-dimensional simulation. III. Modeling and evaluation of the breast ductal network. *Med. Phys.*, 30(7):1914.

Bakic, P. R., Zhang, C., and Maidment, A. D. A. 2011. Development and characterization of an anthropomorphic breast software phantom based upon region-growing algorithm. *Med. Phys.*, 38(6):3165.

Bauer, W. F. 1958. The Monte Carlo method. *J. Soc. Indust. Appl. Math.*, 6(4):438–451.

Berger, M., Hubbell, J., Seltzer, S., Chang, J., Coursey, J., Sukumar, R., Zucker, D., and Olsen, K. 1998. XCOM: Photon Cross Sections Database.

Beutel, J., Kundel, H. L., and Van Metter, R. L., eds 2000. *Handbook of Medical Imaging: Physics and Psychophysics*. SPIE Press, Bellingham, WA.

Bliznakova, K., Bliznakov, Z., Bravou, V., Kolitsi, Z., and Pallikarakis, N. 2003. A three-dimensional breast software phantom for mammography simulation. *Phys. Med. Biol.*, 48(22):3699–3719.

Boone, J. 1999. Glandular breast dose for monoenergetic and high-energy X-ray beams: Monte Carlo Assessment. *Radiology*, 213(1):23–37.

Boone, J. M. 2009. Dose spread functions in computed tomography: A Monte Carlo study. *Med. Phys.*, 36(10):4547.

Boone, J. M., Buonocore, M. H., and Cooper, V. N. 2000a. Monte Carlo validation in diagnostic radiological imaging. *Med. Phys.*, 27(6):1294–1304.

Boone, J. M. and Cooper, V. N. 2000. Scatter/primary in mammography: Monte Carlo validation. *Med. Phys.*, 27(8):1818–1831.

Boone, J. M., Fewell, T. R., and Jennings, R. J. 1997. Molybdenum, rhodium, and tungsten anode spectral models using interpolating polynomials with application to mammography. *Med. Phys.*, 24(12):1863–1874.

Boone, J. M., Lindfors, K. K., Cooper, V. N., and Seibert, J. A. 2000b. Scatter/primary in mammography: Comprehensive results. *Med. Phys.*, 27(10):2408–2416.

Boone, J. M. and Seibert, J. A. 1988. Monte Carlo simulation of the scattered radiation distribution in diagnostic radiology. *Med. Phys.*, 15(5):713–720.

Boone, J. M. and Seibert, J. A. 1997. An accurate method for computer-generating tungsten anode x-ray spectra from 30 to 140 kV. *Med. Phys.*, 24(11):1661–70.

Carton, A.-K., Bakic, P., Ullberg, C., Derand, H., and Maidment, A. D. A. 2011. Development of a physical 3D anthropomorphic breast phantom. *Med. Phys.*, 38(2):891.

Chen, B., Shorey, J., Saunders, R. S., Richard, S., Thompson, J., Nolte, L. W., and Samei, E. 2011. An anthropomorphic breast model for breast imaging simulation and optimization. *Acad. Radiol.*, 18(5):536–46.

Cunningham, I., Westmore, M., and Fenster, A. 1994. A spatial-frequency dependent quantum accounting diagram and detective quantum efficiency model of signal and noise propagation in cascaded imaging systems. *Med. Phys.*, 21:417–427.

Fredenberg, E., Danielsson, M., Stayman, J. W., Siewerdsen, J. H., and Aslund, M. 2012. Ideal-observer detectability in photon-counting differential phase-contrast imaging using a linear-systems approach. *Med. Phys.*, 39(9):5317–35.

Freed, M., Miller, S., Tang, K., and Badano, A. 2009. Experimental validation of Monte Carlo (MANTIS) simulated x-ray response of columnar CsI scintillator screens. *Med. Phys.*, 36(11):4944.

Freed, M., Park, S., and Badano, A. 2010. A fast, angle-dependent, analytical model of CsI detector response for optimization of 3D x-ray breast imaging systems. *Med. Phys.*, 37(6):2593.

Ghetti, C., Borrini, A., Ortenzia, O., Rossi, R., and Ordonez, P. L. 2008. Physical characteristics of GE Senographe Essential and DS digital mammography detectors. *Med. Phys.*, 35(2):456.

Glick, S. J. and Gong, X. 2006. Optimal spectra for indirect detector breast tomosynthesis. In Flynn, M. J. and Hsieh, J., eds, *Proc. SPIE*, 6142, 61421 L–61421 L–9.

Gong, X., Glick, S. J., Liu, B., Vedula, A. A., and Thacker, S. 2006. A computer simulation study comparing lesion detection accuracy with digital mammography, breast tomosynthesis, and cone-beam CT breast imaging. *Med. Phys.*, 33(4):1041.

Hu, Y.-H. and Zhao, W. 2011. The effect of angular dose distribution on the detection of microcalcifications in digital breast tomosynthesis. *Med. Phys.*, 38(5):2455.

Hubbell, J. H. and Seltzer, S. M. 2012. Tables of X-Ray Mass Attenuation Coefficients and Mass Energy-Absorption Coefficients, National Institute of Standards and Technology, Gaithersburg, MD, url: http://www.nist.gov/pml/data/xraycoef/index.cfm, accessed date: 26/04/12.

Johns, P. C. and Yaffe, M. J. 1987. X-ray characterisation of normal and neoplastic breast tissues. *Phys. Med. Biol.*, 32(6):675–695.

Krupinski, E. A., Roehrig, H., and Yu, T. 1995. Observer performance comparison of digital radiograph systems for stereotactic breast needle biopsy. *Acad. Radiol.*, 2(2):116–22.

Lau, B. A., Reiser, I., Nishikawa, R. M., and Bakic, P. R. 2012. A statistically defined anthropomorphic software breast phantom. *Med. Phys.*, 39(6):3375–3385.

Li, C. M., Segars, W. P., Tourassi, G. D., Boone, J. M., and Dobbins, J. T. 2009. Methodology for generating a 3D computerized breast phantom from empirical data. *Med. Phys.*, 36(7):3122.

Lubberts, G. 1968. Random noise produced by x-ray fluorescent screens. *JOSA*, 58(11):1475–1483.

Lubinsky, A., Whiting, B., and Owen, J. 1987. Storage phosphor system for computed radiography: Optical effects and detective quantum efficiency (DQE). *Proc. SPIE*, 767, 167–177.

Ma, A., Gunn, S., and Darambara, D. 2009. Introducing DeBRa: A detailed breast model for radiological studies. *Phys. Med. Biol.*, 54:4533.

Mainprize, J. G., Bloomquist, A. K., Kempston, M. P., and Yaffe, M. J. 2006. Resolution at oblique incidence angles of a flat panel imager for breast tomosynthesis. *Med. Phys.*, 33(9):3159.

Mainprize, J. and Yaffe, M. 2010. Cascaded analysis of signal and noise propagation through a heterogeneous breast model. *Med. Phys.*, 37(10):5243–5250.

Man, B. D. and Basu, S. 2004. Distance-driven projection and backprojection in three dimensions. *Phys. Med. Biol.*, 49(11):2463–2475.

Nishikawa, R. M. and Yaffe, M. J. 1990. Model of the spatial-frequency-dependent detective quantum efficiency of phosphor screens. *Med. Phys.*, 17:894–904.

Nishikawa, R. M., Yaffe, M. J., and Holmes, R. B. 1989. Effect of finite phosphor thickness on detective quantum efficiency. *Med. Phys.*, 16:773–780.

O'Connor, J., Das, M., Didier, C., Mah'D, M., and Glick, S. 2010. Development of an ensemble of digital breast object models. *Digit. Mammogr.*, 6136:54–61.

Park, S., Badal, A., Young, S., and Myers, K. J. 2012. A mathematical framework for including various sources of variability in a task-based assessment of digital breast tomosynthesis. *Proc. SPIE*, 8313, 83134S.

Pokrajac, D. D., Maidment, A. D. A., and Bakic, P. R. 2012. Optimized generation of high resolution breast anthropomorphic software phantoms. *Med. Phys.*, 39(4):2290–2302.

Que, W. and Rowlands, J. A. 1995. X-ray imaging using amorphous selenium: Inherent spatial resolution. *Med. Phys.*, 22(4):365–374.

Rabbani, M., Shaw, R., and Van Metter, R. 1987. Detective quantum efficiency of imaging systems with amplifying and scattering mechanisms. *JOSA A*, 4(5):895–901.

Reiser, I. and Nishikawa, R. M. 2010. Task-based assessment of breast tomosynthesis: Effect of acquisition parameters and quantum noise. *Med. Phys.*, 37(4):1591–1600.

Reiser, I., Nishikawa, R. M., Sidky, E. Y., Chinander, M. R., and Seifi, P. 2007. Development of a model for breast tomosynthesis image acquisition. *Proc. SPIE*, 6510, 65103D–65103D–8.

Reiser, I. S., Nishikawa, R. M., and Lau, B. A. 2009. Effect of non-isotropic detector blur on microcalcification detectability in tomosynthesis. *Proc. SPIE*, 7258, 72585Z–72585Z–10.

Richard, S. and Siewerdsen, J. H. 2008. Comparison of model and human observer performance for detection and discrimination tasks using dual-energy x-ray images. *Med. Phys.*, 35(11):5043.

Rogers, D. W. O. 2006. Fifty years of Monte Carlo simulations for medical physics. *Phys. Med. Biol.*, 51(13):R287–R301.

Saunders, R. S. and Samei, E. 2003. A method for modifying the image quality parameters of digital radiographic images. *Med. Phys.*, 30(11):3006.

Sechopoulos, I., Suryanarayanan, S., Vedantham, S., D'Orsi, C. J., and Karellas, A. 2007a. Scatter radiation in digital tomosynthesis of the breast. *Med. Phys.*, 34(2):564.

Sechopoulos, I., Suryanarayanan, S., Vedantham, S., D'Orsi, C., and Karellas, A. 2007b. Computation of the glandular radiation dose in digital tomosynthesis of the breast. *Med. Phys.*, 34(1):221.

Segars, W. P. 2001. Development and Application of the New Dynamic Nurbs Based Cardiac-Torso (NCAT) Phantom. PhD thesis, University of North Carolina, Chapel Hill.

Segars, W. P. and Tsui, B. M. W. 2009. MCAT to XCAT: The evolution of 4-D computerized phantoms for imaging research. *Proc. IEEE*, 97(12):1954–1968.

Sempau, J., Acosta, E., Baro, J., Fernandez-Varea, J. M., and Salvat, F. 1997. An algorithm for Monte Carlo simulation of the coupled electron-photon transport. *NIM B*, 132:377–390.

Sempau, J., Fernandez-Varea, J. M., Acosta, E., and Salvat, F. 2003. Experimental benchmarks of the Monte Carlo code PENELOPE. *NIM B*, 207:107–123.

Shaheen, E., Marshall, N., and Bosmans, H. 2011. Investigation of the effect of tube motion in breast tomosynthesis: Continuous or step and shoot? *Proc. SPIE*, 7961, 79611E–79611E–9.

Sharma, D., Badal, A., and Badano, A. 2012. HybridMANTIS: A CPUGPU Monte Carlo method for modeling indirect x-ray detectors with columnar scintillators. *Phys. Med. Biol.*, 57(8):2357–2372.

Siddon, R. 1985. Fast calculation of the exact radiological path for a three-dimensional CT array. *Med. Phys.*, 12:252.

Sidky, E. Y., Pan, X., Reiser, I. S., Nishikawa, R. M., Moore, R. H., and Kopans, D. B. 2009. Enhanced imaging of microcalcifications in digital breast tomosynthesis through improved image-reconstruction algorithms. *Med. Phys.*, 36(11):4920.

Siewerdsen, J. H., Antonuk, L. E., El-Mohri, Y., Yorkston, J., Huang, W., Boudry, J. M., and Cunningham, I. A. 1997. Empirical and theoretical investigation of the noise performance of indirect detection, active matrix flat-panel imagers (AMFPIs) for diagnostic radiology. *Med. Phys.*, 24(1):71–89.

Siewerdsen, J. H., Antonuk, L. E., El-Mohri, Y., Yorkston, J., Huang, W., and Cunningham, I. A. 1998. Signal, noise power spectrum, and detective quantum efficiency of indirect-detection flat-panel imagers for diagnostic radiology. *Med. Phys.*, 25(5):614–28.

Siewerdsen, J., Daly, M., Bakhtiar, B., Moseley, D., Richard, S., Keller, H., and Jaffray, D. 2006. A simple, direct method for x-ray scatter estimation and correction in digital radiography and cone-beam CT. *Med. Phys.*, 33:187.

Swank, R. K. 1973. Calculation of modulation transfer functions of x-ray fluorescent screens. *Appl. Opt.*, 12(8):1865–1870.

Tward, D. J. and Siewerdsen, J. H. 2008. Cascaded systems analysis of the 3D noise transfer characteristics of flat-panel cone-beam CT. *Med. Phys.*, 35(12):5510.

Van Metter, R. and Rabbani, M. 1990. An application of multivariate moment-generating functions to the analysis of signal and noise propagation in radiographic screen-film systems. *Med. Phys.*, 17(1):65–71.

Vedantham, S., Karellas, A., Suryanarayanan, S., Albagli, D., Han, S., Tkaczyk, E., Landberg, C. et al. 2000. Full breast digital mammography with an amorphous silicon-based flat panel detector: Physical characteristics of a clinical prototype. *Med. Phys.*, 27(9):558.

Wu, G., Mainprize, J. G., Boone, J. M., and Yaffe, M. J. 2009a. Evaluation of scatter effects on image quality for breast tomosynthesis. *Med. Phys.*, 36(10):4425.

Wu, G., Mainprize, J. G., Boone, J. M., and Yaffe, M. J. 2009b. Evaluation of scatter effects on image quality for breast tomosynthesis. *Med. Phys.*, 36(10):4425.

Wu, T., Moore, R. H., Rafferty, E. A., and Kopans, D. B. 2004. A comparison of reconstruction algorithms for breast tomosynthesis. *Med. Phys.*, 31(9):2636.

Zhang, C., Bakic, P. R., and Maidment, A. D. A. 2008. Proceedings of SPIE. In *Medical Imaging 2008: Visualization, Image-Guided Procedures, and Modeling*, pp. 69180 V–69180 V–10. University of Pennsylvania, SPIE.

Zhao, B. and Zhao, W. 2008a. Imaging performance of an amorphous selenium digital mammography detector in a breast tomosynthesis system. *Med. Phys.*, 35(5):1978.

Zhao, B. and Zhao, W. 2008b. Three-dimensional linear system analysis for breast tomosynthesis. *Med. Phys.*, 35(12):5219.

Zhao, B., Zhou, J., Hu, Y.-H., Mertelmeier, T., Ludwig, J., and Zhao, W. 2009. Experimental validation of a three-dimensional linear system model for breast tomosynthesis. *Med. Phys.*, 36(1):240.

Zhao, W., Ji, W. G., Debrie, A., and Rowlands, J. A. 2003. Imaging performance of amorphous selenium based flat-panel detectors for digital mammography: Characterization of a small area prototype detector. *Med. Phys.*, 30(2):254.

Zhao, W. and Rowlands, J. A. 1997. Digital radiology using active matrix readout of amorphous selenium: Theoretical analysis of detective quantum efficiency. *Med. Phys.*, 24(12):1819–33.

Zhou, J., Zhao, B., and Zhao, W. 2007. A computer simulation platform for the optimization of a breast tomosynthesis system. *Med. Phys.*, 34(3):1098.

System design

Image reconstruction

7 Filtered backprojection-based methods for tomosynthesis image reconstruction

Thomas Mertelmeier

Contents

7.1 INTRODUCTION

Tomosynthesis is characterized by incomplete data acquisition since projections are taken only from a limited angular range around the object. It is a special kind of limited-angle tomography. The typical number of projections is between 7 and 70. In addition, two adjacent single-projection views are usually separated by a relatively large angle increment compared to computed tomography (CT). The typical angle increments are 1–3° for medical applications. This incompleteness will inherently lead to artifacts since the inverse problem is ill-posed and cannot be solved exactly (Natterer 1986). To alleviate those artifacts, special reconstruction methods are required that go beyond the standard CT reconstruction algorithms.

There are principally two broad classes of algorithms to reconstruct the image of the object from the projections: algebraic and analytical algorithms. In the former, the object and the data acquisition process is considered to be discrete, resulting in the system equation. The solution of the problem, the image, is estimated by solving the system equation of the problem numerically, which means the iterative solution of a large system of equations is looked for. These iterative reconstruction algorithms may be based on models, that is, models for the data acquisition or for the object. In analytical reconstruction algorithms, the measurement process is assumed to be ideal and continuous given by the Radon transform (Radon 1917), and the system equation is solved analytically prior to the discretization of the solution. However, for the incomplete data problem, only approximate solutions exist and the effort is put on finding the appropriate approximations. In this chapter, we address the analytical path of solving the inverse problem for tomosynthesis.

7.2 FILTERED BACKPROJECTION FOR TWO-DIMENSIONAL CT

The best example to illustrate the principle of filtered backprojection (FBP) is two-dimensional (2-D) CT in parallel projection geometry (Kak and Slaney 1988). The Radon transform maps the object distribution $f(\mathbf{r})$, $\mathbf{r} = (x, y) \in \mathrm{R}^2$ into the set of line integrals through the object

$$\mathrm{R}f(\theta, s) = R_\theta(s) = \int_{\mathbf{r}\cdot\theta=s} f(\mathbf{r})\, d^2r = \int_{\theta\perp} f(s\theta+t)\, dt, \quad (7.1)$$

where θ is a unit vector in R^2, $\theta = (\cos\theta, \sin\theta)$, and s is the distance of the line from the origin, defined by $\mathbf{r}\cdot\theta = s$ (Figure 7.1).

A line integral, also called projection value, is given by Beer's attenuation law

$$\int_{\mathrm{line}} f(t)\, dt = -\log(I_1/I_0) \quad (7.2)$$

describing the attenuation of the radiation with the initial intensity I_0 and attenuated intensity I_1 along the line.

By 2-D Fourier transformation in polar coordinates (ρ, θ), it can be shown that Equation 7.1 can be inverted as

$$f(\mathbf{r}) = \int_0^\pi \left[F_1^{-1}\left(|\rho|\right) * R_\theta(f)d\theta \right]\Big|_{s=\mathbf{r}\cdot\theta}. \quad (7.3)$$

This is the basic equation for FPB. The object at $\mathbf{r}$ is calculated by back projecting, that is, summing up all line integrals passing through the point $\mathbf{r}$, and filtered by the inverse one-dimensional (1-D) Fourier transform of $|\rho|$. Because of its shape in frequency space, $|\rho|$ is called the ramp filter. It takes into account that the sampling density increases with decreasing distance from the

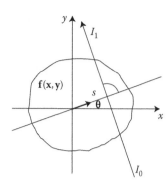

Figure 7.1 The object distribution f(x, y) is mapped to the set of line integrals.

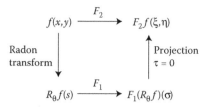

Figure 7.2 The central slice theorem provides a relationship between the 1-D Fourier transform of a projection and a cut through the 2-D Fourier transform of the object function.

origin. Without the filtering step of the projection data with the ramp filter function, the point spread function would not be point-like but smeared over a certain region according to $1/r$.

Using the fact that in the Fourier domain the convolution is the product in the spatial domain and vice versa, the filtering step can also be performed in frequency space.

For ideal, that is, completely and continuously sampled, and noiseless projection data, the 2-D object function can be exactly reconstructed from all its line integrals by means of Equation 7.3. The reconstruction filter is the ramp filter that inverts the blurring caused by the sampling and the backprojection. For practical reasons, the object and the sampling scheme have to be discretized. Employing a finite number of projections, a band-limited object function can be reconstructed with a Nyquist frequency of $1/(2d)$, d being the detector element size. In this case, the ramp filter is given by $|\nu|$ rect (νd) (rect(x): $= 1$ for $|x| \leq \frac{1}{2}$ and 0 otherwise) up to the Nyquist frequency. The spatial domain realization of this filter is called the Ramachandran–Lakshminarayanan kernel (Ramachandran and Lakshminarayanan 1971). However, it emphasizes the high frequencies, and therefore noise. To suppress the high-frequency noise, usually, this filter is apodized and modified, for example, such as the Shepp–Logan kernel that is characterized by $|\nu| \cdot |\sin(\pi \nu d)/(\pi \nu d)| \cdot$ rect (νd) (Shepp and Logan 1974).

An interesting relation between the projections and the data in frequency space is given by the Fourier slice or central slice theorem (Barrett and Swindell 1981). This theorem states that the 1-D Fourier transform of a projection characterized by the angle θ is equivalent to a section of the 2-D Fourier transform of the object f:

$$F_1(R_\theta f)(\sigma) = F_2 f(r, \theta)(\sigma, \tau = 0) \tag{7.4}$$

(see Figure 7.2). In this equation, σ and τ are the coordinates in frequency space in a Cartesian coordinate system, where σ is the frequency axis belonging to the s-axis in spatial domain.

Thus, this section of $F_2 f$ is defined by the angle θ that characterizes the projection. The Fourier slice theorem is particularly helpful to understand the tomosynthetic data acquisition structure in frequency space.

7.3 FPB FOR TOMOSYNTHESIS

7.3.1 TOMOSYNTHESIS SAMPLING

Tomosynthesis is a kind of limited-angle tomography, that is, data are acquired only from a limited angular range with respect to the object. In this section, we will analyze the data acquisition and the implication on the point spread function for the special case of linear tomosynthesis. In this case, the x-ray tube moves relative to the object on a circular arc or on a linear trajectory. Owing to the limited angle from which the projection data are acquired, the object is incompletely sampled. The data are missing with the impact that an exact reconstruction of the object is not possible. As a consequence, artifacts will be inevitable. This incompleteness goes far beyond the incompleteness of a circular sampling of a three-dimensional (3-D) object over 360° that would allow approximate reconstruction with the Feldkamp algorithm (Feldkamp et al. 1984), and in the central plane of the circle, even an exact reconstruction is possible. Thus, for tomosynthesis reconstruction, special procedures have to be applied to alleviate the missing data artifacts.

In the approximation of parallel beam sampling, according to the central slice theorem, data on planes on a double cone in frequency space would be acquired (Figure 7.3). The parallel beam

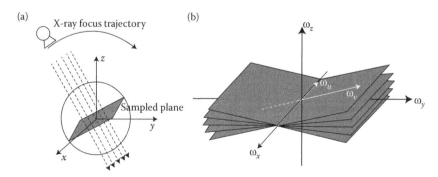

Figure 7.3 Data acquisition in Fourier's space in parallel beam approximation. (a) For the tube position at angle φ ($-\alpha \leq \varphi \leq \alpha$), the indicated sampling plane is created. (b) Moving over an arc from angle $-\alpha$ to α, the Fourier space is sampled on planes in a double-wedge domain. The relation between the object frequency space (ω_x, ω_y, ω_z) and the projection frequency space (ω_u, ω_v) is indicated.

condition is approximately valid for a large tube–object distance. The smaller the tomosynthesis angle α, the smaller the region where data are acquired.

The fact that data are missing has some impact on image quality, which can be seen when looking at the point spread function.

7.3.2 TOMOSYNTHESIS POINT SPREAD FUNCTION AND TRANSFER FUNCTION

Let us consider an acquisition geometry as shown in Figure 7.4. The tube moves on an arc from angle $-\alpha$ to $+\alpha$ around the pivoting point. The situation is treated as a 2-D problem with scan in the direction of the y-axis and translational invariance in the x-direction.

Then the point spread function or response function of one point in space for this acquisition scheme and backprojection is given by (Haerer et al. 2002)

$$h_P(\mathbf{r}) = \text{const.} \int_C \int_{-\infty}^{\infty} \delta(\mathbf{r} - s\mathbf{e}_t)\, ds\, dt. \qquad (7.5)$$

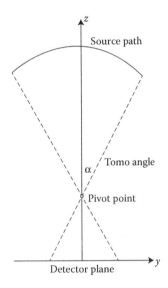

Figure 7.4 Scan geometry for linear tomosynthesis with scan motion in the y–z plane.

The integral over s represents one projection ray in the direction of the unit vector $\mathbf{e}_t$. The source path C is parameterized by $t \in R$ (R = set of real numbers); hence, the integral over C is the backprojection of all projection rays passing through the point $\mathbf{r}$. For equiangular sampling, the integration may be performed in cylindrical coordinates (x, r, φ), which leads to

$$h_P(\mathbf{r}) = \frac{1}{2\alpha} \int_C \int_{-\infty}^{\infty} \delta(\mathbf{r} - s\mathbf{e}_\varphi)\, ds\, d\varphi$$

to yield finally the point spread function using $r^2 = y^2 + z^2$ as

$$h_P(x, y, z) = \frac{1}{2\alpha\sqrt{y^2 + z^2}} \delta(x) \quad \text{for } y/z < \tan\alpha$$
$$= 0 \qquad\qquad\qquad \text{otherwise} \qquad (7.6)$$

Figure 7.5a shows this point spread function h_P.

From the point spread function h_P, it becomes obvious that the information is spread out over all slices. The image of a point becomes a linear structure in neighboring slices with decreasing intensity for increasing distance.

The transfer function is obtained by a Fourier transformation of Equation 7.6 in the sampled region:

$$H_P(\boldsymbol{\omega}) = \frac{1}{2\alpha\sqrt{\omega_y^2 + \omega_z^2}} \qquad (7.7)$$

with the 3-D frequency vector $\boldsymbol{\omega} = (\omega_x, \omega_y, \omega_z)$.

In the approximation for $\omega_z \ll \omega_y$, in the sampled double-wedge-shaped region

$$H_P(\boldsymbol{\omega}) \approx \frac{1}{2\alpha\,|\,\omega_y\,|} \qquad (7.8)$$

is obtained, which can be generally written as

$$H_P(\boldsymbol{\omega}) \approx \frac{1}{2\alpha\,|\,\omega_y\,|}\,\text{rect}\!\left(\frac{\omega_z}{\omega_y\tan\alpha}\right) \qquad (7.9)$$

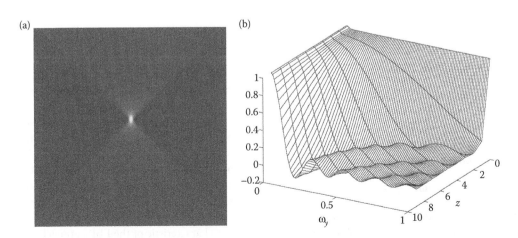

Figure 7.5 (a) Point spread function of linear tomosynthesis with equiangular sampling in the y–z plane. The tomosynthesis half-angle is 45°. (b) Slice transfer function (z is given in arbitrary units). (These graphs were provided by M. Zellerhoff, 1998. Personal communication, Siemens AG, 1998.)

The inverse function of Equation 7.8 is proportional to the ramp filter known from CT, which is the reconstruction filter of the backprojection. Thus, there is a low-pass filter in the scan direction, which leads to the blurring of the simple backprojection reconstruction.

By a 1-D Fourier transformation in the z-direction (denoted as F_z), the slice transfer function is defined as

$$H_{\text{slice}}(\omega_x, \omega_y | z) = F_z^{-1} H_p. \tag{7.10}$$

This slice transfer function measures how structures are transmitted to a slice of distance z. Hence, the slice transfer function is

$$H_{\text{slice}}(\omega_x, \omega_y | z) = \text{sinc}(z\omega_y \tan \alpha). \tag{7.11}$$

This function is plotted in Figure 7.5b demonstrating that the transfer to the neighboring slices and thus the slice thickness depends on the spatial frequency ω_y in the scan direction.

7.3.3 FILTER DESIGN

FBP algorithms are the standard reconstruction methods in CT. For tomosynthesis, fundamental work has been carried out by Grant (1972) who analyzed the transfer function and described the blurring process of tomography as a filtering process. Edholm et al. developed a method called ectomography (Edholm et al. 1980) defining filtering steps for circular scanning to reduce blurring artifacts and to define a slice thickness. Matsuo et al. gave an analysis of the transfer function in frequency space and calculated the point spread function (Matsuo et al. 1993). On the basis of this preceding work, Lauritsch and Härer (1998) formulated an FBP procedure for circular tomosynthesis involving several filtering steps. The reconstruction method described in Lauritsch et al. (1998) for 2-D circular sampling in x- and y-coordinates (circular tomosynthesis) has been adapted to mammographic linear sampling geometry (Haerer et al. 2002, Mertelmeier et al. 2006). For the mammography data acquisition geometry, the tube motion on an arc over the object is treated in parallel beam approximation as a linear sampling path in y-orientation with varying magnification. The parallel beam approximation for designing the reconstruction filters is appropriate since the associated inaccuracies are small compared to the effects induced by the tomosynthetic sampling, which is inherently incomplete.

Starting from the system equation

$$G(\boldsymbol{\omega}) = H(\boldsymbol{\omega}) \cdot F(\boldsymbol{\omega}), \tag{7.12}$$

with F being the Fourier transform of the object, H the system modulation transfer function (MTF), G the Fourier transform of the reconstructed image data set, and $\boldsymbol{\omega}$ the spatial frequency vector in 3-D Fourier space, we assume that the MTF can be split into a filter function H_{filter} and a projection–backprojection part H_p

$$H(\boldsymbol{\omega}) = H_{\text{filter}}(\boldsymbol{\omega}) \cdot H_p(\boldsymbol{\omega}). \tag{7.13}$$

The filter function shall invert the projection–backprojection transfer function as derived in Section 7.3.2 and, since the inversion problem cannot be solved exactly, provides the flexibility to tune image characteristics and to alleviate artifacts.

An appropriate filter function in Fourier space for linear sampling in y-orientation may be chosen as

$$H_{\text{filter}}(\omega_y, \omega_z) = H_{\text{spectrum}}(\omega_y) \cdot H_{\text{profile}}(\omega_z) \cdot H_{\text{inverse}}(\omega_y, \omega_z). \tag{7.14}$$

H_{inverse} inverts the MTF H_p of the projection–backprojection process in the frequency region that is accessible to tomosynthetic data collection. For equiangular sampling, H_p is given by Equation 7.7 (Section 7.3.2). Thus, H_{inverse} is proportional to a ramp-type filter in the sampled region. With realistic noisy data, the ramp filter is known for emphasizing noise that can be regularized by an appropriate spectral filter. For the spectral filtering H_{spectrum}, we may choose a von Hann ("Hanning") window

$$\begin{aligned} H_{\text{spectrum}}(\omega_y) &= 0.5\left(1 + \cos\left(\frac{\pi\omega_y}{A}\right)\right) \quad \text{for } |\omega_y| < A, \\ &= 0 \quad\quad\quad\quad\quad\quad\quad \text{elsewhere} \end{aligned} \tag{7.15}$$

with parameter A (A > 0) to suppress high frequencies and thereby noise.

At this stage, after inversion, even after applying a spectral filter in ω_y, the ω_z-border of the sampled region in Fourier's space (cf. Figure 7.6) still represents a sharp step function. This discontinuity will create a corresponding ringing in the spatial domain, increasing the out-of-plane artifacts already present from the incomplete sampling of tomosynthesis. The third filter part $H_{\text{profile}}(\omega_z)$ (see Equation 7.14) can be used to improve this behavior. It is usually called "slice profile function" or "slice thickness filter." It controls the spatial slice thickness, more precisely, the spatial slice sensitivity profile (Lauritsch et al. 1998) and leads to a suppression of out-of-plane artifacts typical for tomosynthesis.

The slice thickness filter in its simplest form may be designed in a similar way as the spectral filter, for example, as

$$\begin{aligned} H_{\text{profile}}(\omega_z) &= 0.5\left(1 + \cos\left(\frac{\pi\omega_z}{B}\right)\right) \quad \begin{aligned}&\text{for}|\omega_z| < B \quad \text{and}\\ &|\omega_z| < \tan(\alpha)|\omega_y|.\end{aligned} \\ H_{\text{profile}}(\omega_z) &= 0 \quad\quad\quad\quad\quad\quad\quad \text{elsewhere.} \end{aligned} \tag{7.16}$$

The parameter B (B > 0) controls the cutoff frequency in ω_z and thus, via inverse Fourier transformation of H_{profile}, also the slice width of the slice sensitivity function in object space. The incomplete sampling generally prevents from obtaining a constant slice thickness but generates out-of-plane artifacts. Figure 7.6 illustrates how these artifacts can be largely reduced for spatial frequencies ω_y above a certain lower bound of the frequency. If the cutoff on the z-frequency scale is inside the sampling region, a slice thickness is well defined. For small y-frequencies, that is, for $\omega_y < \omega_{z\text{-max}}/\tan(\alpha)$, however, the slice thickness increases with decreasing y-frequencies. In tomosynthesis, the slice thickness cannot be held constant throughout the entire data range, but rather depends on the frequency content of the object.

One characteristic of the FBP approach described here is the suppression of low frequencies leading to an image impression dominated by edge enhancement and looking differently than a

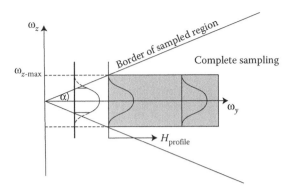

Figure 7.6 The slice thickness filter function $H_{profile}$ (ω_z) ensures a constant depth resolution over a wide range of spatial frequencies.

mammogram (Zhao et al. 2009). Figure 7.7 compares different reconstruction methods based on backprojection. Here, a slice out of the reconstructed data set of a human subject in mediolateral oblique (MLO) position is shown. The result of backprojection without filtering is shown in Figure 7.7a exhibiting blurring artifacts well known from conventional tomography. These can be avoided by ramp-like filtering at the cost of increased noise due to amplification of high frequencies (Figure 7.7b). The reconstruction results of additional spectral and slice thickness filtering are demonstrated in Figure 7.7c where it can be appreciated that spatial resolution and local contrast are vastly improved.

An alternative filter mechanism was proposed in Kunze et al. (2007) and Ludwig et al. (2008). The filter was derived from the simultaneous iterative reconstruction technique (SIRT) leading to a more realistic density representation of the tissue typical for mammograms (Figure 7.7d). This can be attributed to a small nonzero component in the filter function. The first results look promising; however, an evaluation with clinical data still has to be undertaken.

For the backprojection step in FBP reconstruction, it is of essence to correctly take into account the geometry of data acquisition. This can be accomplished in the backprojection step either by directly incorporating the geometry of the acquisition for each projection or by employing projection matrices for the acquisition geometry (Wiesent et al. 2000). With the latter procedure, any sampling geometry with varying magnification factors can be handled. The projection matrices can be determined by calibrating the system with the help of a marker phantom or can be estimated by an online measurement procedure.

7.3.4 ARTIFACTS

Since the inverse problem of tomosynthesis cannot be solved exactly, artifacts are inevitable. The most important artifacts are caused by the missing data due to the small-acquisition angular range and the associated filtering across the borders between the sampled region and the region in Fourier's domain that is not sampled.

The point spread function (Figure 7.5) explains that the image of a localized object extends to neighboring slices. As an example, a calcification in a mammogram located in one specific plane and imaged sharply in that plane is replicated in other planes producing ghost-like images (Figure 7.8). These artifacts are called out-of-plane artifacts. The intensity of out-of-plane artifacts and their range depends on the object contrast and size. The larger the contrast, the larger is the artifact strength. Large objects create long-range artifacts because there is little information in the vertical (z) direction for small in-plane spatial frequency leading to a large physical slice width for these large objects.

Owing to the missing data, filtering creates the so-called overshoot artifacts with black rims in scan direction around high-contrast objects (Figure 7.9). Also, these artifacts appear stronger for increasing object size.

Another type of artifact is caused if not enough projections are taken to reconstruct the image. These artifacts are well known from CT and are most frequently called streak artifacts. They occur mostly at the edges of large high-contrast objects such as bones or metal. Therefore, they appear not severe in breast imaging as the breast may be considered a relatively homogenous object.

On the one hand, tomosynthesis shows some artifacts caused by the missing data. On the other, these artifacts are well known and can even be used in the reading process. For example, all objects being blurred in one slice are located out of this slice and not in focus. The overshoot artifact emphasizes small objects and thus may facilitate detection.

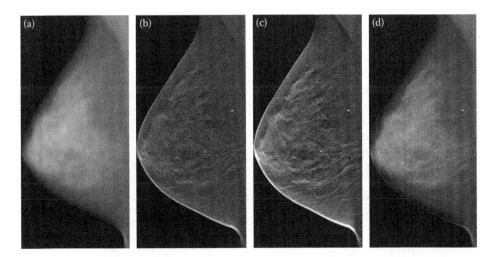

Figure 7.7 Slice 27 mm above the object table of a breast in MLO position. (a) Simple backprojection, (b) backprojection with ramp-like filter, (c) FPB with spectral and slice thickness filter, and (d) FPB with breast density filter. (Projections data courtesy of Dr. I. Andersson, Malmo University Hospital, Malmo, Sweden.)

Image reconstruction

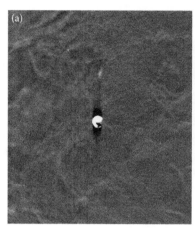

Figure 7.8 A large calcification in a mammogram sharply displayed in slice 31 (a) is replicated as an out-of-plane artifact in slice 45. The slice separation is 1 mm. The image size is approximately 46 × 52 mm. In the left image, the overshoot artifact is exhibited as well.

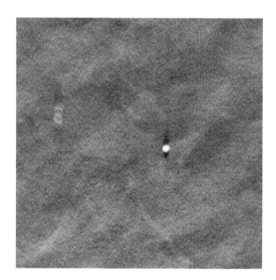

Figure 7.9 Overshoot artifacts manifest themselves as black rims in the scan direction around high-contrast objects such as a calcification. At the left side of the picture, a calcification located in a different slice is visible by its out-of-plane artifact. (The image size is ~36 × 36 mm.)

7.4 CONCLUSION

The FPB algorithm is a versatile and powerful reconstruction method for the tomosynthetic imaging problem providing high spatial and high contrast resolution. The filters can be tuned to minimize artifacts originating from the missing data characteristic for tomosynthesis, and can be adapted to the imaging task. FPB enables fast reconstruction as it is an analytic algorithm, which can be implemented very efficiently.

REFERENCES

Barrett HH, Swindell W. 1981. *Radiological Imaging*, Vol. 2. Academic Press, New York.

Edholm P, Granlund G, Knutsson H, Petersson C. 1980. Ectomography—A new radiographic method for reproducing a selected slice by varying thickness. *Acta Radiol.* 21: 433–442.

Feldkamp LA, Davis LC, Kress JW. 1984. Practical cone-beam algorithm. *Opt. Soc. Am.* 1: 612–619.

Grant DG. 1972. Tomosynthesis: A three-dimensional radiographic imaging technique. *IEEE Trans. Biomed. Eng.* 19: 20–28.

Haerer W, Lauritsch G, Zellerhoff M. 2002. Method for reconstructing a three-dimensional image of an object scanned in the context of a tomosynthesis, and apparatus for tomosynthesis. US Patent No. 6442288.

Kak AC, Slaney M. 1988. *Principles of Computerized Tomographic Imaging*, IEEE Press, New York.

Kunze H, Haerer W, Orman J, Mertelmeier T, Stierstorfer K. 2007. Filter determination for tomosynthesis aided by iterative reconstruction techniques. In: *9th International Meeting on Fully Three-Dimensional Image Reconstruction in Radiology and Nuclear Medicine*, pp. 309–312, Lindau, Germany.

Lauritsch G, Haerer W. 1998. A theoretical framework for filtered backprojection in tomosynthesis. *Proc. SPIE* 3338: 1127–1137.

Ludwig J, Mertelmeier T, Kunze H, Härer W. 2008. A novel approach for filtered backprojection in tomosynthesis based on filter kernels determined by iterative reconstruction techniques. In Krupinski E, ed. *Lecture Notes in Computer Science 5116, Digital Mammography, 9th International Workshop*, IWDM 2008, pp. 612–620, Springer-Verlag, Berlin, Heidelberg.

Matsuo H, Iwata A, Horiba I, Suzumura N. 1993. Three-dimensional image reconstruction by digital tomosynthesis using inverse filtering. *IEEE Trans. Med. Imag.* 12: 307–313.

Mertelmeier T, Orman J, Haerer W, Kumar MK. 2006. Optimizing filtered backprojection reconstruction for a breast tomosynthesis prototype device. *Proc. SPIE* 6142: 61420F1.

Natterer F. 1986. *The Mathematics of Computerized Tomography*, Teubner, Stuttgart.

Radon J. 1917. Über die Bestimmung von Funktionen durch ihre Integralwerte längs gewisser Mannigfaltigkeiten. *Ber. Verh. Sächs. Akad. Wiss. Leipzig*, 69: 262–277.

Ramachandran GN, Lakshminarayanan AV. 1971. Three-dimensional reconstruction from radiographs and electron micrographs: Application of convolutions instead of Fourier transforms. *Proc. Nat. Acad. Sci. USA* 68: 2236–2240.

Shepp LA, Logan BF. 1974. The Fourier reconstruction of a head section. *IEEE Trans. Nucl. Sci.* 21: 21–43.

Wiesent K, Barth K, Navab N, Durlak P, Brunner TM, Schuetz O, Seissler W. 2000: Enhanced 3-D reconstruction algorithm for C-arm systems suitable for interventional procedures. *IEEE Trans. Med. Imag.* 19: 391–403.

Zhao B, Zhou J, Yue-Houng H, Zhao W, Mertelmeier T, Ludwig J. 2009. Experimental validation of a three-dimensional linear system model for breast tomosynthesis. *Med. Phys.* 36: 240–251.

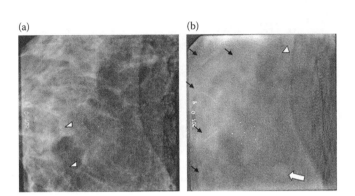

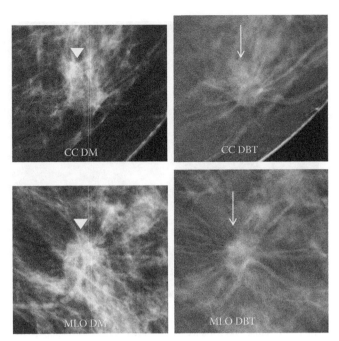

Figure 12.1 (a) DM and (b) DBT images of an American College of Radiology quality assurance phantom imbedded in a mastectomy specimen which has a heterogeneously dense breast composition. Two speck groups (arrowheads) are visible in the DM image (a), while the DBT image at the plane of interest reveals a third speck group (arrowhead), as well as five bar inclusions (arrows) and 1 mass inclusion (block arrow). (Courtesy of Andy Smith, PhD, Hologic, Inc.)

Figure 12.2 Cropped craniocaudal (CC) and medio-lateral oblique (MLO) digital mammography (DM) and digital breast tomosynthesis (DBT) images demonstrating an invasive ductal carcinoma seen as a vague asymmetry (arrowhead) with possible architectural distortion on DM and corresponding conspicuous mass with spiculated margins (arrow) on DBT.

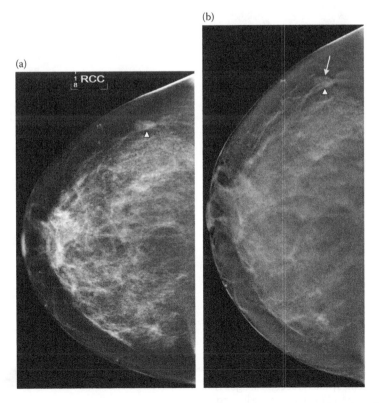

Figure 12.3 (a) Craniocaudal DM and (b) DBT images of a trial subject with a benign intra-mammary lymph node visualized as a nonspecific mass on DM (arrowhead) and a definitely benign intra-mammary lymph node (arrowhead) due to the presence of a fatty hilum (arrow) recognizable on DBT.

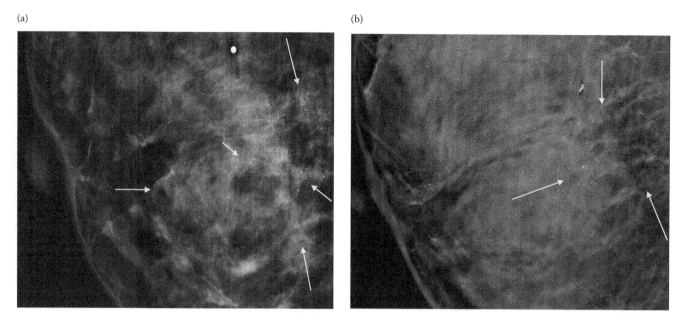

Figure 12.4 (a) DM image demonstrating malignant calcifications (arrows) in a segmental distribution versus (b) DBT thin slice reconstruction image of the same abnormality which shows clustered calcifications only.

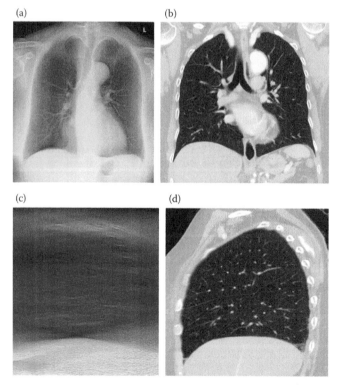

Figure 13.1 (a) A coronal tomosynthesis section image and (b) a coronal CT image at the tracheal level of the patient showing good agreement in the depicted anatomy. In (c), a sagittal reconstruction of the right lung from the coronal tomosynthesis section images is presented to illustrate the low depth resolution due to the limited angular range used. In (d), a sagittal CT image at a similar position is shown for comparison.

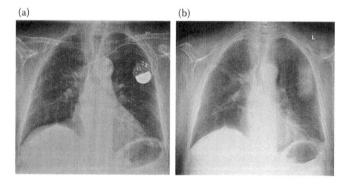

Figure 13.2 (a) A PA chest radiograph of a patient with a pacemaker and (b) a coronal tomosynthesis section image at the tracheal level. The low depth resolution of tomosynthesis results in the pacemaker showing up as an artifact in all section images, thus making the detection of pathology almost impossible.

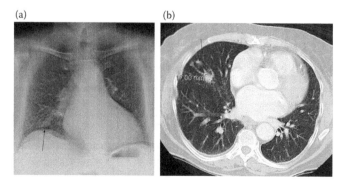

Figure 13.3 A coronal tomosynthesis section image (a) showing the presence of motion artifacts, reducing the visibility of the nodule clearly visible in the corresponding axial CT (b).

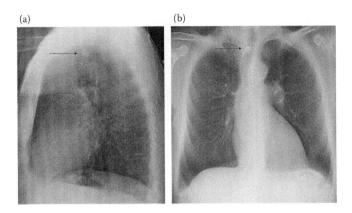

Figure 13.4 (a) A lateral chest radiograph and (b) a coronal tomosynthesis section image at the tracheal level. The calcified lymph node posterior to the trachea on the lateral chest radiograph is depicted as a possible foreign body within the trachea in the tomosynthesis section image.

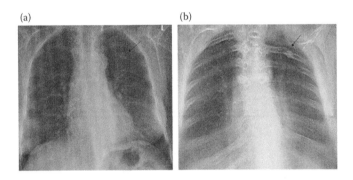

Figure 13.5 (a) A PA chest radiograph and (b) a coronal tomosynthesis section image at the level of the posterior parts of the ribs. The suspected opacity in the chest radiograph is clearly identified as a healed rib fracture on the tomosynthesis section image.

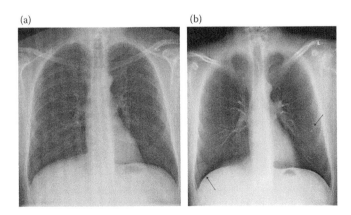

Figure 13.6 (a) A PA chest radiograph and (b) a coronal tomosynthesis section image at the hilum level. The tomosynthesis section image clearly shows two nodules, which are more difficult to discern on the chest radiograph.

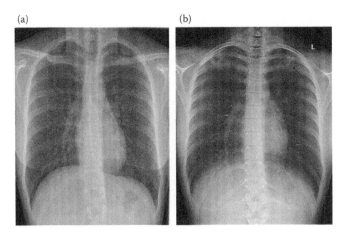

Figure 13.7 (a) A PA chest radiograph and (b) a coronal tomosynthesis section image at the spinal level. The tomosynthesis section image clearly shows a cavity in the left upper lobe, which is more difficult to discern on the chest radiograph.

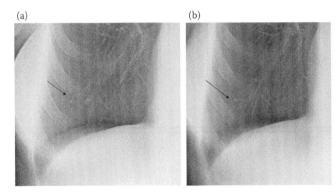

Figure 13.8 (a) A close-up of a chest tomosynthesis section image containing a 5-mm nodule in the middle lobe and (b) the corresponding image at the follow-up examination after 2 years. The images reveal no apparent nodule growth.

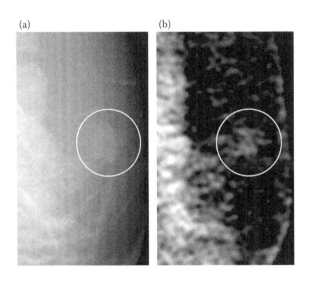

Figure 15.2 Examples of zoomed CE-DBT images of a 25-mm ductal carcinoma *in situ* lesion (circled) for a 69-year-old woman. Total energy image (a) and energy subtraction image (b) from a photon counting spectral imaging system at 120 s following contrast agent injection. Slice thickness was 3 mm. (Images courtesy of Dr. Florian Schmitzberger.)

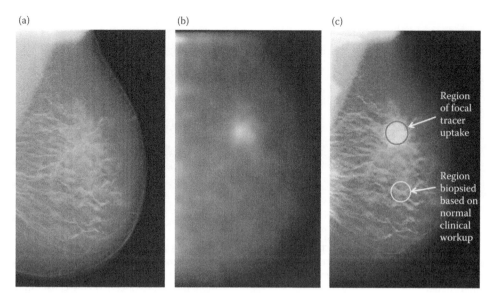

Figure 15.3 Examples of fused DBT and MBI images for a 52-year-old woman. (a) DBT slice at 1 mm thick. (b) Corresponding gamma tomosynthesis image at the same depth. (c) Merged sections from (a) and (b). The enhancing region was biopsy-confirmed ductal carcinoma *in situ* (arrow). (Images courtesy of Dr. Mark Williams, University of Virginia.)

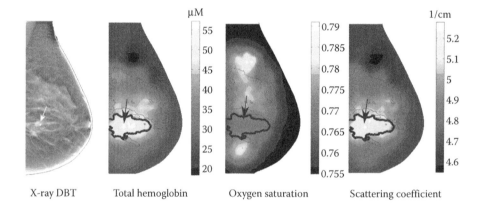

Figure 15.5 Registered DBT slice (far left) and optical images for (left to right) total hemoglobin concentration, blood oxygen saturation, a scattering coefficient for a 45-year-old woman with a 2.50-cm invasive ductal carcinoma (arrow on DBT image and outlined in optical images). (Figure courtesy of Dr. Qianqian Fang, Massachussetts General Hospital.)

8 Iterative image reconstruction design for digital breast tomosynthesis

Emil Y. Sidky

Contents

8.1 INTRODUCTION

This chapter on iterative image reconstruction (IIR) for digital breast tomosynthesis (DBT) is meant to be a quick start guide for implementing such algorithms for the DBT system. The goal, here, is to have the reader be able to implement an IIR algorithm for DBT, which can be used with actual data from a DBT scanner. The emphasis is on algorithm simplicity. Following the simplicity philosophy, we prioritize the issues for implementing IIR for DBT and concentrate only on the top two: data insufficiency and truncated iteration. The focus is on

IIR algorithm design, and while we necessarily develop only few algorithms, we point out that there are many other good options that simply cannot all be covered in a whole book much less a single chapter. Toward the end of the chapter, after the reader has some exposure to basic IIR, a short review of current IIR research for DBT will be presented.

The motivation of this style of presentation is that the relevant literature on IIR is enormous. Design of IIR algorithms can potentially go deep into physical systems and object modeling [1,2], objective assessment of image quality [3] (see Chapter 10), and optimization theory and algorithms [4–6].

Furthermore, research articles on IIR tend to go into specific details derived from various combinations of these general topics so that it can be difficult to grasp the big picture or to separate implementation details from general concepts. The simplifying theme of this chapter is that we focus on the main challenges of DBT.

8.1.1 DATA INSUFFICIENCY

The DBT scan consists of only 10–50 projection views over a limited angular range, which is at most 90°. Relative to computed tomography (CT), such a scan is limited both in angular scanning range and in view-angle sampling density. Even under ideal conditions of computer simulation, it is not possible to recover an accurate image without very strong prior knowledge of the subject. Because sampling insufficiency outweighs all other factors, we use basic line-integration models for the data and do not consider more sophisticated physical modeling for the exposition of IIR for DBT. This is not to say such considerations are unimportant, but rather sampling is more important and should be considered first.

8.1.2 TRUNCATED ITERATION

In developing an image reconstruction algorithm, a mathematical model is set up, which relates the underlying object function to the available data and, in some cases, which includes some prior knowledge of the object. After setting up such a model, an algorithm is derived that solves this model. For IIR, such models often take the form of an optimization problem, where some mathematical quantity is either minimized or maximized, respecting certain constraints of the system. For the DBT system, it is challenging to develop algorithms that solve the relevant model. Even if an IIR algorithm is derived that is mathematically guaranteed to converge to the solution of the DBT model, in practice, it will be necessary to terminate the iteration well short of convergence—the truncated iteration problem. This strong limitation results from the size of realistic DBT systems and mathematical properties common to any DBT model. This nonideality must also take high priority in the design of IIR algorithms for DBT.

The presentation in this chapter is very much along the lines of starting from the ground and working up. In this way, the reader does not need much fore knowledge of image reconstruction to understand the application of IIR to DBT. Along the way, we motivate various concepts, such as the use of optimization theory. The main text focuses on algorithm design and when alternatives in the design come up, we take the simplest option (in terms of computer coding) and refer to other work for more sophisticated and potentially better solutions. We emphasize that in no way do we attempt to make claims of optimality of any of the IIR algorithms. Indeed, application of IIR to DBT is still in its infancy, and at this stage it is more important to characterize the various IIR approaches and methods.

The chapter is divided into the following sections. We start by describing the theoretical background for IIR in DBT in Section 8.2, explaining the discrete-to-discrete imaging model used and the use of optimization theory to solve this model implicitly. Section 8.3 covers fundamental algorithms used for solving optimization problems relevant to IIR. In Section 8.4, we illustrate in detail the application of the basic IIR algorithms to a 2D DBT computer simulation. In Section 8.5, based on what we learn from the simulations, we present a modified IIR algorithm to address the specific needs of DBT. In Section 8.6, we present the application of the discussed IIR algorithms to a clinical DBT data set. Finally, in Section 8.7, we describe how incorporating IIR algorithms into a DBT system differs from filtered back-projection (FBP) and present a brief summary of research efforts in IIR for DBT.

8.2 DISCRETE-TO-DISCRETE IMAGING MODEL FOR DBT AND ITS IMPLICIT SOLUTION

From the point of view of tomographic image reconstruction, the major hurdles for DBT IIR stem from sampling insufficiency: limited angular scanning range and low angular sampling rate. These issues outweigh inconsistency due to physical factors; as a result, for the algorithms developed in this chapter, we eschew detailed physical imaging models and employ the standard monochromatic approximate model for transmitted x-ray intensity, neglecting x-ray scatter, and partial volume averaging. The basic physical model for x-ray transmission imaging is the Beer–Lambert law [7]:

$$I = I_0 \exp\left(-\int_0^\infty dt\, \mu(t)\right), \tag{8.1}$$

where I_0 is the incident x-ray intensity along a ray passing through the subject, I is the transmitted x-ray intensity along this ray, and $\mu(t)$ is the spatially varying linear x-ray attenuation map with t indicating location along the ray. Computing $-\ln(I/I_0)$ isolates the integral in the exponential function, which when written out in terms of a ray originating at x-ray source location $\vec{s}$ and pointing in direction $\hat{\theta}$ yields the data function g for the imaging model:

$$g(\vec{s}, \hat{\theta}) = \int_0^\infty dt f[\vec{s} + t\hat{\theta}], \tag{8.2}$$

where f represents the three-dimensional object function, in this case the x-ray linear attenuation coefficient map μ. This continuous-to-continuous imaging model is known as the x-ray or divergent-beam transform, used in deriving the FBP algorithm.

The difficulty with FBP is that for the present case of DBT we do not have sufficient data to make use of the available analytic inverses to Equation 8.2. The data insufficiency has two aspects: actual data models for digital imaging are continuous-to-discrete and g is measured only at discrete values of $\vec{s}$ and $\hat{\theta}$, and insufficient range coverage in each of these variables. The sampling insufficiency is particularly acute with respect to the x-ray source location, where the scanning arc is quite limited and the angular spacing between views can be large. As shown, however, in Chapter 7, one can ignore the mathematical restrictions and still design image reconstruction algorithms, of practical value, around FBP.

8.2.1 DISCRETE-TO-DISCRETE IMAGING MODEL

The approach for IIR is to solve Equation 8.2 implicitly on a digital computer, and in order to do so the imaging model needs to be converted to a finite linear system of the form

$$\mathbf{g} = X\mathbf{f}, \tag{8.3}$$

where $\mathbf{g}$ is a vector of the estimates of the ray-integration samples, comprising the sinogram; $\mathbf{f}$ is a vector of image expansion coefficients, comprising the image; and X is the system matrix obtained by specifying the image expansion elements and the sample locations for Equation 8.2. This discrete-to-discrete model, on the one hand, directly accounts for discrete sampling of the digital imaging system on the data side, but on the other hand, it introduces approximation by discretizing the object function. At this point, we comment that this linear system for an actual DBT scanner is enormous and as a result its direct inversion is not practical. This fact by itself forces us to consider only implicit solution of Equation 8.3 where only iterative algorithms are available. But before getting into the details of IIR, we must fully specify what is the matrix X. We first address the sampling locations, and then discuss the image expansion.

As we aim to demonstrate IIR design on a clinical data set provided by D. Kopans at Massachusetts General Hospital (MGH) and acquired on a GE prototype DBT system, we specify the scanner geometry and sampling for this particular scanner. The scanner configuration and properties are described in Ref. [8]. As shown in Figure 8.1, the breast is compressed to a thickness of 3–8 cm on a carbon-fiber tray protecting the fixed, flat-panel detector. The x-ray source is moved on an arc, centered on point $h = 21.7$ cm above the detector, and with radius $R = 44.3$ cm. The detector is composed of an array of 1800×2304 bins with width 100 microns, and its physical dimensions are $W = 180.0$ mm $\times L = 230.4$ mm. The number of projections is 11, and they are approximately equally spaced along the 50° arc. We use the term "in-plane" to refer to xy-planes, parallel to the detector, and the term "depth" to refer to the z-direction, perpendicular to the detector. The source position follows the arc described by

$$\vec{r}_0(s) = (0, R\sin s, R\cos s), \tag{8.4}$$

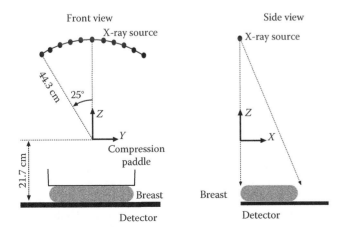

Figure 8.1 Configuration of the GE prototype DBT system.

and the detector bin locations are described by

$$\vec{d}(u,v) = (u, v - L/2, -h). \tag{8.5}$$

The unit vector $\hat{\theta}$ is now a function of the variables s, u, v:

$$\hat{\theta}(s,u,v) = \frac{\vec{d}(u,v) - \vec{r}_0(s)}{|\vec{d}(u,v) - \vec{r}_0(s)|}. \tag{8.6}$$

The actual sampling locations, indexed by (i,j,k), are obtained by evaluating $\vec{r}_0(s_i)$, where

$$s_i = -25° + i5° \quad i = 0, 1, 2, \ldots, 10, \tag{8.7}$$

and $\vec{d}(u_j, v_k)$, where

$$\begin{aligned}
u_j &= j0.1 \quad j = 0, 1, 2, \ldots, j_{max}, \\
v_k &= k0.1 \quad k = 0, 1, 2, \ldots, k_{max},
\end{aligned} \tag{8.8}$$

where the units for u_j and v_k are millimeters. The values of j_{max} and k_{max} depend on subject size and consequently vary from case to case.

Having specified the data sampling, we now turn to the image representation. While many alternatives appear in the literature [2,9,10], we use voxels for the present algorithm description. Again, for DBT, the data sampling insufficiency is the main issue. Substituting the voxelized image representation into the continuous model, Equation 8.2, and evaluating g at the sample points specifies the line-intersection form of the discrete x-ray transform and consequently specifies X in Equation 8.3.

For x-ray CT IIR, it is common to use cubic voxels, or to consider potentially voxel aspect ratios up to a factor of two. But for DBT, with its unusual sampling, it is standard practice to employ voxels with a large aspect ratio, where the depth of the voxels are typically 10 times their in-plane width. For performing IIR with the GE prototype system, for example, voxels are typically 0.1 mm, in-plane, and 1.0 mm, depth. The in-plane voxel width is chosen to be the same as the detector resolution.

8.2.2 NEED FOR OPTIMIZATION THEORY

In performing image reconstruction, one immediately thinks of inverting the imaging model, which is in fact not precisely what is happening in IIR. It is more accurate to describe IIR as an attempt to model the projection data, and the obtained image voxel values are to be interpreted as model parameters, which hopefully also prove useful in forming an image. The distinction becomes especially clear for DBT. It is unlikely that any actual DBT data set is consistent with the linear equation set in Equation 8.3; namely, there is in general no vector $\mathbf{f}$ that would correspond to a measured data set $\mathbf{g}$. Of course one could blame noise or other physical factors, but there is an even more basic modeling error. Even within the ideal theoretical world using computer-simulated phantoms involving images with simple geometric shapes, such as spheres or ellipsoids, we incur modeling error because of the assumption that the underlying x-ray attenuation map is truly voxelized. As a result, data generated even under ideal conditions will often yield no image solution for Equation 8.3. The use of the elongated voxels for DBT IIR could potentially exacerbate this issue.

Looking at IIR from the perspective of data modeling, Equation 8.3 is no longer interpreted directly as the imaging model. Rather, it is taken as an expression for a data model, and in order to obtain an image estimate, we need to find the image coefficients $\mathbf{f}$ that yield a data estimate $\mathbf{g}$ that is in some sense *close* to the available projection data $\mathbf{g}_{\text{data}}$. Seeking an image estimate in this fashion is suggestive of optimization, where a distance expression between $\mathbf{g}$ and $\mathbf{g}_{\text{data}}$ is written down and the voxel coefficients $\mathbf{f}$ are sought, which minimize this distance.

The first thing we will watch for in the presentation below is if the data are adequately modeled; if they are not, it is unlikely that the model parameters, the voxel coefficients, will yield an accurate image. Even if the data are modeled accurately by Equation 8.3, there is still a potential problem, which may interfere with forming an accurate image. Generally, problems arise for image formation when there are more model parameters than data or if the data do not depend strongly on the model parameters. For the specific linear data model, Equation 8.3, this latter issue relates directly to the condition of the matrix X. If X has a nontrivial null-space, consisting of all $\mathbf{f}_{\text{null}}$ such that $X\mathbf{f}_{\text{null}} = 0$, there is ambiguity in the reconstructed image. Many images could yield the same data estimate. Similarly, if X is poorly conditioned, there are $\mathbf{f}$, where $X\mathbf{f}$ is small. Such $\mathbf{f}$ can be difficult to recover in the presence of model error. To some extent, the effects of a null-space and ill-conditioning can be addressed within the optimization framework by imposing constraints on $\bar{f}$, such as nonnegativity or a maximum value. The effect of the ill-conditionedness of $\mathbf{X}$ is one of the two themes running through this IIR algorithm presentation because it is a consequence of the limited angular range scanning and view-angle undersampling.

To summarize, to begin our IIR design, we need (1) to derive an update step that minimizes some data fidelity measure, (2) to verify that we do indeed model the data, and (3) to investigate if we also obtain an accurate image. If we have an accurate data model but the resulting image quality is poor, we can then modify the basic IIR design to incorporate prior information that improves image quality.

8.3 DATA FIDELITY UPDATE FORMULAS

For an IIR algorithm, the first component needed is an update step that will decrease the distance between the data estimate. While there are many such update steps, see Refs. [3,7], we will limit the discussion to three, which have different origins and exhibit different behaviors: gradient descent (GD) on the Euclidean data-error distance, the maximum-likelihood-expectation-maximization (MLEM) for a Poisson noise model, and the algebraic reconstruction technique (ART). Understanding these three data-error reduction steps provides a decent background for understanding many of the other update steps, which appear in the research literature. The update steps considered here take the form

1: $n \leftarrow 0$
2: initialize $\mathbf{f}_0$
3: **repeat**
4: $\mathbf{f}_{n+1} \leftarrow F(\mathbf{f}_n, \mathbf{g})$
5: $n \leftarrow n + 1$
6: **until** $n \geq N$

where n is the iteration index, $\mathbf{f}_0$ is the initial image estimate, and N is the total number of iterations taken. The function F at line 4 takes different forms for the three update steps. For GD, it is an additive step; for MLEM, it is a multiplicative step; and for ART, this line is replaced by another "inner" loop of less expensive (computationally) additive steps.

8.3.1 GRADIENT DESCENT

Conceptually, GD on the Euclidean data-error is the simplest update step. This data-error is given by

$$D_{\text{Euclidean}}(X\mathbf{f}, \mathbf{g}) = \| X\mathbf{f} - \mathbf{g} \|_2 \equiv \sqrt{\sum_j |(X\mathbf{f})_j - g_j|^2}, \quad (8.9)$$

where j is an index for the individual ray projections g_j of the data set $\mathbf{g}$. Because minimizing $D_{\text{Euclidean}}(X\mathbf{f}, \mathbf{g})$ is equivalent to minimizing its square, formally we employ an update step aimed at solving the following minimization:

$$\mathbf{f}^* = \arg\min_{\mathbf{f}} \frac{1}{2} \left\{ D_{\text{Euclidean}}(X\mathbf{f}, \mathbf{g}) \right\}^2, \quad (8.10)$$

where $\mathbf{f}^*$ denotes the minimizer, and the factor of 1/2 in the objective function is for convenience in that it simplifies, slightly, the expression for the objective function gradient. Note that multiplying the objective function by a positive constant will not affect its minimizer. To derive an update step, we need to decide what direction to go, and how far to go in that direction. For GD, the direction is along the negative gradient of the objective function at the current estimate $\mathbf{f}_n$. The expression for the gradient is

$$\nabla \left(\frac{1}{2} \left\{ D_{\text{Euclidean}}(X\mathbf{f}_n, \mathbf{g}) \right\}^2 \right) = X^T (X\mathbf{f}_n - \mathbf{g}), \quad (8.11)$$

where the T superscript indicates matrix transpose and X^T is commonly known as back-projection. The corresponding update step takes the form

$$\mathbf{f}_{n+1} \leftarrow \mathbf{f}_n - \alpha_n X^T (X\mathbf{f}_n - \mathbf{g}), \quad (8.12)$$

where α_n is a scalar expressing how far to go along the direction of the negative gradient. More needs to be said about X^T and α_n, and the discussion on these quantities is heavily influenced by the fact that we desire IIR algorithms, which are useful at low iteration numbers—well short of true minimization of Equation 8.10.

The gradient expression in Equation 8.11 represents the back-projection of the current data residual. If the current image estimate $\mathbf{f}_n$ solves the linear model, this data residual will be zero and consequently the gradient expression $X^T(X\mathbf{f}_n - \mathbf{g})$ will also be a zero. In this case, the GD update will not alter the image.

Note that this will also be the case if X^T is replaced by different back-projector, B. Much use of this fact is made in the literature for two reasons: often complex forward projectors are employed for which the corresponding matrix transpose is difficult to implement, and a popular choice of discrete back-projector B is a discretization of the continuous back-projection, the adjoint of Equation 8.2. The use of an unmatched back-projector B can, however, complicate conditions for algorithm convergence [11], and it adds algorithmic degrees of freedom, which may not be so useful for DBT, because the undersampling of DBT is the main issue. Hence, for the remainder of this chapter, we employ only the true transpose X^T as the back-projector.

As for the step length parameters α_n, common options are to choose a constant independent of n, or a decaying sequence with a diverging cumulative sum. Because we are interested in IIR algorithms that can be truncated early, it makes more sense to make as much descent progress as possible at each iteration. Deriving this step length comes from noting that the objective function is a multidimensional quadratic function, and its functional dependence in the search direction is therefore a one-dimensional quadratic, which can be analytically minimized. (Performing this derivation makes a nice homework exercise. *Hint:* Substitute the right-hand side of Equation 8.12 into the objective function of Equation 8.10 to obtain a quadratic function of the parameter α_n. Take the derivative with respect to α_n and set it equal to zero.) Let $\mathbf{g}_n = X\mathbf{f}_n$ represent the current data estimate and $\mathbf{d}_n = X^T(X\mathbf{f}_n - \mathbf{g})$; the desired step size becomes

$$\alpha_n = \frac{(X\mathbf{d}_n)^T(\mathbf{g}_n - \mathbf{g})}{(X\mathbf{d}_n)^T(X\mathbf{d}_n)}. \quad (8.13)$$

Putting it all together, we have the first IIR algorithm pseudocode in Algorithm 1. Aside from the model parameters specifying X, the free parameters of this algorithm are the initial image vector $\mathbf{f}_0$ and number of iterations N. (This GD implementation, though straightforward, can be made more computationally efficient. For IIR, the expensive computations are projection, i.e., multiplication by the system matrix X, and back-projection, i.e., multiplication by X^T. The implementation in Algorithm 1 has two projection operations, one at line 4 and one at line 6, and one back-projection at line 5. This entails a total of three matrix-vector multiplications per loop. It turns out that one of the projections can be eliminated from the loop, so that it is only necessary to compute one projection and one back-projection per loop. (Homework: Rewrite the pseudocode in Algorithm 1 so that only one projection and one back-projection are performed per iteration.)

Algorithm 1: N-steps of gradient descent on the Euclidean data-error

1: $n \leftarrow 0$
2: initialize $\mathbf{f}_0$
3: **repeat**
4: $\quad \mathbf{g}_n \leftarrow X\mathbf{f}_n$
5: $\quad \mathbf{d}_n \leftarrow X^T(\mathbf{g}_n - \mathbf{g})$

6: $\quad \mathbf{h}_n \leftarrow X\mathbf{d}_n$
7: $\quad \alpha_n \leftarrow \dfrac{\mathbf{h}_n^T(\mathbf{g}_n - \mathbf{g})}{\mathbf{h}_n^T\mathbf{h}_n}$
8: $\quad \mathbf{f}_{n+1} \leftarrow \mathbf{f}_n - \alpha_n\mathbf{d}_n$
9: $\quad n \leftarrow n + 1$
10: **until** $n \geq N$

8.3.2 MAXIMUM-LIKELIHOOD-EXPECTATION-MAXIMIZATION

The next update step of interest derives from MLEM applied to an uncorrelated Poisson noise model (for the remainder of the text, we refer to this update step simply as MLEM) for the data. We do not present its derivation, but there are plenty of references on this for the interested reader [1,2,12,13]. The update step for MLEM is

$$\mathbf{f}_{n+1} \leftarrow \mathbf{f}_n \left(X^T \frac{\mathbf{g}}{X\mathbf{f}_n} \right) \Big/ (X^T \mathbf{1}). \quad (8.14)$$

In this case, the image is updated by a multiplicative factor, which is the back-projection of the ratio between the data and the current data estimate normalized by system matrix summed over the data indexes. This normalization is computed by back-projecting a sinogram vector with all components set to one. MLEM applies to linear systems such as Equation 8.3 when the system matrix and data are nonnegative. From the form of this update step, it is clear that any zero values for the image estimate $\mathbf{f}_n$ remain at zero because the update is multiplicative. In implementing an algorithm based on this update step, one must take care to avoid a potential divide-by-zero. Such an issue can be avoided by proper modeling with system matrix X (recall the first goal of an IIR algorithm is to model the data). For example, if the image array upon which the volume image is reconstructed is smaller than the object support, projection with X yields zero data estimates for rays passing along the periphery of the subject, which can lead to a divide-by-zero issue. Accounting for these properties of the update step, we present a basic MLEM implementation in Algorithm 2.

Algorithm 2: N-steps of MLEM. In initializing f0, select values greater than zero. For example, all voxels can be initialized to 1. In setting up the image array, ensure projection of the image array with all voxels set to one leads to a strictly positive data estimate at each detector bin

1: $n \leftarrow 0$
2: initialize $\mathbf{f}_0$
3: **repeat**
4: $\quad \mathbf{f}_{n+1} \leftarrow \mathbf{f}_n \left(X^T \dfrac{\mathbf{g}}{X\mathbf{f}_n} \right) \Big/ (X^T \mathbf{1})$
5: $\quad n \leftarrow n + 1$
6: **until** $n \geq N$

The question we now address is why choose MLEM over the GD approach mentioned previously. There may be some motivation in terms of physics if the Poisson noise model applies, but DBT being an x-ray transmission modality, the Poisson noise model is not appropriate for the measurements when converted to the line-integral form [14]. Also, the interest here is for IIR algorithms truncated well short of convergence, meaning that the maximum likelihood image is likely not going to be achieved. Nevertheless, MLEM does have some nice properties. It can be shown that the MLEM update decreases the Kullback–Leibler (KL) divergence between the available and estimated data. This divergence is

$$D_{KL}(\mathbf{g}, X\mathbf{f}) = \sum_i [X\mathbf{f} - \mathbf{g} + \mathbf{g}\ln(\mathbf{g}/X\mathbf{f})]_i. \tag{8.15}$$

This function is called a divergence, not a distance, because the scalar value of the function depends on the order of the vector arguments, that is, $D_{KL}(X\mathbf{f}, \mathbf{g}) \neq D_{KL}(\mathbf{g}, X\mathbf{f})$ in general. The potential advantages for IIR are: first, that the MLEM update enforces image nonnegativity without the need to explicitly implement this constraint, and second, the KL divergence has a milder increase with larger data discrepancy than the quadratic dependence of Equation 8.10. This latter property is known in statistics as robust fitting [5], which addresses the issue that least-squares fitting is sensitive to outliers because of the quadratic objective function. With the milder increase of the KL divergence, the fitting objective function has less of a contribution from outliers.

8.3.3 ALGEBRAIC RECONSTRUCTION TECHNIQUE

The final update we consider here is the ART, which was introduced to the imaging community by Herman and collaborators [15,16]. This method is also known as the Kazcmarz algorithm [17] for solving a linear system such as Equation 8.3. In terms of the generic optimization pseudocode at the beginning of this section (Section 8.3), line 4 takes and image estimate $\mathbf{f}_n$ and yields an image estimate $\mathbf{f}_{n+1}$, which is hopefully closer to data agreement. For ART, line 4 is replaced by an inner loop that processes the data sequentially:

1: $\mathbf{f}_{n+1,1} \leftarrow \mathbf{f}_n$
2: $M \leftarrow$ size $(\mathbf{g})$
3: $m \leftarrow 1$
4: **repeat**
5: $\quad \mathbf{f}_{n+1,m+1} \leftarrow \mathbf{f}_{n+1,m} + \beta_n \dfrac{g_m - \mathbf{f}_{n+1,m} \cdot X_m}{X_m \cdot X_m}$
6: $\quad m \leftarrow m + 1$
7: **until** $m > M$
8: $\mathbf{f}_{n+1} \leftarrow \mathbf{f}_{n+1,M}$

Note that $\mathbf{f}_n$ goes in at line 1 of this pseudocode snippet and $\mathbf{f}_{n+1}$ is obtained at line 8. In this pseudocode, the symbol $\cdot$ is used for a vector dot product, that is, $\mathbf{a} \cdot \mathbf{b} \equiv \mathbf{a}^T\mathbf{b}$. The sensitivity vector X_m is the mth row of the system matrix X. Thus, it has the same size as an image vector and when dotted with the image estimate, it yields an estimate of the datum g_m. The parameter β_n is the relaxation parameter, which should be chosen in the interval $(0,2)$ and is often selected to be a decreasing sequence with n such that $\sum_n \beta_n = \infty$. This update step may appear computationally more expensive, but this inner loop is comparable in computational effort to a single pair of projection and back-projection operations. This update step is interesting for a couple of reasons: it is a

special case of a general projection strategy called projection-onto-convex-sets (POCS), which provides an alternative picture [4] for IIR algorithms than optimization theory, and this type of algorithm, where the data are processed sequentially, is known to yield useful images relatively quickly. Both these points are particularly relevant for designing IIR algorithms that will be terminated after tens of iterations. Before discussing both these points in more detail, we comment on the degrees of algorithmic freedom. As alluded to earlier, there is choice in selecting the sequence β_n, but for this chapter, we consider only static choices $\beta_n = \beta$. The sequential processing of the data, in general, leads to different results for different access orders. In particular, there may be some advantage to randomly chosen access orders [18]. Again, in this chapter, we do not explore this degree of freedom and consider only monotonic access order in the data array index.

The ART picture can be illustrated nicely in considering linear systems with two independent variables. First, consider the following system:

$$\begin{pmatrix} X_{11} & X_{12} \\ X_{21} & X_{22} \end{pmatrix} \begin{pmatrix} f_1 \\ f_2 \end{pmatrix} = \begin{pmatrix} g_1 \\ g_2 \end{pmatrix}. \tag{8.16}$$

Solving this system with the ART algorithm has an intuitive picture illustrated in Figure 8.2. Both rows of this linear system are equations describing a line. The intersection of the lines is the solution of the linear system. ART involves making an initial guess at the location of the solution, then alternately projecting the current solution estimate onto each line. One step of the inner ART loop involves projecting onto one line, and one step of the ART outer loop involves a cycle of projecting onto all (in this case both) lines. For the POCS interpretation, we note that each line in Figure 8.2 is a convex set. (A convex set is set where for any two points belonging to it, all points in between also belong to it). The POCS picture is useful because other convex constraints can be added to the projection cycle. Nonnegativity, for example, specifies that the desired solution lies in the first quadrant, a convex set, of the Figure 8.3. This figure illustrates how the nonnegativity projection can be added to the POCS cycle.

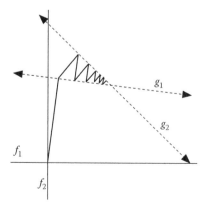

Figure 8.2 Schematic of the ART cycle. The lines labeled g_1 and g_2 correspond to the top and bottom rows of the linear system in Equation 8.16. The solution of this linear system in the f_1, f_2 plane is the point where the two lines intersect. Starting with the point (0,0), the illustrated ART trajectory (the thick line) is obtained by alternately projecting the (f_1,f_2) estimate onto the g_1 and g_2 lines. The trajectory is seen to converge to the intersection of these lines.

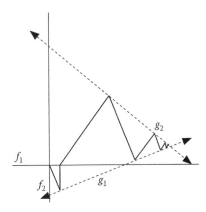

Figure 8.3 Schematic of the ART cycle (see Figure 8.2) including a nonnegativity projection. The lines labeled g_1 and g_2 are different than those in Figure 8.2, and accordingly the ART trajectory is altered. We note that the nonnegativity projection affects the ART trajectory only at the second leg, because thereafter the (f_1, f_2) estimate remains in the positive quadrant of the f_1, f_2 plane.

The POCS picture is also useful for gaining intuition on three important properties of the linear system Equation 8.3, which are relevant for DBT: under-determinedness, ill-conditionedness, and linear system inconsistency. For our simple two variable system, direct under-determinedness means there are fewer equations than unknowns, so in this case we would have only one equation, shown in Figure 8.4. Note that for this special system, ART terminates with a single projection, and the resulting solution depends on the starting guess. A related property that behaves similarly is ill-conditionedness. With the two variable model, the condition of the system is related to the angle between the two lines. A smaller angle leads to worse conditioning. The ART cycle shown in Figure 8.4 illustrates what happens with a small angle between the two lines. The cycle makes slow progress toward the line intersection, the solution of the system. In light of the desire

to have IIR algorithms with few iterations, it can be seen that ill-conditionedness behaves similarly to under-determinedness. The ART cycles move the solution estimate very slowly away from the result of the initial projection cycle. As a result, early termination of ART will yield an estimate, which depends on the initial guess. (The next homework problem involves understanding the well-conditioned case. A perfectly conditioned system, here, occurs when the lines intersect at 90°. Pick an initial guess for this two variable system and graphically draw the first cycle. How many ART outer loops are needed for convergence? Does the result depend on initial guess?)

For the simple two variable system, inconsistency can only be shown with an over-determined system; namely, consider the situation where there are three equations and two unknowns. Figure 8.5 shows both consistent cases, where the three lines intersect at a single point, and inconsistent cases, where pairs of lines intersect but all three do not. For the inconsistent case, there is no solution, and ART algorithm will cycle indefinitely for $\beta_n = 1$; hence, one of the motivations for introducing decay of the relaxation parameter is to force convergence to a single result for the inconsistent case. From the linear systems perspective, DBT is almost always both ill-conditioned and inconsistent, and depending on the choice of voxel size it may even be strictly under-determined. The combination of under-determined and inconsistent cannot be easily visualized in a low-dimensional example, except for the pathological case where the 2D system consists of two parallel, nonoverlapping lines. (Homework 3: (A) Consider what the effect of noise would be on the consistent-case plot of Figure 8.5. (B) An alternative way to deal with inconsistency is to exploit the general POCS picture by altering the convex sets to form a different, consistent system. This is done by fattening each of the lines corresponding to g_1, g_2, and g_3 into strips so that there exists a point common to the three convex sets. Introduce a strip thickness parameter δ and modify the ART cycle accordingly. *Hint*: Projecting a point A onto a convex set entails finding the point in the convex set closest to A.)

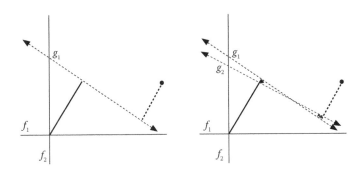

Figure 8.4 Left, a schematic of the ART cycle (see Figure 8.2) for an under-determined system: two variables f_1 and f_2, and one datum g_1; and right, a schematic of the ART cycle for an ill-conditioned system: two lines corresponding to data g_1 and g_2, which are nearly parallel. The thick, solid line is the ART trajectory for the zero initial estimate, and the thick, broken line is the ART trajectory for the initial estimate indicated by the solid circle. For the under-determined system on the left, any point along the line labeled g_1 is a valid solution. The ART solution depends on the initial estimate for (f_1, f_2). Note that for this special circumstance of a single equation in the linear system, ART converges in one step. For the ill-conditioned system on the right, the alternating projections make very little progress per cycle. While the ART algorithm must eventually converge to the intersection of lines g_1 and g_2, early truncation can leave the image estimate far from this solution.

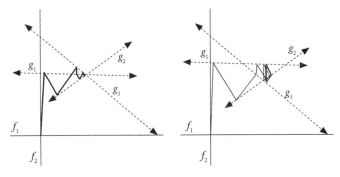

Figure 8.5 Schematic of the ART cycle (see Figure 8.2) for an over-determined, consistent (left), and inconsistent (right) system. For the consistent system on the left, the three lines intersect at a single, common point, and the ART cycle converges to this intersection. The inconsistent system on the right is seen graphically as three lines that do not have a common intersection, where all three lines meet. In this case, the ART cycle does not converge to a single point. Rather, the convergence of ART tends to a limit cycle (a triangle for the illustrated trajectory). One reason for allowing the relaxation parameter β_n to tend to zero is to obtain convergence to a single point for inconsistent linear systems.

To end this section on POCS, we state the algorithm that we will later use to illustrate its application to DBT IIR. The algorithm shown in Algorithm 3 consists of an ART cycle and nonnegativity projection. The latter projection appears in the pseudocode as $pos(\cdot)$, and its effect is to simply threshold any negative values in the image estimate to zero. The $pos(\cdot)$ operator is used to get a nonnegative image estimate, which is reasonable for x-ray tomographic imaging, and later in demonstrating the algorithms we will compute different data-error terms such as the KL divergence, which can be problematic when there are negative values in the image array.

Algorithm 3: *N*-outer steps of POCS algorithm, consisting of an ART cycle together with a nonnegativity projection. For the application of this algorithm, we consider only fixed β_n, because we will operate only at low iteration number where the solution cycling from data inconsistency is not an important issue

1: $M \leftarrow \text{size}(\mathbf{g})$
2: $n \leftarrow 0$
3: initialize $\mathbf{f}_0$
4: **repeat**
5:    $m \leftarrow 1$
6:    $\mathbf{f}_{n+1,1} \leftarrow \mathbf{f}_n$
7:    **repeat**
8:       $\mathbf{f}_{n+1,m+1} \leftarrow \mathbf{f}_{n+1,m} + \beta_n \dfrac{g_m - \mathbf{f}_{n+1,m} \cdot \mathbf{X}_m}{\mathbf{X}_m \cdot \mathbf{X}_m}$
9:       $m \leftarrow m + 1$
10:   **until** $m > M$
11:   $\mathbf{f}_{n+1} \leftarrow pos(\mathbf{f}_{n+1,M})$
12:   $n \leftarrow n + 1$
13: **until** $n \geq N$

In the next section, we will take a break from algorithm development and see how these three basic pseudocodes work for simulated DBT data. In illustrating some fundamental issues of these IIR algorithms, it will be clear that we need some control over image regularity (think smoothness) and then we will build on the ART/POCS algorithm to introduce this additional control.

8.4 IIR FOR DBT WITH COMPUTER SIMULATED DATA

In demonstrating IIR for DBT with simulated data, we utilize two types of data sets: ideal consistent data and ideal data with a mismatched model. Ideal consistent data are generated from a pixelized computer phantom using the same projection matrix X as what is used in the IIR algorithm. The purpose of performing such tests is to understand how the IIR algorithms behave under ideal conditions. This test is particularly relevant for IIR with truncated iteration and limited data problems such as DBT, where either factor likely prevents exact recovery of the phantom even under ideal consistent data conditions.

The next experiments with ideal data use a mismatched model, where the projection data are generated from analytic projection of simple geometric shapes such as ellipsoids, cones, rods, and boxes. The IIR data model, however, still

employs pixels, which are incapable of exactly representing the object and hence the simulated projection data. This issue of pixelization may not seem that important when considering volume images as a whole, but it does have a significant impact on image quality in terms of texture and visualization of small objects—microcalcifications.

Based on what we learn from DBT simulations, we return in Section 8.5 to IIR algorithm development and present a modified version of POCS that is more suited for the DBT scanning configuration. Finally, in Section 8.6, we apply the discussed IIR algorithms to actual DBT data, which include the issues of imperfect modeling together with data inconsistencies due to all of the unaccounted-for physical factors present in an actual projection DBT data set. The two computer-simulated studies will be performed in 2D, because the main issues with DBT sampling can be captured in a 2D model and a complete image is easier to display for this setting. Of course, IIR with actual DBT data will be performed in a typical 3D display volume.

8.4.1 COMPUTER-SIMULATED DATA, MATCHED MODEL

With the DBT configuration specified in Figure 8.1, the detector lies in an x,y-plane, and the x-ray source trajectory is an arc embedded in an y,z-plane at $x = 0$. For this set of simulations, we consider only the 2D system corresponding to this $x = 0$ plane; the x-ray source trajectory is the same, but now the detector is a single strip of detector bins along the line specified by $x = 0$ and $z = -h$. The test phantom is composed of the pixelized image shown in Figure 8.6, where there are structures that loosely model microcalcifications, masses, and vessels. The data dimensions, as specified above, are 11 projections taken on a 1024-bin strip

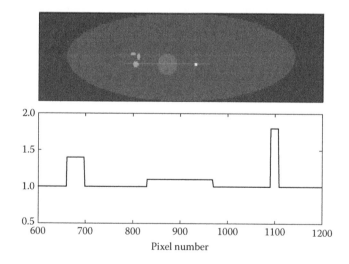

Figure 8.6 Two-dimensional compressed breast phantom embedded in a depth plane, parallel to the y and z axes from the 3D DBT geometry described in Figure 8.1. In keeping with this geometry, the horizontal axis is labeled y and the vertical z. The top panel shows the gray-scale image and the line, not part of the phantom, indicates the location for the phantom profile shown in the bottom panel. The units for the gray scale are chosen so that the background tissue is one and the abscissa of the plot is given in units of pixels, which are chosen to be 100 microns in the y-direction. This phantom is pixelized so that perfectly consistent simulation data can be generated by applying the system matrix X to this phantom.

detector with each bin having a width of 100 microns. The pixel width in y, along the detector, is also 100 microns and the depth in z is 1 mm. The image array is 60×1024 pixels, and the physical dimension is 6×10.24 cm. We note immediately that the posed imaging problem is under-determined because there are a total of 11,264 (11 views $\times$ 1024 bins) data samples and 61,480 unknown pixel coefficients. For this DBT configuration, the under-determinedness is compounded by the limited angular sampling, which further worsens the conditioning of the system matrix X. We forge ahead and apply the GD, EM, and POCS implementations discussed in the previous section. Because we are dealing with an ill-posed system and because we will be truncating the iteration short of convergence, we expect the results of each of these implementations to depend on starting image, so we must specify what is the starting image in each case. For GD and POCS, we initialize the image pixels to zero, and for EM, we initialize all pixels to one.

8.4.1.1 Do we model the data?

Recall that for the implicit estimation of IIR algorithms the first task is to model the data, which in principle we should be able to do exactly in this ideal situation. Data-error summary metrics are plotted as a function of iteration number for GD, EM, and POCS in Figure 8.7. The metrics used are root mean square error (RMSE), which is proportional to the Euclidean data distance; the KL data divergence Equation 8.15, normalized to the magnitude of the test data, $\|\mathbf{g}\|_2$; and a data-error merit function $\|D_y(\mathbf{g} - Xf)\|_2$, which computes the data-error after performing numerical differentiation in the y-direction, normalized by $\|D_y\mathbf{g}\|_2$. This numerical differentiation is encoded in the matrix D_y. The last metric is interesting because it is more sensitive to discontinuities (edges) in the image estimate, and most of the information content of tomographic images takes the form of such discontinuities. Each of these metrics tells a slightly different story, because they all put different weighting on the individual errors at each detector bin. These plots also illustrate the fact that even though the GD and EM algorithms are designed to reduce the Euclidean and KL data divergences, respectively, both algorithms in fact reduce each of the shown data-error metrics for this ideal situation. The reason for this is clear because in each case the algorithm is driving the estimated data Xf toward the available data $\mathbf{g}$. And for this artificial example, we have set up the problem such that we know there exists an image, namely, the phantom, where the

data and estimated data are equal. This behavior will also be seen for the inconsistent model case, when the iteration number is truncated.

Interestingly, the POCS implementation drives both the RMSE and y-derivative-RMSE metrics rapidly toward zero even though it is not designed from the point of view of generating guaranteed descent steps for any data-error metric. The basic MLEM algorithm, although it has an anecdotal reputation of being slow, actually shows the most rapid convergence on the KL data-divergence metric, for which it was designed, on this particular data set. (As an aside, in discussing IIR algorithms, one often hears that algorithm X is slow or that algorithm Y is faster than algorithm X. Making such statements is of course important because there is often a very restricted budget for performing algorithmic iterations, but one must also be aware that such statements can rarely be made in an absolute sense. They tend to be valid within the narrow confines of the particular experiment performed. This goes for the present text as well. While we will often not repeat the phrase "for this particular set of conditions," this should be implicitly understood for general statements made about algorithm speed or efficiency.) The speed at which the POCS algorithm closes the gap between estimated data and DBT data, as measured by the RMSE metrics, makes it an interesting algorithm for IIR algorithms with severe truncation of the iteration number. This convergence speed, we point out, is not limited to POCS; rather, it is a more general observation of many IIR algorithms that process the data sequentially: for example, simultaneous-ART (SART) processes the data projection-by-projection [7] and various ordered-subset (OS) strategies [2,19] can be applied to MLEM to accelerate reduction of the data divergence. Noting the generality of the sequential-processing strategy, we will return exclusively to POCS for two reasons: (1) we want to present few algorithms in great detail, and (2) an effective strategy for regularization with truncated iteration IIR fits neatly into the POCS picture. We do stress, however, that we do not advocate favoring one algorithm over another exclusively; the purpose of this chapter is to illustrate IIR algorithm design for DBT.

To obtain a more detailed sense of the convergence of the estimated data to the data, we plot the data profile for the central projection along with the corresponding estimate for different iteration numbers in Figure 8.8. Indeed, we observe that the estimated data does approach the projection of the phantom in each case. The rate of convergence of the estimated data to the input data is also important to note because of the truncated

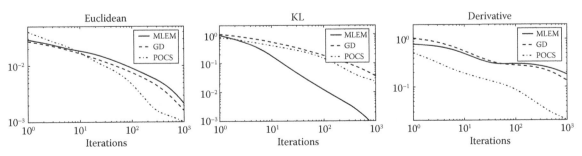

Figure 8.7 Various data residuals for the MLEM, GD, and POCS algorithms as a function of iteration number. The KL and Euclidean data residuals are normalized to the magnitude of the test data vector, and the derivative data residual is normalized to the magnitude of the test data derivative. Note that for this particular system, POCS exhibits faster convergence to data consistency as measured by the Euclidean and derivative data residuals. The convergences rates are relevant because IIR in practice is employed with truncated iteration.

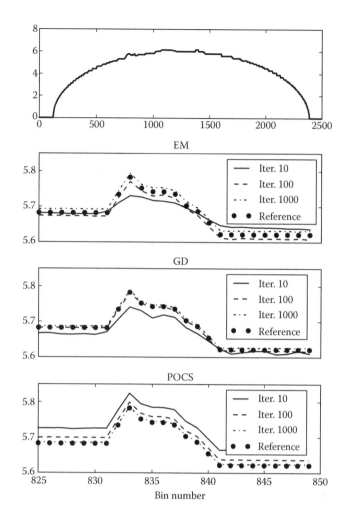

Figure 8.8 Top panel: projection of the pixelized compressed breast phantom at the central view. Note the stair-casing in this profile is due to the phantom pixelization. The other panels show convergence, for the three algorithms of study, of the data estimate to the data at a small segment of the projection containing the small objects on the left of the phantom in Figure 8.6.

iteration issue. Again, these results depend on all the particulars of the linear data model, and changing the system matrix X will in general alter the rate of convergence; thus, it is paramount to perform such tests for any modification to the data model.

In fact, it is often a good idea to select an unrealistically coarse image array, which generally leads to a system matrix X with better conditioning and consequently faster convergence of the estimated data to the input data. Observing this numerical convergence is an important check on the IIR program and to some extent the theory. IIR algorithms can be difficult to debug; observe in Figure 8.7 that very different algorithmic steps can lead to a descent direction for the same data-error term. Likewise, a programming error in the computation of, say, the GD step can still lead to a descent direction for the Euclidean data-error. But it is less likely that the same error will consistently lead to a descent direction for all data estimates until numerical convergence.

8.4.1.2 Do we recover the phantom?

For this matched model case, the three update steps are guaranteed to converge the estimated data to the input data. This fact, however, does not translate necessarily to the model

parameters—the reconstructed image. The main reason is that for DBT, different model parameters (images) can often lead to the same data to within numerical error of the computer. And this issue will only worsen for the unmatched data model or real data situations. (Homework problem: what property of the system matrix X is most relevant to the issue of different images yielding the same projection data?) For each of the simulations described above, we show the corresponding images at the last computed iteration in Figure 8.9. In comparing with the input phantom, the most obvious difference is the poor resolution recovery along z. The reconstructed images also generally have quite some shift in the background gray level, resulting from the fact that the images are not confined to the object support in the z-direction. It is interesting to observe that the reconstructed images and phantom can be so different, yet yield similar projection data. Performing IIR with this ideal matched data model allows us to characterize image artifacts due to the ill-conditionedness of X alone. We will see similar artifacts in images shown below where other data model nonidealities are considered; thus, at least qualitatively, we will be able to separate artifacts due to ill-conditioning and those due to mismatch of the data model.

8.4.2 COMPUTER-SIMULATED DATA, UNMATCHED MODEL

We address now the unavoidable issue of having a mismatch between the data model and the input data. There are many sources of model mismatch, but perhaps one of the most important stems from discretization of the image. Scanned subjects will not be exactly represented by pixels, and no matter what pixel values are arrived at through the IIR algorithm, there will be error in the estimated projection data. It would seem that such error could be alleviated by decreasing pixel size, but doing so increases the size of the image vector while the data vector remains at the same size—resulting in poorer conditioning of X. In this subsection, we will not fully explore this trade-off because enough experience with IIR in DBT has been gained that it is common practice to use voxels with a high aspect ratio. Instead we take an in-depth look isolating this issue for DBT, particularly addressing the question on whether it makes sense to employ voxels with a width of 100 microns and a depth of 1 mm to represent microcalcifications with a dimension of 200–500 microns. Furthermore, we examine whether the depth-blurred reconstructed images are impacted by model mismatch, or if this factor is overwhelmed by the limited angular range of the DBT scanning configuration.

For the unmatched model, there is a greater distinction between the point of view that Equation 8.3 is an imaging model for the DBT scanner and the point of view that the image vector $\mathbf{f}$ are model parameters for the DBT projection data $\mathbf{g}$. If one attempts to substitute unmatched data into the left-hand side of Equation 8.3, no $\mathbf{f}$ can be found such that $X\mathbf{f}$ yields the data. Instead, understanding $\mathbf{g}$ in Equation 8.3 as a data estimate, dependent on the model parameters $\mathbf{f}$, makes more sense. The values of $\mathbf{f}$ are sought so as to minimize a distance between the estimated and input data.

For the simulations in this section, we employ a similar computer phantom, shown in Figure 8.10, as the previous section except that the various structures used to generate the data consist of continuous ellipses. For such shapes, analytic expressions for

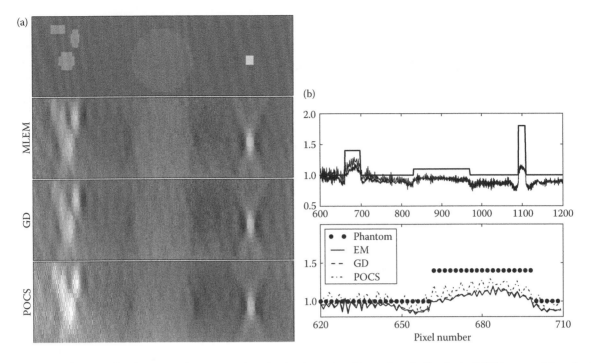

Figure 8.9 (a) Region-of-interest (ROI) images of reconstructions by the three algorithms of study. Top is the ROI blow-up of the phantom in Figure 8.6, and the other three ROI images show the thousandth iteration of each algorithm in a gray-scale window of [0.8,1.2]. (b) The corresponding profiles for quantitative comparison on two different y intervals (in the top panel of this figure, the curves from the different algorithms have the same line style because they are difficult to distinguish at this scale). That the data-error is small and the image-error is large results from the ill-conditionedness of X.

line integrals traversing the phantom are available. To further increase the realism of the modeled projection data, some accounting of partial area averaging is included by subdividing each detector bin into four subbins and averaging over their corresponding line-integrations. This additional averaging is not included in the system matrix X, used for IIR. The issue of data

Figure 8.10 (a) ROI of the continuous version of the compressed breast phantom and (b) ROI of the discrete compressed breast phantom as shown in Figure 8.6. For the present continuous version, no pixelization of the geometric shapes is performed and a series of four microcalcification-like objects replace the single square object of the pixelized phantom. The image is magnified so that all the microcalcifications can be seen. The data generated from the continuous phantom take advantage of analytically known line-integrals through ellipses and, in order to simulate partial-area averaging, each data sample represents the average over four rays originating at the x-ray source and terminating at equally spaced locations within each detector bin.

mismatch due to the pixel representation impacts imaging of small objects more greatly than large structures, and accordingly the small high-contrast object, from the discrete phantom, is removed and replaced by four smaller circles with diameters 100, 200, 400, and 800 microns. Again, the pixel dimensions are 1000 microns in depth and 100 microns in width.

8.4.2.1 Do we model the data?

Having introduced inconsistency between the data model used for generating the data and the one used in the IIR algorithm, the "do we model the data?" question becomes more interesting. Of particular interest are the data relevant for the small objects mimicking microcalcifications, where it is clear that image discretization is insufficient to faithfully represent the object; the image pixel depth is 1 mm and the circular microcalcification objects are smaller than this. The data-error metrics are plotted for each algorithm in Figure 8.11. The resulting curves are interesting in that they look very much like the previous consistent-case study in Figure 8.7. Only the y-derivative-RMSE reveals a significant difference between two studies. For the present case, where the data are generated from continuous ellipses, the y-derivative-RMSE is larger at all iterations than the previous case with the discrete phantom. Also, the MLEM algorithm shows a markedly nonmonotonic decrease of the y-derivative-RMSE. This particular merit function for data fidelity reveals a difference because it heavily weights data discrepancies for rays that are tangent to edge discontinuities in the phantom, located at the transition from one tissue type to another or the borders of the phantom itself. One expects the effect of object pixelization to be the most pronounced at these edges.

Image reconstruction

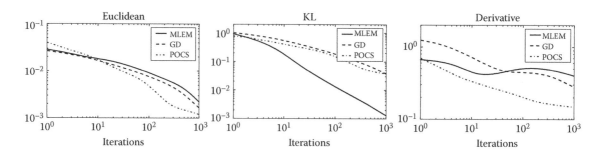

Figure 8.11 Same as Figure 8.7, except that the modeled input data are generated from a continuous test phantom including partial area averaging.

To examine in more detail the question of fitting the data, we present zoomed-in profiles of the estimated projection data in Figure 8.12. By the one-thousandth iteration, each of the three algorithms shows good agreement between the estimated data profile and the continuous projection of the phantom. This agreement is perhaps not surprising for the central projection

where the projection axis is normal to the detector, and the pixels are being projected along the pixel depth direction. But the project profiles are also accurate for the most oblique projection, for which the 1 mm pixel depth is potentially more of an issue. That the DBT projection of the small 100 micron circle is representable by an image array composed of substantially larger

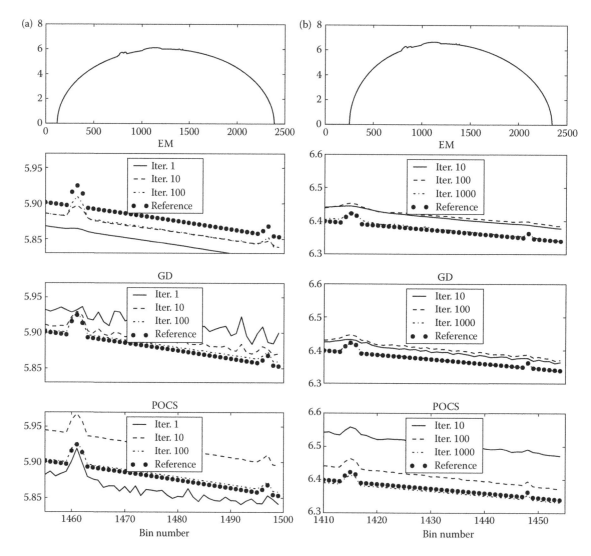

Figure 8.12 Same plots as shown in Figure 8.8 except that the data are derived from a continuous integration including partial-area averaging. The blow-up regions of the projection profiles focus on the projection of the smallest two microcalcification objects. The profile comparisons are shown for the central projection (a), and the last projection (b), where the x-ray source is most oblique to the detector normal vector. The profiles of the oblique projection are shown for larger iteration numbers, because they converge more slowly than the profiles in the central projection.

pixels seems a bit surprising, but we note that multiple pixels along the ray measurements can be contributing to these profile, and we need to examine the reconstructed images to see what is going on. Before going on, however, we comment on algorithm efficiency because we are interested in truncated IIR for DBT.

In the profiles presented in Figure 8.12, we observe that the POCS algorithm captures the structure of the data profiles at earlier iterations than GD or MLEM. This coincides with the more rapid convergence of y-derivative-RMSE for POCS. While the structure convergence is quick for POCS, there is clearly a slower convergence of a background shift relative to the projections from the continuous data model. For visualizing clinically relevant structures, however, deviations in the background gray level are not that crucial.

8.4.2.2 Do we recover the phantom?

The reconstructed images from the continuous data model for this DBT simulation are shown in Figure 8.13. The first general observation is that the image quality from the inconsistent, continuous model is quite similar to that of the consistent, discrete model in Figure 8.9. Looking at the microcalcification objects on the right side of Figure 8.13, as expected there is substantial blurring in the z-direction. We also note that the second smallest object is barely visible and the smallest object is not seen at all in the reconstructed images. That these objects

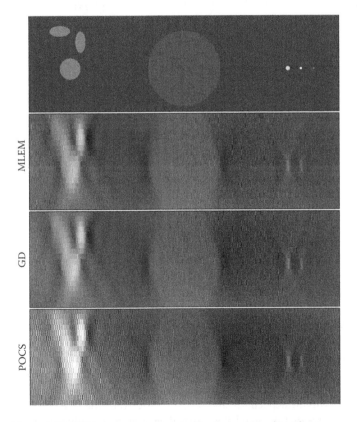

Figure 8.13 ROI images reconstructed by the various algorithms under study at 1000 iterations. Note that many of the depth blurring artifacts are similar to what was seen in the case of image reconstruction with a matched, discrete data model in Figure 8.9. Even though the projections of all four microcalcifications are well modeled by the estimated data, the corresponding images in this figure indicate only a hint of the second smallest object and no sign of the smallest.

are clearly seen in the estimated projections (see Figure 8.12) and not visible in the images is a consequence of the limited angular scanning of DBT. The DBT scan does not contain the side projections needed to concentrate the energy of small objects in the depth direction.

At this point we have examined three different update steps, which form the basis of most IIR algorithms for DBT. The 2D simulations have been performed using an ideal pixelized phantom and a more realistic continuous phantom to generate projection data. We could, of course, continue to add more realism in the data model and include additional physical factors in order to see how these basic data fidelity processing steps handle these inconsistencies. And in fact we will address, later in this section, IIR with actual DBT data. But before going on, it is important to make a couple of observations on the results so far in order to see if there are some simple steps that can be taken to improve DBT image quality. As we have seen so far, the POCS algorithm, with its sequential image estimate updates, reveals structure in the reconstructed images more quickly than GD or MLEM. Because we are interested in IIR with truncated iteration, we focus exclusively on modifying POCS.

8.5 AN ALGORITHMIC INTERLUDE: MODIFYING POCS FOR DBT

In examining the POCS image in Figure 8.13, we see that the image appears quite noisy. Keeping in mind that presented DBT images lie in planes parallel with the detector, corresponding to horizontal lines in the shown images, the speckle noise is detrimental to DBT image quality particularly in visualizing small objects. To mitigate these image artifacts, we need to understand their origin. Because similar artifacts appear in the GD and MLEM images, they are not exclusive to POCS. The simulation leading to the images in Figure 8.13 models both limited angular range scanning and a mismatched data model. In order to see which factor causes the image noise artifacts, we go back to the results of Figure 8.9, for which the data model and IIR system model match. We see that the same noisy artifacts are seen in Figure 8.9; thus these artifacts originate from the limited angular range scanning.

To address this undersampling artifact, we need an additional control knob on the algorithm that regularizes the image by introducing an objective function, which increases with variations between neighboring pixels. To implement this regularization, it is important to take advantage of the fact that the origin of the noise is undersampling. Traditionally, roughness terms are introduced to smooth out image noise originating from data noise, and as a result such terms are interpreted as a penalty, entailing a trade-off between image regularity and data fidelity. Here, the use of a roughness objective function is different. We are using image roughness to break the degeneracy between the multiple solutions of our under-determined imaging model. Thus, it is possible to find images with greater regularity that those seen in Figures 8.9 and 8.13 without sacrificing data fidelity.

In terms of optimization, the difference between these two uses of image roughness terms is formulated in the following way.

Introducing a roughness penalty, one can include another term in Equation 8.10:

$$\mathbf{f}^{*} = \arg\min_{\mathbf{f}} \left\{ \frac{1}{2} D_{\text{Euclidean}} (X\mathbf{f}, \mathbf{g})^2 + \gamma R(\mathbf{f}) \right\}, \qquad (8.17)$$

where the function $R(\cdot)$ is some function that measures image roughness, and γ is a regularization parameter that controls the strength of the smoothing. By combining both data fidelity and image roughness into one objective function, γ parametrizes a trade-off between the two terms. Larger γ smooths the image and increases the gap between the available and estimated data.

The optimization governing our use of a roughness term for an under-determined linear system takes the form of a constrained minimization:

$$\mathbf{f}^{*} = \arg\min_{\mathbf{f}} R(\mathbf{f}) \text{ such that } X\mathbf{f} = \mathbf{g}. \qquad (8.18)$$

In this idealized, constrained minimization, there is no parameter in the optimization. The equality constraint admits only images that agree perfectly with the data, and among those, chooses the one that minimizes $R(\mathbf{f})$. This optimization applies only to the ideal simulation with a matched data model. If there is any inconsistency in the data model, no images satisfy the data equality constraint. It turns out that these two optimizations are related, if the limit $\gamma \to 0$ is taken in Equation 8.17, the resulting solution approaches that of Equation 8.18. (Homework: For Equation 8.17, under what conditions is its solution for $\gamma = 0$ the same as the limiting solution for $\gamma \to 0$?) Physically, this correspondence means that the use of a roughness penalty to overcome undersampling should involve small values of γ.

These optimization problems are useful for guiding design of IIR algorithms, but they cannot be used directly for DBT IIR. The primary reason for this is that we are interested in IIR with truncated iteration. As we have seen in Figures 8.7 and 8.11, IIR algorithms for DBT can take hundreds of iterations to converge simple optimizations involving only data fidelity objective functions. Despite recent advances in large-scale optimization algorithms [20–23], at least this many iterations will be needed to solve variants of Equations 8.17 and 8.18. With current computational technology and typical DBT data sizes, IIR algorithms need to be restricted to numbers of iterations on the order of ten. Accordingly, we modify the POCS implementation in Algorithm 3 in a way that is motivated by Equation 8.18.

Algorithm 3 reduces the distance between the data estimate and the available projection data while maintaining nonnegative image estimates. Additional GD steps on $R(\mathbf{f})$ can be inserted into this algorithm to reduce the roughness objective function. The step-size of these GD steps can be controlled in such a way that the net change of the POCS and GD on $R(\mathbf{f})$ steps still leads to a reduction of the data-error, $D_{\text{Euclidean}}(X\mathbf{f}, \mathbf{g})$. In this way, the roughness of the image estimate is reduced while also reducing data-error. The modified pseudocode appears in Algorithm 4.

Algorithm 4: Pseudocode for an adaptive steepest descent-projection-onto-convex-sets (ASD-POCS) algorithm, consisting of alternating POCS and GD on $R(\mathbf{f})$ steps. The algorithmic parameters N_{GD} and r_{GD} are set to specific values in the pseudocode, leaving the parameters β and N as nearly orthogonal control parameters. The parameter β controls image roughness; lower values of β decreases image roughness. The parameter N controls data fidelity; data fidelity increases with N

1: $M \leftarrow \text{size}(\mathbf{g})$
2: $n \leftarrow 0$
3: initialize $\beta \in (0,1]$, N
4: $\mathbf{f}_0 \leftarrow 0$; $N_{\text{GD}} \leftarrow 10$; $r_{\text{GD}} \leftarrow 0.75$
5: **repeat**
6:    $m \leftarrow 1$; $\mathbf{f}_{n+1,1} \leftarrow \mathbf{f}_n$
7:    **repeat**
8:       $\mathbf{f}_{n+1,m+1} \leftarrow \mathbf{f}_{n+1,m} + \beta \dfrac{g_m - \mathbf{f}_{n+1,m} \cdot \mathbf{X}_m}{\mathbf{X}_m \cdot \mathbf{X}_m}$
9:       $m \leftarrow m + 1$
10:    **until** $m > M$
11:    $\mathbf{f}'_{n+1} \leftarrow pos(\mathbf{f}_{n+1,M})$; $d_{\text{POCS}} = \|\mathbf{f}'_{n+1} - \mathbf{f}_n\|_2$
12:    $m \leftarrow 1$; $\mathbf{f}'_{n+1,1} \leftarrow \mathbf{f}'_{n+1}$
13:    **repeat**
14:       $\mathbf{f}'_{n+1,m+1} \leftarrow pos\left(\mathbf{f}'_{n+1,m} - (r_{\text{GD}})^p d_{\text{POCS}} \dfrac{\nabla R(\mathbf{f}'_{n+1,m})}{\|\nabla R(\mathbf{f}'_{n+1,m})\|_2} \right)$
15:       where p is the smallest nonnegative, integer such that $R(\mathbf{f}'_{n+1,m+1}) \leq R(\mathbf{f}'_{n+1,m})$
16:       $m \leftarrow m + 1$
17:    **until** $m > N_{\text{GD}}$
18:    $\mathbf{f}_{n+1} \leftarrow \mathbf{f}'_{n+1,M}$; $\mathbf{f}_{\text{GD}} = \mathbf{f}_{n+1} - \mathbf{f}'_{n+1}$; $d_{\text{GD}} = \|\mathbf{f}_{\text{GD}}\|_2$
19:    **if** $d_{\text{GD}} > d_{\text{POCS}}$ **then**
20:       $\mathbf{f}_{n+1} \leftarrow \mathbf{f}'_{n+1} + \mathbf{f}_{\text{GD}}(d_{\text{POCS}}/d_{\text{GD}})$
21:    **end if**
22:    $n \leftarrow n + 1$
23: **until** $n \geq N$

This algorithm alternates a steepest descent (SD, which is equivalent to GD) step on $R(\mathbf{f})$ with POCS, and the step-size on the SD part is controlled adaptively so that a net decrease in the data-error is realized. This type of algorithm is called ASD-POCS [24–26], and the particular implementation depends on whether the goal is to solve a constrained optimization problem accurately or to obtain images of practical utility with truncated iteration. We have the latter goal, and Algorithm 4 follows most closely Ref. [26]. This algorithm up to line 11 is the same as the POCS implementation in Algorithm 3. After this POCS section, a steepest descent loop is inserted at lines 13 through 17. The heart of the steepest descent step is line 14, where ∇R is the pixel-wise gradient of the roughness objective function; d_{POCS} is the maximum step length; and $(r_{\text{GD}})^p$ is a back-tracking factor that ensures that the step length is small enough to actually reduce the roughness of the image estimate; the direction of the negative gradient is only guaranteed to be a descent direction locally where the linear approximation to $R(\mathbf{f})$ is valid. As seen in the pseudocode, the loop parameter N_{GD} is set to ten. There are two reasons for this: (1) at early iterations of ASD-POCS, the POCS step is much larger than the SD step; many SD steps can be taken before causing the data estimate to diverge from the given data, and (2) SD steps are computationally more efficient than the POCS step; multiple SD steps per POCS step do not significantly impact algorithm run time. At line 19, we have included an adaptive control that checks if the change in the image estimate due to the SD loop is larger than that due to POCS. If it is, then the image estimate change due to the SD loop, $\mathbf{f}_{\mathrm{GD}}$, is scaled back so as to not exceed d_{POCS}. This adaptive control causes each iteration n of ASD-POCS to yield a net reduction in the data-error. Empirically, for low iteration number, we have found that this adaptive control has little effect because d_{POCS} initially is much greater than d_{GD}, but we include this line so that the pseudocode maintains the ASD-POCS design for any iteration number.

8.5.1 DO WE MODEL THE DATA? ... AND REDUCE IMAGE ROUGHNESS?

In the present implementation of ASD-POCS, we use the image total variation (TV) as the image roughness metric. The image TV is the sum over the gradient magnitude image, where the gradient is computed by finite differencing. The use of TV for image processing applications was originally promoted in Ref. [27], and interest in image TV has been renewed for sparsity-exploiting IIR where it was noted that exploiting gradient magnitude sparsity can be useful in medical imaging applications [28–30]. Here, we demonstrate ASD-POCS with the function $R(\cdot)$ set to the image TV, and the projection data are generated from the continuous projection model.

To illustrate the global behavior of ASD-POCS, we plot in Figure 8.14 the data-error versus image TV for each iteration, comparing with POCS. Starting with the POCS results, we see that as the iteration number increases the data-error is reduced,

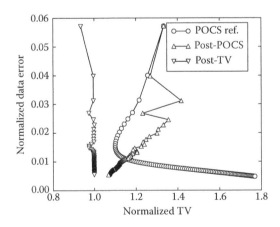

Figure 8.14 As ASD-POCS is designed to lower both data-error and image roughness, the plot has each of these measures on the ordinate and abscissa, respectively. Each symbol of the three curves represents an iteration with the first iteration in each case being the topmost symbol. The normalized data-error is the Euclidean difference of the estimated data and the projection data divided by the magnitude of the projection data. Normalized TV is the measure of image roughness, and it is computed by the ratio of the image estimate TV and the phantom TV. These measures are plotted for ASD-POCS for the image estimates f′, post-POCS, and f, post-TV, see Algorithm 4. The POCS reference is put in the plot to see quantitatively the effect of the additional TV SD loop from the ASD-POCS algorithm.

as expected, but we also see that the image TV initially decreases, then increases at an ever-growing rate. From the shape of the POCS curve, we observe a trade-off between image roughness, measured by TV, and data fidelity. Larger iteration number leads to lower data-error and large image TV. This type of behavior is also a generic feature of the GD and MLEM basic algorithms. As an aside, there is an interesting parallel between these basic update steps as a function of iteration number, and the unconstrained optimization in Equation 8.17, which combines data-fidelity and roughness-penalty terms.

Turning to the ASD-POCS results, the normalized data-error and image TV is shown for the iterates, $\mathbf{f}_n$, and for the intermediate images captured after the POCS loop, $\mathbf{f}_n'$, setting the parameter, $\beta = 1$. The main feature of the evolution of the ASD-POCS iterates, $\mathbf{f}_n$, is that the data-error decreases, while the image TV remains nearly constant, as the iteration number increases. This behavior contrasts with the divergent behavior of the image TV for POCS. As ASD-POCS performs alternating SD and POCS steps, it is also of interest to examine the trajectory of the intermediate image estimates, $\mathbf{f}_n'$. At each iteration, these estimates have a higher TV than $\mathbf{f}_n$, and as the iteration progresses, the gap between $\mathbf{f}_n$ and $\mathbf{f}_n'$ decreases. It is important, however, to be aware of this gap because we are interested in IIR algorithms with truncated iteration and this gap is likely to be nonnegligible. Thus, the iterates, $\mathbf{f}_n'$, may have practical use in some situations. But for the remainder of the chapter, we focus only on the iterates, $\mathbf{f}_n$. Though not shown in the figure, the effect of selecting lower β is to shift the vertical ASD-POCS curve to the left, toward lower TV. The net result is that ASD-POCS affords nearly independent control over image data-error and TV: increasing the iteration number reduces data-error with small changes in TV, and lowering β yields lower image TV. These nearly independent algorithm controls allow the image estimate to be steered to both low data-error and low image TV, which is advantageous for scanning configurations with incomplete data such as DBT.

8.5.2 DO WE RECOVER THE PHANTOM?

The impact of the SD step on the image TV is dramatic on the reconstructed images. Shown in Figure 8.15 are regions of interest (ROIs) of image reconstructed by ASD-POCS at different values of β. In each of the images, the background noisy artifacts seen in Figures 8.9 and 8.13 are eliminated, and as a consequence, the small microcalcification objects are more visible that they are in Figure 8.13. The blurring in the depth direction, however, remains in the ASD-POCS reconstructions. Note also that the fourth, smallest microcalcification is not visible either in any of these images.

In order to have a more quantitative picture of the ASD-POCS results, profiles through the middle two microcalcifications are shown in Figure 8.16. As a reference, the MLEM profile is also shown; MLEM is chosen because it will be used as the reference in the clinical results of the next section. In these plots, we note again that the oscillations seen in the MLEM profile are removed in the ASD-POCS profiles. The microcalcification profiles are also seen to be larger and closer to the phantom for ASD-POCS profiles, compared with the MLEM profile. The results from the basic GD and POCS algorithms, not shown, are similar to that of

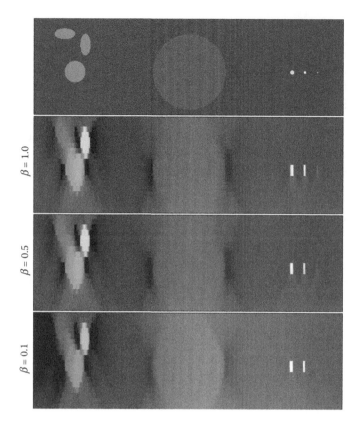

Figure 8.15 Same plot as Figure 8.13 except ROI images are reconstructed by ASD-POCS for different relaxation parameters β at 100 iterations.

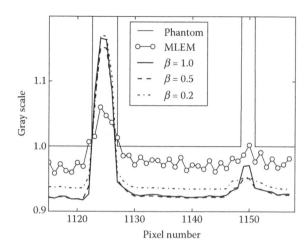

Figure 8.16 Image profiles plotted for the middle two microcalcifications from the phantom and IIR by MLEM and ASD-POCS with different values of β. The low TV images from ASD-POCS can enhance contrast of these objects, but the effect is nonlinear and depends on the object. The gray-scale range of the plot clips the larger values of the microcalcifications in the phantom, which extend up to a value of 2.0.

MLEM modulo a background shift. It is rather remarkable that the shown image profiles, which appear to be quite different, all belong to images that, when projected, match the DBT data very closely. Of course, this phenomenon is well understood and it is a consequence of the nontrivial null-space of the DBT system matrix X, but it is interesting to have a sense of what kind of variation is possible in the reconstructed images while still having their projections match the data. We observe that the constrained minimization strategy, of selecting images with a low TV, constrained to a given, low data-error has the potential to improve visibility of microcalcifications.

Before moving on to IIR with clinical data, it is important to make one more observation that the ASD-POCS profiles show in Figure 8.16. The profiles of the larger microcalcification on the left do not vary much with β but the profiles of the smaller microcalcification on the right do have a strong dependence on β. This difference in behavior underscores the nonlinear nature of IIR; results of IIR algorithms depend not only the data but also the underlying object. Because of this nonlinearity, it is difficult to make general conclusions about IIR algorithms. For example, the ASD-POCS profile shows a contrast twice that of the MLEM result for the microcalcification on the left. This certainly does not mean that the ASD-POCS contrast will always be twice that of MLEM, as seen by visual inspection of Figure 8.13 compared with Figure 8.15. Furthermore, introduction of noise into the data model, for example, could affect the MLEM and ASD-POCS implementations differently so that this contrast ratio may be altered significantly. In general, due to the nonlinearity of IIR, the gap between testing IIR algorithms on simulated data and

using IIR algorithms on clinical data can be large—more so than for FBP algorithms. We return to this discussion in Section 8.7.

8.6 CLINICAL DBT DATA

The previous three studies characterize three basic algorithms—GD, MLEM, and POCS—for IIR in DBT and the last of these studies illustrates a modification to POCS to address artifacts due to the limited angular scan of DBT. These simulation studies can be extended, and IIR algorithms refined, with further controlled investigations on the impact of other important physical factors such as noise, x-ray beam polychromaticity, and scatter. At this point, however, we demonstrate the presented algorithm on clinical DBT data. In the spirit of algorithm design, we start with the basic algorithms first, then show the impact of further modification to one of these basic algorithms. Images are shown for ranges of algorithm parameters to obtain a sense of the image dependence. No claims of the optimal algorithm or parameter settings are made. For clinical DBT, the computation will naturally be fully three-dimensional.

The DBT case reconstructed here consists of a projection data set with 11 views projected onto a detector, which has 1109×2296 bins of 100 micron width, and the geometric configuration was described previously in Section 8.2. This particular case shown in Figure 8.17 contains a malignant mass and microcalcifications. The case was selected in order that the basic image properties could be discussed in terms of structures important for DBT imaging, but the focus of this chapter is confined to IIR. The IIR algorithms, here, are run with much lower iteration numbers than the previous simulations, because the data size for the clinical DBT is much larger. As a result, more discussion focuses on the impact of truncated iteration. In fact, iteration number becomes a parameter of the IIR algorithm.

In the previous simulation, studies we examined how well we modeled the data and recovered the underlying phantom. For the application of IIR DBT clinical data, we do not repeat these

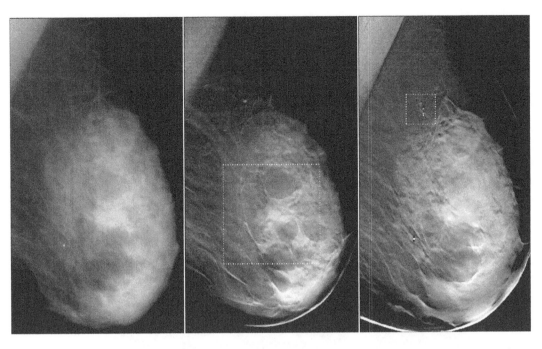

Figure 8.17 Images from a DBT case containing a malignant mass and microcalcifications. The central x-ray projection is shown, left. Reconstructed DBT slices are shown for a slice centered on a malignant mass, middle, and for a slice centered on a cluster of microcalcifications, right. Indicated with the dashed boxes are a 5 × 5 cm ROI containing the malignant mass and a 1 × 1 cm ROI containing the microcalcifications. The displayed DBT reconstructions in the subsequent figures focus on these ROIs, magnified.

assessments. With respect to data modeling, similar conclusions on convergence would be obtained as seen in the previous simulations. In fact, the simulation results provide a sense of how well we can expect the estimated DBT projections from the reconstructed DBT volumes to match the available data. Given that we will be running IIR for 5–20 iterations, we can see from the simulations that we will obtain images that are far from minimizing any of the data-error metrics. With respect to object function recovery, we cannot perform such studies, because the underlying truth is not available. It is for this reason that validation of image reconstruction algorithms is a science in itself and relevant image quality metrics range from being system oriented (see Chapter 9) to being imaging task oriented (see Chapter 10). Owing to the complexity of IIR algorithm validation, we aim here only to characterize the various algorithms by showing selected ROIs obtained for various IIR algorithm parameter settings.

For the presented clinical DBT images, we show results for MLEM, POCS, and ASD-POCS surveying the algorithm parameter values. The MLEM algorithm is initialized by a uniform volume of ones, and the POCS and ASD-POCS algorithms are initialized by zeros. Images obtained by each algorithm are shown at 1, 2, 5, 10, and 20 iterations. In terms of computational load, each of these algorithms is nearly equivalent to performing a projection/back-projection pair per iteration. The POCS and ASD-POCS algorithms are run for $\beta = 1.0, 0.5, 0.2,$ and 0.1 in order of increasing regularization. In the previous simulations for POCS, we did not consider any other parameter setting than $\beta = 1.0$, but here we do so in order to assess the impact of the TV gradient descent loop in ASD-POCS. The figures are organized by the β value, and the MLEM images are repeated in each of these figures because the basic MLEM implementation does not have this algorithmic parameter.

The goal of the DBT image presentation is to reveal the impact of the various algorithm designs and parameters. As the presentation is in hard copy, it is not possible to duplicate clinical DBT viewing, where the observer is free to scroll through slices, to change window and level settings, and to magnify on ROIs. Slices and ROIs, at different length scales, are shown for a malignant mass and microcalcifications. Addressing optimal window/level settings is challenging, and here, these settings are determined by the image values of the last displayed iteration. The gray values are analyzed for their mean and standard deviation. The gray-scale window for all images in a particular iteration sequence is centered on the computed mean and the width of the window is set to six times the computed standard deviation.

8.6.1 DBT CLINICAL IMAGES: MALIGNANT MASS

The first series of images shown in Figures 8.18 through 8.21 shows a 5 × 5 cm ROI from a DBT volume centered on a malignant mass. The MLEM series is repeated in each of the figures, and initially the MLEM image is quite smooth with additional detail filling in as the iteration number increases. This type of behavior is rather generic for MLEM and many other IIR algorithms, and in practice iteration truncation is often used as a form of image regularization. Because of the ubiquity of MLEM for IIR in many tomographic modalities, we use these images as a reference for POCS and ASD-POCS.

The POCS series of images do differ as a function of β. Starting with the $\beta = 1.0$ series shown in Figure 8.18, one notes the conspicuous cross-hatching artifacts seen in each of the panels. These artifacts correspond to the "noise artifacts" seen in the 2D simulations shown in Figure 8.13. The character of these artifacts is more apparent in Figure 8.18 because these images are in a plane parallel to the detector, as is standard in DBT, while

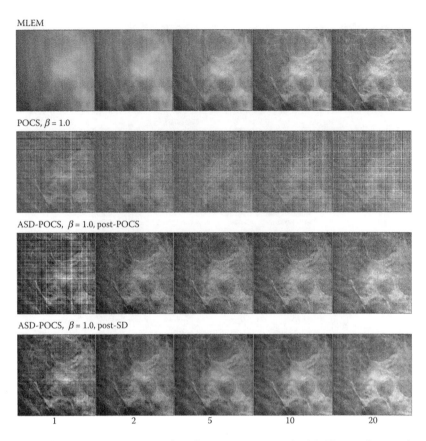

Figure 8.18 Reconstructed images of the malignant mass ROI for relaxation parameter $\beta = 1.0$. The numbers under each column of images indicate the iteration number of the respective IIR algorithm.

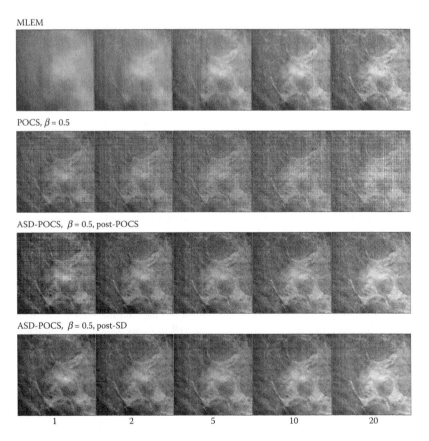

Figure 8.19 Reconstructed images of the malignant mass ROI for relaxation parameter $\beta = 0.5$. The numbers under each column of images indicate the iteration number of the respective IIR algorithm.

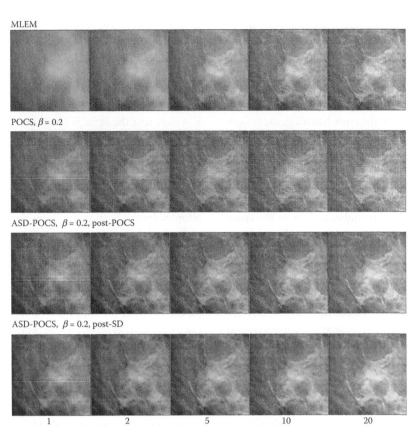

Figure 8.20 Reconstructed images of the malignant mass ROI for relaxation parameter $\beta = 0.2$. The numbers under each column of images indicate the iteration number of the respective IIR algorithm.

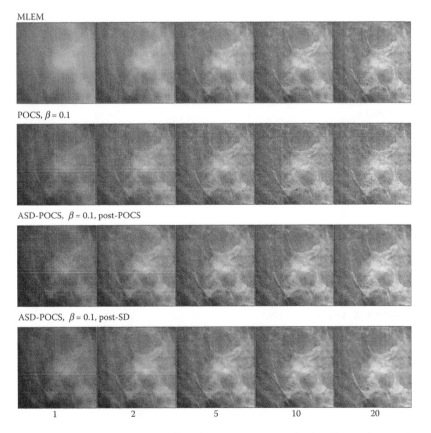

Figure 8.21 Reconstructed images of the malignant mass ROI for relaxation parameter $\beta = 0.1$. The numbers under each column of images indicate the iteration number of the respective IIR algorithm.

the 2D simulations necessarily show planes perpendicular to the detector, which is unusual for DBT. These artifacts will also eventually appear in the images from our MLEM implementation at larger iteration numbers, because the projector/back-projector pairs are the same for all algorithms. It is possible that these artifacts can be mitigated by alternate projector/back-projector implementations [31]. Aside from the cross-hatching, the image at one iteration shows remarkable detail, roughly equivalent to 10 or 20 iterations of MLEM. This acceleration is the generic IIR feature, alluded to earlier, that OS-type approaches where the image is sequentially updated by subsets of the data results in substantial acceleration of IIR. The remaining images in the POCS $\beta = 1.0$ series show increasing noise, aside from the cross-hatching artifact. For POCS, decreasing β reduces the cross-hatching artifact, and it also slows down the appearance of image detail. For example, looking at the POCS $\beta = 0.1$ series in Figure 8.21, the details in the image for the first iteration are blurred relative to the first iteration of POCS with larger β.

Turning now to the ASD-POCS algorithm, both the $\mathbf{f}'$, post-POCS, and $\mathbf{f}$, post-SD, are shown, because at these low iteration numbers, the gap between these two iterates can be substantial (see Figure 8.14). Starting with ASD-POCS for $\beta = 1.0$, we note that the first iteration, post-POCS, is identical to the first panel of the corresponding POCS series. This, of course, is to be expected because ASD-POCS alternates POCS and SD loops. Remarkably, the post-SD image at the first iteration of ASD-POCS substantially reduces the cross-hatching artifact. While some image smoothing occurs, there is still much image detail

in this first iteration of ASD-POCS—roughly corresponding to the fifth iteration of the image from MLEM. Progressing in iteration number, detail fills in rapidly and the gap between the post-POCS and post-SD images closes. However, it is clear that the post-SD are slightly more regularized than the post-POCS images. In comparing the $\beta = 1.0$ POCS and ASD-POCS series, it is clear that the extra SD loop in ASD-POCS effectively removes the cross-hatching artifact stemming from DBT undersampling without compromising visual image resolution. In going to the subsequent figures with lower β, we make the same observations but note that the ASD-POCS series becomes increasingly regular with β. The fine-tuning on image regularity afforded by the β parameter is needed due to the rapid appearance of image detail inherent to POCS-based IIR algorithms.

8.6.2 DBT CLINICAL IMAGES: MICROCALCIFICATIONS

The next series of images shown in Figures 8.22 through 8.25 shows a small ROI, 1×1 cm, centered on a microcalcification cluster. Many of the same observations can be made for the microcalcification images as the previous set. We do note, however, that generally the IIR algorithm design and parameters have a larger effect on this ROI's visual qualities due to its smaller size.

Of particular interest for this series of images is the microcalcification contrast. We saw in the 2D simulations presented in Figures 8.13 and 8.16 that there is a large potential to increase microcalcification contrast with ASD-POCS IIR design over the other basic IIR algorithms. Indeed, we do note that the ASD-POCS images show obvious contrast enhancement over the basic

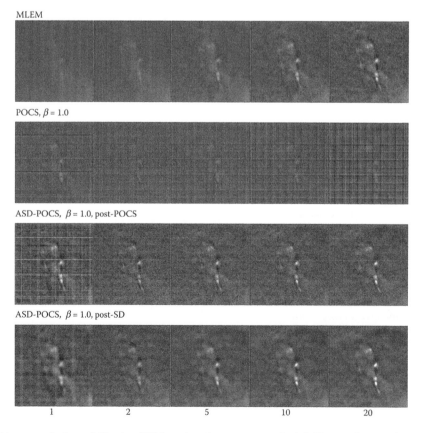

MLEM

POCS, $\beta = 1.0$

ASD-POCS, $\beta = 1.0$, post-POCS

ASD-POCS, $\beta = 1.0$, post-SD

1　　　2　　　5　　　10　　　20

Figure 8.22 Reconstructed images of microcalcification ROI for relaxation parameter $\beta = 1.0$. The numbers under each column of images indicate the iteration number of the respective IIR algorithm.

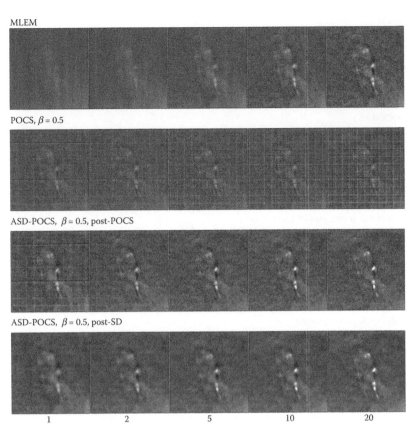

Figure 8.23 Reconstructed images of microcalcification ROI for relaxation parameter $\beta = 0.5$. The numbers under each column of images indicate the iteration number of the respective IIR algorithm.

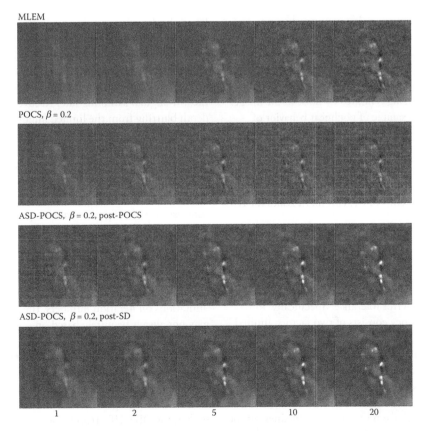

Figure 8.24 Reconstructed images of microcalcification ROI for relaxation parameter $\beta = 0.2$. The numbers under each column of images indicate the iteration number of the respective IIR algorithm.

Image reconstruction

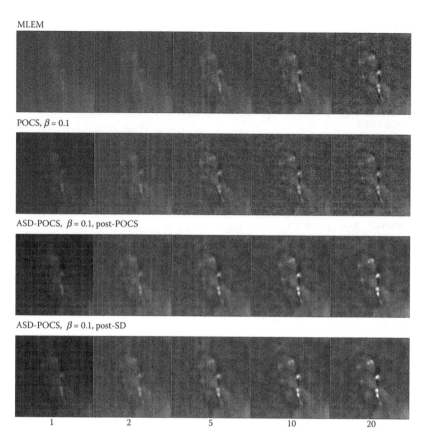

Figure 8.25 Reconstructed images of microcalcification ROI for relaxation parameter $\beta = 0.1$. The numbers under each column of images indicate the iteration number of the respective IIR algorithm.

POCS and MLEM implementations as iteration number increases. Interestingly, examining the most heavily regularized ASD-POCS image, $\beta = 0.1$, 20th iteration, post-SD, the background tissue is clearly overregularized yet the microcalcification contrast is extremely high. This property of ASD-POCS is useful particularly because microcalcification imaging is limited by noise in the image. Again, this is another specific example of nonlinear behavior of IIR algorithms as the background is highly smoothed but the microcalcifications themselves are not blurred.

8.6.3 CLINICAL DBT IMAGES SUMMARY

We summarize the results in terms of the two main issues for IIR in DBT.

8.6.3.1 Data insufficiency due to limited angular range scan

Generically, the data insufficiency issue means that there are multiple images that agree with the DBT data, that is, that solve Equation 8.3 with a low data-error tolerance. Specifically, in simulation, the data insufficiency consists of two factors: limited scanning angular range and a low angular sampling rate. The former factor causes the familiar depth blurring of DBT and the latter results in streaks in the image, which appeared as noise in the 2D simulations and cross-hatching in the images from clinical DBT volumes.

Because these artifacts result from under-determinedness of the linear data model, Equation 8.3, the algorithm design to address these artifacts uses an image regularizing objective function to break the degeneracy of images in the solution space. Instead of having a trade-off between image regularity and data fidelity, the ASD-POCS algorithm design aims at both low image roughness and data-error. Using the image TV as the regularizer for ASD-POCS effectively eliminates the cross-hatching artifact due to the low angular sampling rate. Recovery of contrast due to the depth blurring from the limited angular range appears to be more difficult. Results from the illustrated ASD-POCS implementation shows that some contrast recovery is possible for high-contrast objects, which is important for imaging microcalcifications in DBT. Looking back at the clinical DBT results, the ASD-POCS algorithm with the TV regularizer hardly has any impact on the contrast in the malignant mass images, while there is an obvious enhancement of contrast in the microcalcification images.

8.6.3.2 Truncated iteration

The number of iterations an IIR algorithm takes is an important practical consideration for clinical DBT. Because of the present limitations on iteration numbers, the design of IIR algorithms may not always be purely based on optimization. For example, on the topic of image regularization, one can perform descent steps on a combined objective function containing data fidelity and roughness penalty terms, but it is difficult to control the balance between these terms when the IIR algorithm is truncated well short of converging to the minimum of this objective function. In fact, for DBT IIR, truncation of the iteration of data-error descent algorithms, such as POCS, GD, or MLEM, itself is the most common form of imposing image regularity. We see in the results from our MLEM implementation applied to clinical DBT data that we can obtain regularized images by selecting results from lower iteration number without explicitly including regularization in the IIR formulation. This strategy also has the advantage that it involves only a single run of the IIR algorithm.

In IIR for DBT with explicit image regularization, the algorithm design has to take into account truncated iteration. Explicit regularization involves descent steps of at least two different objective functions, data fidelity and image roughness. The step lengths in reducing each objective function have to be controlled to ensure progress on each term at each iteration. The illustrated ASD-POCS implementation in Algorithm 4 represents one realization of this strategy. In the clinical DBT image series, the practical benefit of explicit regularization control afforded by ASD-POCS is that images with high detail are obtained at earlier iteration numbers than those of the basic MLEM implementation, and because of the explicit regularization control the ASD-POCS images do not have the cross-hatching artifacts seen in the POCS results.

8.7 SUMMARY OF IIR FOR DBT

This chapter has covered the basic building blocks for IIR with an eye toward its application to DBT. Design of IIR algorithms is generally based upon some form of optimization, most commonly, the fundamental image update steps are derived from a descent direction for a data fidelity objective function. In adapting IIR methods to DBT, the main issues to consider are the limited angular sampling of the DBT configuration and truncated iteration.

8.7.1 ALGORITHM DESIGN OF IIR COMPARED WITH FBP

The focus of this chapter has been on fundamental aspects of IIR design, and it is instructive to summarize IIR design issues with their FBP counterpart, for which there is more experience. For implementation of IIR, the only issue that is fully appreciated is algorithm efficiency; FBP involves only one back-projection while IIR entails multiple projection/back-projection operations. Beyond this obvious difference, there are other important implementation issues for IIR that originate from the fact that IIR algorithms are designed to solve data models implicitly and IIR algorithms process the DBT projection data in a nonlinear way.

The implicit nature of IIR means such algorithms aim to estimate the projection data. The image voxels are just one set of parameters needed to model the data. This view of IIR design affords incredible flexibility as more detailed physical models can be employed than image reconstruction based on direct inversion of a data model such as FBP. There are still large design differences between IIR and FBP even when both algorithms are based on the x-ray transform. In this chapter, for example, the data model used for developing IIR is just the discretized form of the x-ray transform. Because IIR aims to model the data, a nearly complete representation of the reconstruction volume is needed. This means that the whole volume must be reconstructed together, while FBP affords reconstruction of ROIs. Also, changing the resolution of the IIR volume representation can have unpredictable effects because the data model is being altered,

while the FBP volume can be computed at any voxel size—the true FBP resolution being limited only by the data sampling.

The nonlinear nature of IIR entails design differences from FBP. For example, in the results for the ASD-POCS algorithm, the nonlinear smoothing of clinical images allows for increased contrast of microcalcifications. But with this potential image quality gain, come many new design challenges. Validation of IIR algorithms can only convincingly be performed on clinical DBT data sets or a close simulation. Because the resulting image quality depends on the subject generating the data, phantom testing has limited use for validation and traditional image quality metrics have limited use. For example, one characterization of FBP is through the use of various spatial and contrast resolution phantoms, but these phantoms are often piecewise constant even if they have challenging, small or low-contrast objects in them. The use of the TV objective function in IIR favors piecewise constant subjects, and, as a result, such phantoms may give an overoptimistic assessment of resolution and contrast afforded by the TV-based IIR algorithm. Phantom testing still has a role within IIR development, but it is more for proof-of-principle as is done in this chapter. We see a marked improvement in the image quality of the phantom simulations in switching from POCS to ASD-POCS. But the only conclusion from that comparison is that seeking a low TV image that agrees with the data is a good strategy for that particular simulated object. Validation of this IIR algorithm requires further testing with clinical DBT data and the enhancement of microcalcification contrast seen there is more significant.

The nonlinear nature of IIR results in more complicated algorithm parameter dependence. Changing filter properties, for example, in FBP has fairly generic and understood effects on image quality. Changing IIR algorithm parameters, however, does not have directly predictable effects, and these effects in general vary according to the properties of each data set. Consider the array of images reconstructed from clinical DBT data presented in Section 8.6, where ROI images are presented for different algorithms and parameter settings, and suppose that one of the ROIs is selected to be optimal for a given imaging task. There is no guarantee that the same algorithm and parameter setting will be optimal for another DBT data set even for the same imaging task. The nonuniformity of parameter setting dependence compounds the computational efficiency issue because multiple reconstructions need to be performed at different parameter settings.

While there has been much progress in understanding IIR algorithm development for DBT, there still remains much research effort needed to address practical issues of IIR in order to squeeze the maximum information out of DBT projection data sets for its clinical application.

8.7.2 RESEARCH LITERATURE ON IIR IN DBT

The three basic update steps described in Section 8.3 provide a basic understanding of the main part of IIR algorithms, but many more update steps appear in the literature. Also, this chapter has not considered any physics in the data model beyond Beer's law, which leads to the x-ray transform in Equation 8.2 and its discrete form in Equation 8.3. Much research effort on IIR in DBT is devoted to improved physics modeling by considering realistic noise models and energy dependence of x-ray attenuation.

Comparison of various image reconstruction algorithms, including IIR, for DBT has been performed in Refs. [8,32]. One of the first applications of IIR to DBT by Wu et al. [33] employed an MLEM algorithm designed for the Poisson noise model on the transmitted x-ray intensity by Lange and Fessler [14]. A penalized maximum likelihood [1], which includes noise modeling and an edge-preserving regularization penalty, was applied to DBT by Das et al. [34]. X-ray beam polychromaticity and noise modeling were both taken into account in IIR by Chung et al. [35]. Other research articles on IIR in DBT have exploited a combination of image processing techniques with IIR to help reduce DBT artifacts and enhance microcalcification conspicuity [36–39]. While the application of image processing methods to reconstructed images is not new, incorporating such steps in an IIR loop can boost algorithm robustness and data fidelity because the image estimate is constrained by the available data.

As can be seen by the simulation and clinical results of this chapter, there is tremendous variability in the appearance of DBT images even within the basic modeling presented. This variability ultimately arises from the need to truncate the iteration number and undersampling of DBT. As IIR for DBT develops further and these basic issues become better characterized, we can expect that image quality will be further improved as more physics modeling is included along with the developed image processing techniques.

ACKNOWLEDGMENTS

The author thanks D. Kopans and R. Moore for providing the clinical DBT data used for this chapter. The author would also like to point out that much progress is enabled by the availability of numerous, free, open source software tools for scientific computing, and the author's work and preparation of this chapter have benefited tremendously from these tools [40,41]. In particular, all images and plots in this chapter have been generated using Matplotlib [41] (matplotlib.org), a publicly available Python library for scientific computing created by John Hunter (1968–2012). The work was supported in part by National Institutes of Health (NIH) R01 grants CA158446, CA120540, and EB000225. The contents of this chapter are solely the responsibility of the author and do not necessarily represent the official views of the National Institutes of Health.

REFERENCES

1. J. A. Fessler, Statistical image reconstruction methods for transmission tomography, in *Handbook of Medical Imaging*, M. Sonka and J. M. Fitzpatrick, Eds., SPIE Press, Bellingham, WA, 2000, vol. 2 of *Medical Image Processing and Analysis*, pp. 1–70.
2. J. Qi and R. M. Leahy, Iterative reconstruction techniques in emission computed tomography, *Phys. Med. Biol.*, 51(15), R541–R578, 2006.
3. H. H. Barrett and K. J. Myers, *Foundations of Image Science*, John Wiley & Sons, Hoboken, NJ, 2004.
4. P. L. Combettes, The foundations of set theoretic estimation, *IEEE Proc.*, 81, 182–208, 1993.
5. S. P. Boyd and L. Vandenberghe, *Convex Optimization*, Cambridge University Press, Cambridge, UK, 2004.
6. J. Nocedal and S. Wright, *Numerical Optimization, 2nd ed.*, Springer, New York, 2006.
7. A. C. Kak and M. Slaney, *Principles of Computerized Tomographic Imaging*, IEEE Press, New York, 1988.
8. T. Wu, R. H. Moore, E. A. Rafferty, and D. B. Kopans, A comparison of reconstruction algorithms for breast tomosynthesis, *Med. Phys.*, 31, 2636, 2004.
9. M. H. Buonocore, W. R. Brody, and A. Macovski, A natural pixel decomposition for two-dimensional image reconstruction, *IEEE Trans. Biomed. Eng.*, 28(2), 69–78, 1981.
10. S. Matej and R. M. Lewitt, Practical considerations for 3-D image reconstruction using spherically symmetric volume elements, *IEEE Trans. Med. Imag.*, 15(1), 68–78, 1996.
11. G. L. Zeng and G. T. Gullberg, Unmatched projector/backprojector pairs in an iterative reconstruction algorithm, *IEEE Trans. Med. Imag.*, 19, 548–555, 2000.
12. A. P. Dempster, N. M. Laird, and D. B. Rubin, Maximum likelihood from incomplete data via the EM algorithm, *J. Royal Stat. Soc. B*, 39, 1–38, 1977.
13. L. A. Shepp and Y. Vardi, Maximum likelihood reconstruction for emission tomography, *IEEE Trans. Med. Imag.*, 1, 113–122, 1982.
14. K. Lange and J. A. Fessler, Globally convergent algorithms for maximum a-posteriori transmission tomography, *IEEE Trans. Imag. Proc.*, 4, 1430–1438, 1995.
15. R. Gordon, R. Bender, and G. T. Herman, Algebraic reconstruction techniques (ART) for three-dimensional electron microscopy and X-ray photography, *J. Theor. Biol.*, 29, 471–481, 1970.
16. G. T. Herman, *Image Reconstruction from Projections*, Academic, New York, 1980.
17. F. Natterer, *The Mathematics of Computerized Tomography*, J. Wiley & Sons, New York, 1986.
18. H. Guan and R. Gordon, A projection access order for speedy convergence of ART (algebraic reconstruction technique): A multilevel scheme for computed tomography, *Phys. Med. Biol.*, 39, 2005–2022, 1994.
19. H. Erdoğan and J. A. Fessler, Ordered subsets algorithms for transmission tomography, *Phys. Med. Biol.*, 44, 2835–2851, 1999.
20. A. Beck and M. Teboulle, Fast gradient-based algorithms for constrained total variation image denoising and deblurring problems, *IEEE Trans. Imag. Proc.*, 18, 2419–2434, 2009.
21. A. Chambolle and T. Pock, A first-order primal-dual algorithm for convex problems with applications to imaging, *J. Math. Imag. Vis.*, 40, 120–145, 2011.
22. T. Pock and A. Chambolle, Diagonal preconditioning for first order primal-dual algorithms in convex optimization, in *International Conference on Computer Vision (ICCV 2011)*, Barcelona, Spain, 2011, pp. 1762–1769.
23. T. L. Jensen, J. H. Jørgensen, P. C. Hansen, and S. H. Jensen, Implementation of an optimal first-order method for strongly convex total variation regularization, *BIT Num. Math.*, 52, 329–356, 2012.
24. E. Y. Sidky, C.-M. Kao, and X. Pan, Accurate image reconstruction from few-views and limited-angle data in divergent-beam CT, *J. X-Ray Sci. Tech.*, 14, 119–139, 2006.
25. E. Y. Sidky and X. Pan, Image reconstruction in circular cone-beam computed tomography by constrained, total-variation minimization, *Phys. Med. Biol.*, 53, 4777–4807, 2008.
26. E. Y. Sidky, X. Pan, I. S. Reiser, R. M. Nishikawa, R. H. Moore, and D. B. Kopans, Enhanced imaging of microcalcifications in digital breast tomosynthesis through improved image-reconstruction algorithms, *Med. Phys.*, 36, 4920–4932, 2009.
27. L. I. Rudin, S. Osher, and E. Fatemi, Nonlinear total variation based noise removal algorithms, *Physica D*, 60, 259–268, 1992.
28. A. H. Delaney and Y. Bresler, Globally convergent edge-preserving regularized reconstruction: An application to limited-angle tomography, *IEEE Trans. Imag. Proc.*, 7(2), 204–221, 1998.

29. M. Persson, D. Bone, and H. Elmqvist, Total variation norm for three-dimensional iterative reconstruction in limited view angle tomography, *Phys. Med. Biol.*, 46, 853–866, 2001.

30. X. Pan, E. Y. Sidky, and M. Vannier, Why do commercial CT scanners still employ traditional, filtered back-projection for image reconstruction? *Inv. Prob.*, 25, 123009, 2009.

31. B. De Man and S. Basu, Distance-driven projection and backprojection in three dimensions, *Phys. Med. Biol.*, 49, 2463–2475, 2004.

32. Y. Zhang, H.-P. Chan, B. Sahiner, J. Wei, M. M. Goodsitt, L. M. Hadjiiski, J. Ge, and C. Zhou, A comparative study of limited-angle cone-beam reconstruction methods for breast tomosynthesis, *Med. Phys.*, 33, 3781–3795, 2006.

33. T. Wu, A. Stewart, M. Stanton, T. McCauley, W. Phillips, D. B. Kopans, R. H. Moore, J. W. Eberhard, B. Opsahl-Ong, L. Niklason, and M. B. Williams, Tomographic mammography using a limited number of low-dose cone-beam projection images, *Med. Phys.*, 30, 365–380, 2003.

34. M. Das, H. C. Gifford, J. M. O'Connor, and S. J. Glick, Penalized maximum likelihood reconstruction for improved microcalcification detection in breast tomosynthesis, *IEEE Trans. Med. Imag.*, 30, 904–914, 2011.

35. J. Chung, J. G. Nagy, and I. Sechopoulos, Numerical algorithms for polyenergetic digital breast tomosynthesis reconstruction, *SIAM J. Imag. Sci.*, 3, 133–152, 2010.

36. T. Wu, R. H. Moore, and D. B. Kopans, Voting strategy for artifact reduction in digital breast tomosynthesis, *Med. Phys.*, 33, 2461–2471, 2006.

37. Y. Zhang, H.-P. Chan, B. Sahiner, Y.-T. Wu, C. Zhou, J. Ge, J. Wei, and L. M. Hadjiiski, Application of boundary detection informabon in breast tomosynthesis reconstruction, *Med. Phys.*, 34, 3603–3613, 2007.

38. Y. Zhang, H.-P. Chan, B. Sahiner, J. Wei, C. Zhou, and L. M. Hadjiiski, Artifact reduction methods for truncated projections in iterative breast tomosynthesis reconstruction, *J. Comp. Assist. Tomo.*, 33, 426–435, 2009.

39. Y. Lu, H.-P. Chan, J. Wei, and L. M. Hadjiiski, Selective-diffusion regularization for enhancement of microcalcifications in digital breast tomosynthesis reconstruction, *Med. Phys.*, 37, 6003–6014, 2010.

40. T. E Oliphant, Python for scientific computing, *Comp. Sci. Eng.*, 9(3), 10–20, 2007.

41. J. D. Hunter, Matplotlib: A 2D graphics environment, *Comp. Sci. Eng.*, 9, 90–95, 2007.

Image reconstruction

Section IV

System performance

9 Fourier-domain methods for optimization of tomosynthesis (NEQ)

Ying (Ada) Chen, Weihua Zhou, and
James T. Dobbins III

Contents

9.1 INTRODUCTION

Tomosynthesis and computed tomography (CT) belong to the technology of tomographic imaging, which demonstrates the important features over conventional projection radiography (Dobbins and Godfrey 2003). The tomographic imaging technology enables three-dimensional reconstruction of objects with depth resolution. It improves the conspicuity of structures by removing the ambiguities caused by overlapping tissues. In the late 1990s, tomosynthesis research was reignited as a result of several technological advancements (Dobbins and Godfrey 2003): the invention of digital flat-panel detectors that are capable of producing high-quality digital images with rapid readout rates and the high-performance computation that enables tomosynthesis reconstruction and image processing. Digital tomosynthesis has been investigated and applied to various medical imaging clinical applications, including chest imaging, joint imaging, angiography, dental imaging, liver imaging, and breast imaging (Sklebitz and Haendle 1983, Dobbins 1990, Sone et al. 1995, Suryanarayanan et al. 2000, Warp et al. 2000, Badea et al. 2001, Godfrey and Dobbins 2002, Godfrey et al. 2001, 2003, 2006, Duryea et al. 2003, Maidment et al. 2006, Rakowski and Dennis 2006, Zhang et al. 2006, Bachar et al. 2007, Mertelemeier et al. 2006, Wu et al. 2011).

With the advancement of tomosynthesis research, many medical imaging manufacturers are actively engaged in designing digital tomosynthesis prototype. Manufacturers are eager to optimize their designs to improve their images. Although the conventional physical measurement techniques of image quality metrics such as linearity, signal-to-noise ratio, modulation transfer function (MTF), and noise power spectrum (NPS) (Dobbins 2000, Godfrey et al. 2006) can be applied to characterize the tomosynthesis image, it is urgent and essential to develop the

appropriate strategies to compare and optimize tomosynthesis systems and image reconstruction algorithms.

There are various imaging parameters to evaluate medical imaging systems and reconstruction algorithms. The performance of an imaging device is customarily described in two domains: the spatial domain of Cartesian space and the frequency domain. The reason for this dual description is partly historical. It also has its basis in the fact that there are certain global responses of the system and spatially correlated system-response characteristics. Compared to the traditional spatial domain analysis tools of image quality, the frequency-domain description is frequently referred to as classical systems analysis. The main reason for evaluating frequency-domain response is due to the spatial correlations of the response of the system. The traditional global descriptions in spatial domain would be completely satisfactory to characterize all behaviors of the system, if the system response was truly the same at all points. However, the fact that both signal and noise are spatially correlated to some degree makes it very helpful to consider a frequency-response characterization of imaging systems and reconstruction algorithms (Dobbins 2000).

With the development of computational technologies, Fourier's transform has been researched with fast computation. Signals of objects can be decomposed into the combination of sine waves with different amplitudes, frequencies, and phases to be evaluated in the spatial frequency domain (Nishikawa 2011). In this chapter, we will focus on the optimization of tomosynthesis image reconstruction algorithms and imaging configurations in the spatial frequency domain based on image quality characterization.

The optimization of the imaging configurations and reconstruction algorithms is of great importance to improve tomosynthesis imaging performance. However, several factors play important roles in the optimization tasks and some of them are associated together to some extent. The nonlinearity property of digital tomosynthesis system makes it difficult to evaluate image quality. Therefore, it is essential to find an effective methodology that allows one to optimize tomosynthesis imaging configurations and reconstruction for breast tomosynthesis imaging.

In the spatial frequency domain, MTF, NPS, and noise-equivalent quanta (NEQ(f)) are the main image quality factors that are used frequently to characterize the performance of medical imaging systems and digital detectors. Physical measurements and computational analysis of MTF, NPS, and NEQ are well published in the literature (Dobbins 2000, Samei et al. 2006, Dobbins et al. 2006). In 2006, Godfrey et al. proposed methods of MTF and NPS measurement to quantitatively characterize the selection of optimal acquisition parameters for digital chest tomosynthesis (Godfrey et al. 2006). A series of experiments on impulse response, NPS(f), and NEQ(f) analysis with a prototype tomosynthesis system were published for breast tomosynthesis (Chen et al. 2005, 2006, 2007).

In this chapter, we will introduce the imaging model in tomosynthesis, including imaging configurations and reconstruction algorithms, to facilitate one's understanding. Then the methodologies of MTF, NPS, and NEQ measurement will be explained. Examples of MTF(f), NPS(f), and NEQ(f) characterizations will be illustrated for breast tomosynthesis optimization.

9.1.1 OPTIMIZATION OBJECTIVES

In this section, we will introduce several important imaging parameters that should be considered and optimized for a typical digital tomosynthesis imaging system.

Digital tomosynthesis prototype systems have been developed for various kinds of clinical applications. The specifications and requirements for different applications may vary. To simplify the optimization problem, here, we take the design of digital breast tomosynthesis (DBT) system as an example. DBT imaging system has received much attention. One manufacturer has received the FDA (Food and Drug Administration) approval to market their device in the United States. The DBT imaging method is challenging the traditional technology of standard mammography.

Most current DBT prototype systems reutilize the conventional mammography design with the associated mechanical, electrical, and sensor techniques. The x-ray tube typically rotates along an arc path above the object to acquire projection images at specified positions with limited view angle (VA). This kind of design is called partial iso-centric, as shown in Figure 9.1. In Figure 9.1, the object (i.e., the compressed breast) is located above the detector surface. SID represents the source-to-image distance. The x-ray tube moves above the breast object to acquire multiple projection images. The number of projection images varies from 11 to 49 for different prototype systems. Tomosynthesis reconstruction algorithms are applied to the acquired tomosynthesis projection image data set to reconstruct slice images passing through different depths of the object. In Figure 9.1, a representative reconstruction slice S is shown for illustration.

During a typical tomosynthesis imaging procedure, image acquisition (x-ray source, detector, etc.), image reconstruction

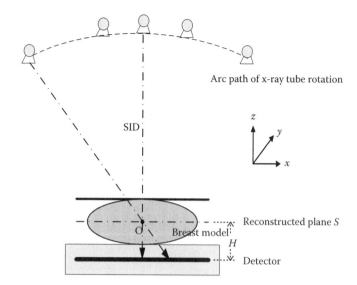

Figure 9.1 Partial iso-centric tomosynthesis imaging configuration.

algorithms, and image display are the main factors that influence the resulting image quality.

With this kind of partial iso-centric breast tomosynthesis system design, one can upgrade the traditional digital mammography device to acquire projection images of tomosynthesis imaging. However, it has also been demonstrated that the rotation of an x-ray tube in the partial iso-centric imaging configuration of DBT design may result in motion blur associated with the x-ray tube's motion (Lalush et al. 2006).

Recently, another breast tomosynthesis imaging design was developed. Fixed multibeam field emission tomosynthesis imaging technique was invented with parallel imaging geometry (Yang et al. 2008, Zhou et al. 2010). The x-ray tubes were developed based on carbon nanotube techniques and were fixed along a line parallel to the detector plane. This system design has great potential to eliminate the motion blur and patients' discomfort associated with partial iso-centric design of typical DBT prototype systems. It is proposed that the image acquisition may also be faster compared with that of other designs.

Figure 9.2 illustrates this new parallel tomosynthesis imaging configuration. One can see that multiple x-ray sources

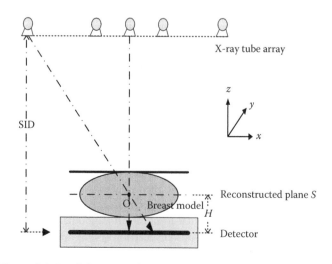

Figure 9.2 Parallel tomosynthesis imaging configuration.

are fixed along a line with this parallel imaging configuration design. No x-ray tube's motion exists. Control signals activate individual x-ray source to make projection images one following another to acquire a whole dataset of tomosynthesis projections.

Currently, both partial iso-centric and parallel tomosynthesis imaging configurations exist in breast tomosynthesis imaging. In other tomosynthesis imaging applications, scientists are developing various designs as well, such as parallel imaging configuration for chest tomosynthesis design and C-arm tomosynthesis for head imaging applications. To compare these different imaging configurations for each clinical application, one needs to select a methodology to optimize the imaging configuration design to provide better resolution. This becomes an important optimization objective for researchers in digital tomosynthesis imaging.

Another key objective in digital tomosynthesis imaging is the optimization and comparison of various tomosynthesis reconstruction algorithms. Tomosynthesis reconstruction algorithms play a significant role in transforming the two-dimensional (2D) projection information into a three-dimensional reconstructed object. An arbitrary number of reconstructed images can be generated with appropriate reconstruction algorithms.

Tomosynthesis and CT share the common theoretical foundation of Radon's integral principles where the pixel intensity on the detector can be considered as the accumulation of photon energy. The main difficulty in developing an ideal tomosynthesis reconstruction algorithm comes from the incomplete sampling in tomosynthesis imaging. In tomosynthesis, only a few limited-angle projection images are available as the foundation to reconstruct the three-dimensional information. Therefore, to improve the solution of this incomplete sampling problem, efforts to optimize reconstruction algorithms never stop.

In summary, with a specific tomosynthesis application and hardware design, the optimization of imaging configurations and image reconstruction algorithms is essential to provide optimal system performance and image resolution. Especially, the imaging configurations typically include several configurable parameters such as the number of projection images, view angle (VA), imaging geometries, and so on. The combinations of these configurable parameters vary with different systems and should be compared and optimized.

Table 9.1 shows the representative imaging configurations and reconstruction algorithms from three major manufacturers in breast tomosynthesis imaging field.

9.1.2 IMAGING CONFIGURATION ANALYSIS OF TOMOSYNTHESIS

During digital tomosynthesis image acquisition, digital detectors are used to record images as discrete arrays with a limited intensity range. The spatial and temporal integral of the image irradiance is recorded. A detailed theory about image formation can be found in Barret and Myers (2004).

In tomosynthesis reconstruction, slices passing through an object will be reconstructed based on a tomosynthesis dataset of x-ray projection images. Digital computers are usually used to compute the reconstruction. It is necessary to represent the actual continuous object as a discrete set of numbers. A common way for the representation of the discrete small elements is pixels or voxels (Barrett et al. 2004).

If the statistical nature of the imaging process is ignored, the mapping from the object o to a single-projection image p can be written as (Barret and Myers 2004).

$$p = h \cdot o \tag{9.1}$$

The mapping operator h can be either linear or nonlinear. The property of homogeneity in linear systems makes it easier to analyze than nonlinear systems. Here, we begin with the assumption of linearity. In Fourier's frequency domain, one can use

$$P = H * O \tag{9.2}$$

to denote the imaging mapping. H is the Fourier transform of h and it represents the transfer function. P and O are the Fourier representations of the projection image p and the object o, respectively.

To simplify the imaging configuration consideration, we extract the linear parallel tomosynthesis imaging system (Figure 9.2) from Figure 9.3. The central point O is located in the reconstructed plane S. We have the projection image on the detector P. Under this assumption, S is a radiopaque plane with

Table 9.1 **Imaging configurations and reconstruction algorithms of DBT**

MANUFACTURER	VA	PROJECTION IMAGE NUMBER	RECONSTRUCTION ALGORITHM
GE (prototype system)	±20° (MGH, Sunnybrook)	15	MLEM/SART
Hologic (commercial system)	±30°	21	SART
	(University of Michigan)		
Siemens (prototype system)	±7.5°	11	FBP
	±22° (Duke, Malmo, SUNY)	25	FBP

Data source: Zhao, W., Zhao, B., and Hu, Y. H. 2011. Latest advances in digital breast tomosynthesis. Retrieved September 24, 2011, from http://www.aapm.org/meetings/amos2/pdf/42–11930–81688–724.pdf.2011.

Note: In the United States, only Hologic's breast tomosynthesis system is approved for clinical use.

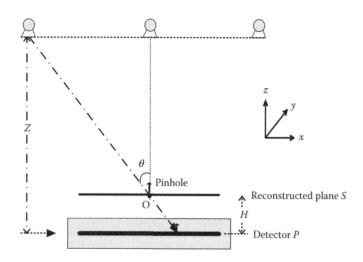

Figure 9.3 Impulse response imaging in tomosynthesis.

a small pinhole O located on the vertical axis. One can find that this input produces a replica of the x-ray source geometry on the detector with a Z-depth-dependent scaling factor (Grant 1972).

The line length of the replica of impulse response on the detector is

$$\Delta = 2rZ \tan \theta = dZ \qquad (9.3)$$

where r is the magnification and θ is half of the total scan angle (called VA in the following). With the Fourier transform of the impulse response function, the transfer function in Equation 9.2 becomes (Grant 1972):

$$H(w_x|Z) = \frac{\sin(w_x dZ/2)}{(w_x dZ/2)} \qquad (9.4)$$

The impulse response function has the properties given below (Grant 1972):

1. The blurring from undesirable planes is basically a linear filtering process.
2. The system's impulse response is a scaled replica of the scan configuration.
3. The position of the impulse response on the detector is Z-depth dependent.
4. The transfer function is a direct quantitative measure of the system's ability to blur undesirable planes and provides a valid method of comparing imaging configurations. It also provides the means of evaluating the effectiveness of a particular imaging configuration before setting up the actual measurement.

For a linear tomosynthesis imaging configuration with N evenly distributed x-ray sources of parallel imaging configurations, the impulse response is simply a series of N infinitesimal points. The corresponding transfer function is extended into (Grant 1972):

$$H^N(w_x|Z) = \frac{1}{N} \frac{\sin N(\delta w_x/2)}{\sin(\delta w_x/2)} \qquad (9.5)$$

The transfer function becomes a series of peaks occurring at harmonics of the sampling frequency.

Godfrey et al. (2006) extended Grant's results and presented the MTF analysis results with varying the VA and the distances to the plane S. They demonstrated that out-of-plane artifacts can be suppressed with increased number of projection images (N).

9.2 MODULATION TRANSFER FUNCTION

The MTF is used to analyze the resolution of the imaging system in the spatial frequency domain. Technically, the "resolution" of a system is the minimum distance of two objects that can be distinguished. In practice, one can simulate a delta function to evaluate the response of the system or algorithm to be investigated. This response is called the point-spread function (PSF). It contains the deterministic spatial-transfer information of the system (Dobbins 2000).

One main problem of image quality degradation in digital imaging is undersampling (Dobbins 2000). It often happens when the image is not sampled sufficiently to represent all spatial frequencies without aliasing. Unfortunately, almost all digital systems are undersampled to some degree. Designers of digital imaging systems face the constraints from limited resources, including image acquisition, computation, storage, network bandwidth, and data transfers. An essential problem is how to compare undersampled digital imaging systems.

The MTF is a useful descriptor of the system spatial response because the stages of system response can be considered as "filters" as described in our linear system analysis in Section 9.1. One can compute the MTF as a 2D Fourier-transform function of spatial frequencies. The composite MTF of a system is the product of the MTFs coming from all individual stages. Since tomosynthesis imaging is a complicated procedure in which both image acquisition and image reconstruction are involved, a simplified MTF measurement methodology is to combine the MTFs from each stage. In this section, we call the MTF due to the image acquisition stage as the system MTF and the MTF due to the image reconstruction stage as reconstruction MTF.

The system MTF is discussed first. Two kinds of MTFs are often used to represent the MTF specifications: the presampling MTF (PMTF) and expectation MTF (EMTF). PMTF characterizes a system's response to a single sinusoid input; EMTF characterizes the approximate response of the system with a broadband input of frequencies, for example, a delta impulse input. It does not accurately describe the response of the system to a single sinusoid (Dobbins 2000). Both the PMTF and EMTF should be reported due to the following reasons: (1) they work together to characterize system response with a single sinusoid or a delta function input on the undersampled system and (2) they are matched over the range of frequencies at which there is no aliasing.

Experimentally, two general physical setups have been used to investigate the MTF specifications: angulated slit and angulated edge (Samei et al. 2006). During the measurement process, the predefined signal is considered as the input to the detector and the system response is read and analyzed. The edge method is typically implemented by using either a translucent edge or an

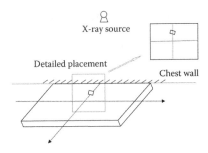

Figure 9.4 Setup of a system MTF measurements.

opaque edge, which is primarily characterized by the difference in its radiolucency (Samei et al. 2006).

In this section, we illustrate a system MTF measurement of a breast tomosynthesis prototype system based on edge measurement. An MW2 technique proposed by Saunders et al. (Saunders and Samei 2003, Saunders et al. 2005) was used for the system MTF measurement with tungsten/rhodium target/filtration spectrum. The system was operated with 28 keV and W/Rh spectrum. A tomosynthesis dataset of 11 projection images with the total 303 mAs was acquired.

Figure 9.4 shows the experimental setup of the system MTF measurement. As shown in Figure 9.4, a 0.1 mm Pt–Ir edge phantom was placed with 1–3° oblique against the pixel axis on the detector surface. Projection images were acquired and regions of interests (ROIs) around the edge area were used to compute the presampled system MTF. Figure 9.5 shows the edge method measured system MTF when the edge center was about 4 cm away from the chest wall and when the x-ray tube was at angles of 0°, ±15°, and ±25°, specifically. There is no big difference in the system MTF for the different angles. The MTF plot of −25° is slightly lower than that of other angular positions. This may be caused by the x-ray tube's motion and velocity difference at the starting position of the −25° location, where the x-ray tube begins the motion. Mainprize et al. reported the measured PMTFs at different incidence angles of 0°, 10°, 20°, 30°, and 45° with a GE Senographe 2000D digital mammography system. Similar results were observed. The MTF of a larger incidence angle of 40° was lower than that of the other angular positions (Mainprize et al. 2006). At larger incidence angle of breast tomosynthesis projections, increased resolution loss may occur.

Compared to the system MTF, the reconstruction MTF can be computed based on computer simulations. The relative reconstruction MTF describes the calculated relative MTF associated with specific algorithm and acquisition parameters. A simulated delta function was used as a standard signal input to test the reconstruction MTF. PSF, edge spread function, and other forms can be characterized to measure the reconstruction MTF. Here, we present an example to illustrate how to calculate reconstruction MTF based on impulse response simulation and PSF to optimize the imaging configuration and algorithms.

Figure 9.6 shows the impulse simulation based on the ray-tracing method. A tomosynthesis projection image dataset of a single delta function at defined height of H above the detector is computer simulated using a ray-tracing method. The delta function (impulse) is projected onto the detector at location P when the x-ray source is located at the specific position as shown in Figure 9.6. If position P falls into a noninteger pixel location, linear interpolations should be considered to distribute the pixel intensity to surrounding pixels. On the basis of this ray-tracing method, a set of tomosynthesis projection images without background can be simulated. Typically, two impulse locations should be considered: the delta function is simulated at a location that is near the chest wall; the delta function is simulated at a location that is away from the chest wall. These two situations represent the structures of interests to be evaluated. One can then apply tomosynthesis reconstruction algorithms to reconstruct the plane where the simulated impulse is located (H distance above the detector in Figure 9.6). The impulse response on this specific reconstruction plane should be evaluated for comparison purposes.

The relative reconstruction MTF (without normalization) can be calculated as the Fourier transform of the impulse response or PSF along the tube's motion direction. Here, we should point out that the reconstruction MTF is affected by the location of the simulated impulse. One should include this concern into the calculation by evaluating the reconstruction MTF for multiple sample locations inside a single pixel range. Ideally, the more sample locations inside a single pixel range, the better the average overall impulse position can be estimated. However, more samples

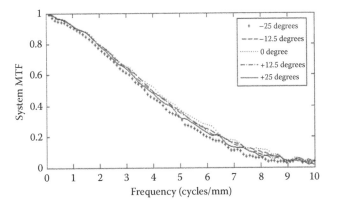

Figure 9.5 System MTF with the edge located about 4 cm away from the chest wall.

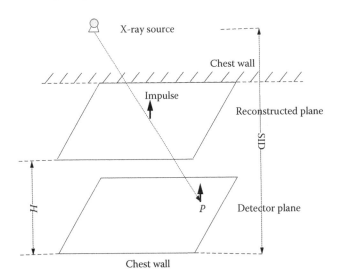

Figure 9.6 Impulse simulation based on ray-tracing method.

System performance

Table 9.2 Nine sample impulse locations inside a pixel when the impulse was near the chest wall and about 4 cm away from the chest wall

SAMPLE LOCATION	NEAR THE CHEST WALL		ABOUT 4 CM AWAY FROM THE CHEST WALL	
	X PIXEL LOCATION	Y PIXEL LOCATION	X PIXEL LOCATION	Y PIXEL LOCATION
#1	2027.50	1023.50	1023.50	1023.50
#2	2028.00	1023.50	1024.00	1023.50
#3	2027.50	1024.00	1023.50	1024.00
#4	2028.00	1024.00	1024.00	1024.00
#5	2028.50	1023.50	1024.50	1023.50
#6	2028.50	1024.00	1024.50	1024.00
#7	2028.50	1024.50	1024.50	1024.50
#8	2028.00	1024.50	1024.00	1024.50
#9	2027.50	1024.50	1023.50	1024.50

will significantly increase the computing time. On the basis of the realistic computation time, in this section, we used nine evenly distributed sample locations inside a pixel to estimate the impulse response and relative reconstruction MTF along the x-ray tube's motion direction. This method of nine sample locations provides an approximate estimation of relative reconstruction MTF. Table 9.2 shows the nine sample locations inside a pixel for impulse response and relative reconstruction MTF calculation. Pixel locations along the x and y axes are shown in Table 9.2. The image size was 2048×2048 with a pixel size of 85 μm.

Figures 9.7 through 9.9 illustrate an example of a reconstruction MTF measurement for the filtered backprojection (FBP) reconstruction algorithm and the imaging configuration: number of projection images $N = 25$, VA = 50°. The FBP algorithm in this example was developed based on a point-by-point backprojection (BP) with the applied Hamming filter $w(i)$:

$$w(i) = 0.5 + 0.5 \times \cos\left(\frac{2\pi \cdot i}{M}\right)$$

where i is the individual frequency bin in the total points of M in the spatial frequency space (Chen 2007).

Figure 9.7 shows the in-plane impulse response of FBP for a simulated breast tomosynthesis configuration of $N = 25$,

VA = 50°. The x and y axes show the pixel location on the reconstruction plane. A pixel region of 40×40 pixels close to the impulse is shown for clarity. Figure 9.7a shows the in-plane response when the impulse was located about 4 cm away from the chest wall and 40.5 mm above the detector. Figure 9.7b shows the in-plane response when the impulse was close to the chest wall (20 pixels from the chest wall) and 40.5 mm above the detector. FBP shows a sharp in-plane impulse response and edge enhancement phenomena for the in-plane reconstruction.

Figures 9.8 and 9.9 illustrate the relative reconstruction MTF of FBP, using the above imaging configuration, for nine sample locations of simulated impulses that are evenly distributed within a pixel. In Figures 9.8 and 9.9, the nine curves follow two trends due to sampling locations inside the pixel. X-axis represents the frequency bins. Y-axis represents the magnitude of Fourier transform of the impulse response at the defined reconstruction plane where the simulated impulse was located (40.5 mm above the detector). The nine calculated relative MTFs from nine different impulses' locations inside a pixel can be averaged to generate the relative MTF for this specific imaging configuration and FBP reconstruction. The reconstruction MTF reflects the applied Hamming low-pass filter by showing the dominated arc appearance in the spatial frequency domain.

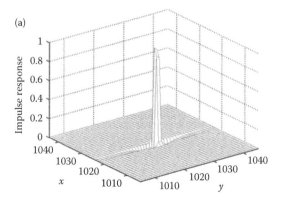
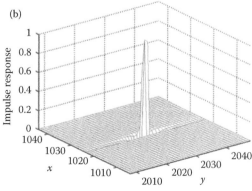

Figure 9.7 In-plane impulse response of FBP with $N = 25$ and VA = 50° partial iso-centric imaging configuration. (a) The impulse was located about 4 cm away from the chest wall and (b) the impulse was located near the chest wall.

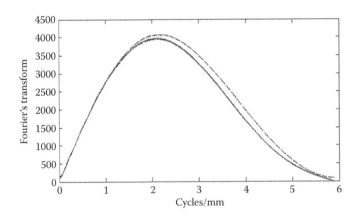

Figure 9.8 Relative reconstruction MTF of nine samples of FBP with N = 25 and VA = 50° partial iso-centric imaging configuration; the impulse was located about 4 cm away from the chest wall.

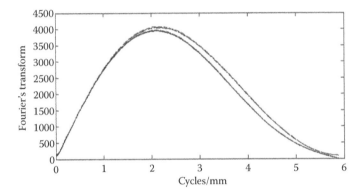

Figure 9.9 Relative reconstruction MTF of nine samples of FBP with N = 25 and VA = 50° partial iso-centric imaging configuration; the impulse was located near the chest wall. The nine curves follow two trends due to sampling locations.

9.3 NOISE POWER SPECTRUM

The NPS is a metrics that is frequently used to characterize the noise properties of an imaging system. The spatial frequency-dependent NPS(f) is defined as the variance per frequency bin of a stochastic signal in the spatial frequency domain (Dobbins 2000). It can be directly computed from the squared Fourier amplitude of 2D imaging data by (Dobbins 2000)

$$
\begin{aligned}
\mathrm{NPS}(u_s, v_t) &= \lim_{M,N \to \infty} (MN\Delta X \Delta Y) < \left| \mathrm{FT}_{st}\left[I(x,y) - \bar{I} \right] \right|^2 > \\
&= \lim_{M,N \to \infty} \lim_{K \to \infty} \frac{MN\Delta X \Delta Y}{K} \sum_{k=1}^{K} \left| \mathrm{FT}_{st}\left[I(x,y) - \bar{I} \right] \right|^2 \\
&= \lim_{M,N,K \to \infty} \frac{\Delta X \Delta Y}{K \cdot MN} \\
&\quad \sum_{k=1}^{K} \left| \sum_{i=1}^{M} \sum_{j=1}^{N} [I(x_i,y_j) - \bar{I}] e^{-2\pi i (u_s x_i + v_t y_j)} \right|^2
\end{aligned}
$$

(9.6)

where $I(x_i,y_j)$ is the image intensity at the pixel location (x_i,y_j) and $\bar{I}$ is the global mean intensity. u and v are the spatial frequencies

conjugate to the x- and y-axis. M and N are the numbers of pixels in the x and y directions of the digital image. ΔX and ΔY are the pixel spacings in the x and y directions, and K is the number of ROIs used for analysis.

In digital imaging, the two main noise sources are photon noise that arises from the discrete nature of photoelectric interactions, and electronic noise from detector amplifiers (Barret and Myers 2004). There are two main types of digital x-ray detectors: energy-integrating detectors that detect the total energy deposited by the incident x-rays, and photon counting detectors that detect individual x-ray quanta as discrete events. Compared with energy-integrating detectors, photon counting detectors tend to exhibit a lower electronic noise power or dark current rate. Scientists recently reported an intrinsic difference in imaging performance between the two types of detectors (Acciavatti and Maidment 2010). In this chapter, we focus on flat-panel energy-integrating detectors, which are widely used by academic researchers and manufacturers of tomosynthesis imaging devices. The difficulties of NPS analysis come from several factors (Dobbins 2000): the finite spatial domain and sampling in Equation 9.6, the static artificial components contained in the measured data, and inherent noise on the detector. Therefore, an appropriate compromise must be found when acquiring data for the NPS measurement.

Since the measured image data are given, there is a trade-off between the number and size of ROIs (Dobbins 2000). The ROI size should contain just enough pixels to adequately demonstrate the structures in the NPS curve. If the NPS curve is smoothly varying with spatial frequency, a very small ROI may be used. If there are spikes in the power spectrum, then the ROI should contain more pixels to have adequate sampling in frequency space so that the shape of the spikes is not adversely impacted. The defect of the detector may also cause inherent stochastic fixed pattern noise (Dobbins 2000).

The different stages of tomosynthesis imaging will bring different noise features. Tremendous efforts have been made to develop the methods for the NPS measurements (Chen et al. 2006, Zhao and Zhao 2008). In Zhao's model, NPS in DBT is affected by the following factors: (1) the aliased NPS of the projection images, including the detector performance; (2) reconstruction algorithms; and (3) the total VA of DBT acquisition. According to the paper, with the increase in total VA, the frequency for three-dimensional NPS is sampled better, especially at low spatial frequency.

A simplified NPS calculation version was proposed by Zhang et al. (2006). For this calculation, a 512 × 512 ROI of NPS can be extracted from a homogeneous area on the reconstructed plane. The region can be divided into multiple 512 × 16-pixel strips and the adjacent strips are overlapped by eight pixels, resulting in 63 samples. For each sample, a line profile of mean values along the sample direction should be calculated and then a second-order polynomial fitting technique can be applied to the mean values to correct the nonuniformities. The resulting line is called as background-corrected noise profile and its one-dimensional (1D) fast Fourier transform (FFT) is taken. The final 1D NPS is estimated by averaging the squared magnitude of the 1D FFT of all samples. The purpose is to compare different algorithms; so, only reconstructed slices will be used and no normalization of the NPS is performed in this method. For this method, Zhang et al. came to the conclusion that (1) the BP method produced much lower NPS level in the reconstructed slice, (2) both the simultaneous algebraic reconstruction technique (SART) and the maximum likelihood expectation maximization (MLEM) methods significantly amplified the noise at all frequencies, and (3) SART and MLEM have very similar NPS behaviors.

In this section, we illustrate an example of mean-subtracted NPS measurement with FBP reconstruction and imaging configuration of the number of projection images $N = 49$, VA = 50°. To mimic the breast tissue equivalent attenuation and scattered radiation, two identical phantom slabs of BR12 (47% water/53% adipose equivalent) for a total of 4 cm thickness were directly placed on the surface plate (detector cover) of a prototype breast tomosynthesis system. Ten identical tomosynthesis sequences of flat images with the phantom slabs on the detector were acquired. The FBP reconstruction algorithm described in Section 9.4 was applied to reconstruct the tomosynthesis images.

Mean-subtracted image reconstruction datasets were analyzed and compared on a defined reconstruction plane at the same location (40.5 mm above the detector) as described in the reconstruction MTF measurement (Section 9.4). A W/Rh spectrum and a 28-kV spectrum with a cumulative tube output of 358 mAs were used. Mean-subtracted NPS is capable to remove fixed pattern noise, including structured noise and system artifacts. The NPS data were generated using 64 ROIs of size 128×128 (Chen 2007). The NPS(f) results can be computed as the average of this 10 mean-subtracted NPS for this specific imaging configuration.

Figure 9.10 illustrates the example of the mean-subtracted NPS measurement of FBP with the above-mentioned specific imaging configuration of 49 projection images and 50° VA. Figure 9.10a represents the 10 individual mean-subtracted NPS; Figure 9.10b represents the averaged mean-subtracted NPS. There was no big variation between 10 individual mean-subtracted NPS for each algorithm. FBP shows an arc-shaped performance due to the applied Hamming filter in the frequency domain.

9.4 NOISE-EQUIVALENT QUANTA (NEQ(F)) AND DETECTIVE QUANTUM EFFICIENCY (DQE(F))

NEQ gives the effective number of quanta used by the imaging device. It is widely recognized as a relevant metrics related to the signal-to-noise ratio in radiographic imaging (Dobbins 2000, Samei et al. 2006).

In the spatial frequency domain, as discussed in Sections 9.4 and 9.5, the MTF describes the signal response of a system at a given spatial frequency and the NPS describes the amplitude variance at a given spatial frequency. The ratio of these quantities, properly normalized, gives information about the maximum available signal-to-noise ratio as a function of spatial frequency.

The spatial frequency-dependent NEQ can be defined as (Dobbins 2000)

$$NEQ(u,v) = \frac{(\text{large area signal})^2 \, MTF^2(u,v)}{NPS(u,v)} \quad (9.7)$$

The DQE is defined as (Dobbins 2000)

$$DQE(u,v) = \frac{NEQ(u,v)}{SNR^2_{\text{incident}}} \quad (9.8)$$

Here, we give an example to illustrate the relative NEQ(f) measurement for FBP and an imaging configuration of $N = 25$ and VA = 50°. The relative NEQ(f) combines the MTF of signal performance and the NPS of noise characteristics. The relative NEQ(f) can be expressed as

$$NEQ_{\text{relative}}(f) = \frac{MTF^2_{\text{recon}}(f) \cdot MTF^2_{\text{system}}(f)}{NPS_{\text{tomo}}(f)} \quad (9.9)$$

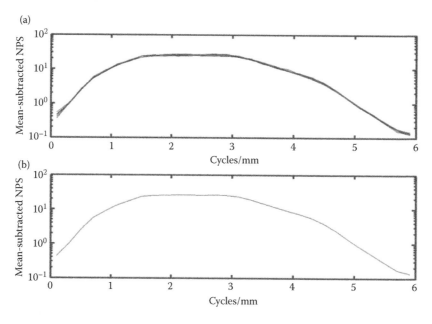

Figure 9.10 Mean-subtracted NPS of FBP with $N = 49$ and VA = 50° partial iso-centric imaging configuration. (a) 10 individual mean-subtracted NPS and (b) averaged mean-subtracted NPS.

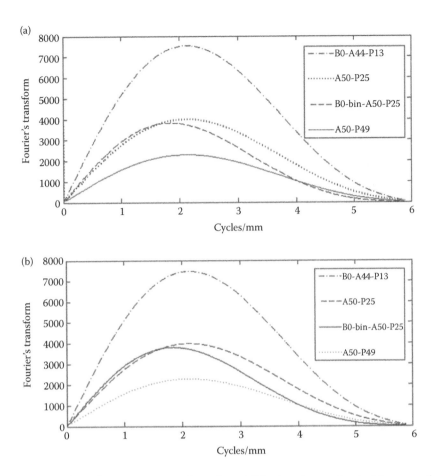

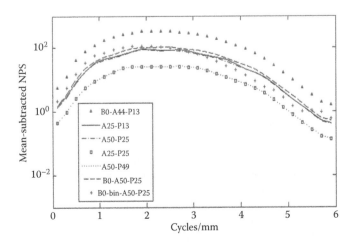

Figure 9.11 RECONMTF$_{relative}$ (*f*) of FBP for different imaging configurations. (a) The impulse was located 4 cm away from the chest wall and (b) the impulse was located near the chest wall.

The MTF$_{recon}$ (*f*) is the relative MTF with the specific image reconstruction algorithm and imaging configuration parameters. The MTF$_{system}$ (*f*) is the measured MTF of the imaging system. The NPS(*f*) is the mean-subtracted NPS on the same reconstruction plane.

Here, we should point out that different tomosynthesis image reconstruction algorithms may result in different gain factors for different imaging configurations. One cannot compare one combination of the imaging configuration and reconstruction algorithm against another purely based on MTF or sub-NPS results. However, the relative NEQ(*f*) analysis combining MTF and NPS together can provide a fair comparison with the same signal and noise inputs.

For each imaging configuration and image reconstruction algorithm, the same inputs, including the same simulated impulse magnitude for relative reconstruction MTF measurement and the same accumulative tomosynthesis sequence exposure level for NPS measurement, should be used in all cases. Therefore, it is feasible to make relative comparisons to evaluate the performance of different algorithms and imaging configurations with the same inputs.

In this section, we illustrate our relative NEQ(*f*) measurement for the FBP reconstruction algorithm and a breast tomosynthesis prototype system. The relative NEQ(*f*) along the tube's motion direction was examined. Figures 9.11 through 9.13 show the RECONMTF$_{relative}$ (*f*), sub-NPS$_{tomo}$ (*f*), and NEQ$_{relative}$ (*f*) results of FBP reconstruction algorithm, respectively. In Figures 9.11 and 9.13, the plots in (a) show results when the

simulated impulse was located 4 cm away from the chest wall and 40.5 mm above the detector and the plots in (b) show results when the simulated impulse was near the chest wall. Figure 9.13 shows the relative NEQ(*f*) results for FBP with several investigated partial iso-centric imaging configurations, respectively.

As shown in Figure 9.13, FBP performed slightly better for nonbinning modes with more projection numbers such as A50-P49 (*N* = 49 projections, VA = 50° angular range) and A50-P25

Figure 9.12 Sub-NPS$_{tomo}$ (*f*) of FBP for different imaging configurations.

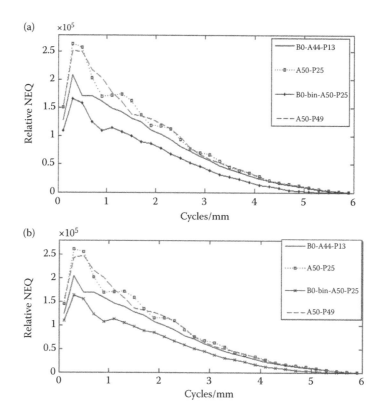

Figure 9.13 NEQ$_{relative}$ (f) of FBP for different imaging configurations. (a) The impulse was located 4 cm away from the chest wall and (b) the impulse was located near the chest wall.

(N = 25 projections, VA = 50° angular range) by providing slightly higher relative NEQ curves. Among the three nonbinning modes, FBP performed worse for B0-A44-P13 (N = 13 projections, VA = 44° angular range, "Bo" means a simple dark current image between each frame). In this situation, FBP suffers from incomplete sampling in the frequency space with a limited number of projection (N = 13) while with a wide angular range of VA = 44°.

FBP has the lowest relative NEQ values with B0-bin-A50-P25 binning modes (N = 25 projections, VA = 50° angular range, "Bo" means a simple dark current image between each frame, and "bin" means the binning mode of the neighboring two pixels) at all frequencies. One possible reason is that the relative RECONMTF was calculated by computer simulation, while the sub-NPS was measured experimentally. During experimental measurements, noise was smoothed by the binning mode. The binning mode provides a trade-off between reducing the resolution by averaging neighboring pixel values and improving the speed of readout.

9.5 CONCLUSIONS

With the rapid development of tomosynthesis imaging for various clinical applications, it is necessary to develop appropriate measurement methodologies to evaluate image quality for optimization of imaging configurations and reconstruction algorithms. A typical digital tomosynthesis system includes several components including image acquisition (x-ray source, detector), image reconstruction, and image display. Every step can influence the signal and noise properties of the imaging system.

Optimized image acquisition techniques are essential to provide better tomosynthesis reconstruction images with different image reconstruction algorithms. None of the imaging acquisition parameters are independent of one another. In this chapter, we discussed a spatial frequency domain method for optimization of tomosynthesis systems. The characterization of image quality using MTF, NPS, and NEQ(f) was introduced and investigated for tomosynthesis imaging.

REFERENCES

Acciavatti, R. J. and Maidment, A. D. A. 2010. A comparative analysis of OTF, NPS, and DQE in energy integrating and photo counting digital x-ray detectors. *Med. Phys.* 37(12): 6480–6495.

Bachar, G., Siewerdsen, J. H., Daly, M. J., Jaffray, D. A., and Irish, J. C. 2007. Image quality and localization accuracy in C-ary tomosynthesis-guided head and neck surgery. *Med. Phys.* 34(12): 4664–4677.

Badea, C., Kolitsi, Z., and Pallikarakis, N. 2001. A 3D imaging system for dental imaging based on digital tomosynthesis and cone beam CT. *Proc. Int. Fed. Med. Biol. Eng.* 2: 739–741.

Barrett, H. H. and Myers, K. J. 2004. *Foundation of Image Science*. John Wiley & Sons. Hoboken, New Jersey.

Chen, Y., Lo, J. Y., and Dobbins, J. T., III. 2005. Impulse response analysis for several digital tomosynthesis mammography reconstruction algorithms. *Proc. SPIE* 5745: 541–549.

Chen, Y., Lo, J. Y., Baker, J. A., and Dobbins, J. T. III. 2006. Noise power spectrum analysis for several digital breast tomosynthesis reconstruction algorithms. *Proc. SPIE* 6142: 1677–1684.

Chen, Y., Lo, J. Y., and Dobbins, J. T., III. 2007a. Importance of point-by-point back projection (BP) correction for isocentric motion in digital breast tomosynthesis: Relevance to morphology of microcalcifications. *Med. Phys.* 34(10): 3885–3892.

Chen, Y., Lo, J. Y., Ranger, N. T., Samei, E., and Dobbins, J. T., III. 2007b. Methodology of NEQ(f) analysis for optimization and comparison of digital breast tomosynthesis acquisition techniques and reconstruction algorithms. *Proc. SPIE* 6510: 65101-I.

Chen, Y. 2007. Digital breast tomosynthesis (DBT)—A novel imaging technology to improve early breast cancer detection: Implementation, comparison and optimization. PhD dissertation, Duke University.

Dobbins, J. T., III. 1990. *Matrix Inversion Tomosynthesis Improvements in Longitudinal x-Ray Slice Imaging*. US Patent #4,903,204. Assignee: Duke University.

Dobbins, J. T., III. 2000. Image quality metrics for digital systems. In *Handbook of Medical Imaging*, Vol. 1. *Physics and Psychophysics*, eds. J. Beutel, H. L. Kundel, and R. L. Van Metter, 161–222. SPIE, Bellingham, WA.

Dobbins, J. T., III. and Godfrey, D. J. 2003. Digital x-ray tomosynthesis: Current state of the art and clinical potential. *Phys. Med. Biol.* 48: 65–106.

Dobbins, J. T., Samei, E., Ranger, N. T., and Chen, Y. 2006. Intercomparison of methods for image quality characterization. II. Noise power spectrum. *Med. Phys.* 33(5): 1466–1475.

Duryea, J., Dobbins, J. T., III, and Lynch, J. A. 2003. Digital tomosynthesis of hand joints for arthritis assessment. *Med. Phys.* 30: 325–333.

Godfrey, D. J., Warp, R. L., and Dobbins, J. T., III. 2001. Optimization of matrix inverse tomosynthesis. *Proc. SPIE* 4320: 696–704.

Godfrey, D. J. and Dobbins, J. T., III. 2002. *Optimization of Matrix Inversion Tomosynthesis via Impulse Response Simulations*. RSNA 88th Scientific Assembly, Chicago, IL.

Godfrey, D. J., Rader, A., and Dobbins, J. T., III. 2003. Practical strategies for the clinical implementation of matrix inversion tomosynthesis. *Proc. SPIE* 5030: 379–390.

Godfrey, D. J., McAdams, H. P., and Dobbins, J. T., III. 2006. Optimization of the matrix inversion tomosynthesis (MITS) impulse response and modulation transfer function characteristics for chest imaging. *Med. Phys.* 33(3): 655–667.

Grant, D. G. 1972. Tomosynthesis: A three-dimensional radiographic imaging technique. *IEEE Trans. Biomed. Eng.* BME-19: 20–28.

Lalush, D. S., Quan, E., Rajaram, R., Zhang, J., Lu, J., and Zhou, O. 2006. Tomosynthesis reconstruction from multi-beam x-ray sources. *Proceedings of 2006 3rd IEEE International Symposium on Biomedical Imaging: Macro to Nano*, Arlington, VA, 1180–1183.

Maidment, A. D., Ullberg, C., Lindman, K., Adelöw, L., Egerström, J., Eklund, M., Francke, T. et al. 2006. Evaluation of a photon-counting breast tomosynthesis imaging system. *Proc. SPIE* 6142: 89–99.

Mainprize, J. G., Bloomquist, A. K., Kempston, M. P., and Yaffe, M. J. 2006. Resolution at oblique incidence angles of a flat panel imager for breast tomosynthesis. *Med. Phys.* 33(9): 3159–3164.

Mertelemeier, T., Orman, J., Haerer, W., and Dudam, M. K. 2006. Optimizing filtered backprojection reconstruction for a breast tomosynthesis prototype device. *Proc. SPIE* 6142: 131–142.

Nishikawa, R. M. 2011. The fundamentals of MTF, Wiener spectra, and DQE. Retrieved September 28, 2011, from http://www.aapm.org/meetings/99AM/pdf/2798-87374.pdf.

Rakowski, J. T. and Dennis, M. J. 2006. A comparison of reconstruction algorithms for C-arm mammography tomosynthesis. *Med. Phys.* 33(8): 3018–3032.

Saunders, R. S. and Samei, E. 2003. A method for modifying the image quality parameters of digital radiographic images. *Med. Phys.* 30: 3006–3017.

Saunders, R. S., Samei, E., Jesneck, J., and Lo, J. Y. 2005. Physical characterization of a prototype selenium-based full field digital mammography detector. *Med. Phys.* 32(2): 588–599.

Samei, E., Ranger, N. T., Dobbins, J. T., and Chen, Y. 2006. Intercomparison of methods for image quality characterization. I. Modulation transfer function. *Med. Phys.* 33(5): 1454–1465.

Sklebitz, H. and Haendle, J. 1983. Tomoscopy: Dynamic layer imaging without mechanical movements. *AJR* 140: 1247–1252.

Sone, S., Kasuga, T., Sakai, F., Kawai, T., Oguchi, K., Hirano, H., Li, F. et al. 1995. Image processing in the digital tomosynthesis for pulmonary imaging. *Eur. Radiol.* 5: 96–101.

Suryanarayanan, S., Karellas, A., Vedantham, S., Glick, S. J., D'Orsi, C. J., Baker, S. P., and Webber, R. L. 2000. Comparison of tomosynthesis methods used with digital mammography. *Acad. Radiol.* 7(12): 1085–1097.

Warp, R. J., Godfrey, E. J., and Dobbins, J. T., III. 2000. Applications of matrix inverse tomosynthesis. *Proc. SPIE* 3977: 376–383.

Wu, Q. J., Meyer, J., Fuller, J., Godfrey, D., Wang, Z., Zhang, J., and Yin, F. F. 2011. Digital tomosynthesis for respiratory gated liver treatment: Clinical feasibility for daily image guidance. *Int. J. Radiat. Oncol. Biol. Phys.* 79(1): 289–296.

Yang, G., Rajaram, R., Cao, G., Sultana, S., Liu, Z., Lalush, D. S., Lu, J., and Zhou, O. 2008. Stationary digital breast tomosynthesis system with a multi-beam field emission x-ray source array. *Proc. SPIE* 6913: 69131A.

Zhang, Y., Chan, H., Sahiner, B., Wei, J., Goodsitt, M. M., Hadjiiski, L. M., Ge, J., and Zhou, C. 2006. A comparative study of limited-angle cone-beam reconstruction methods for breast tomosynthesis. *Med. Phys.* 33(10): 3781–3795.

Zhao, B. and Zhao, W. 2008. Three-dimensional linear system analysis for breast tomosynthesis. *Med. Phys.* 35(12): 5219–5232.

Zhao, W., Zhao, B., and Hu, Y. H. 2011. Latest advances in digital breast tomosynthesis. Retrieved September 24, 2011, from http://www.aapm.org/meetings/amos2/pdf/42-11930-81688-724.pdf.

Zhou, W., Qian, X., Lu, J., Zhou, O., and Chen, Y. 2010. Multi-beam x-ray source breast tomosynthesis reconstruction with different algorithms. *Proc. SPIE* 7622H: 1–8.

System performance

10

Spatial-domain model observers for optimizing tomosynthesis

Subok Park

Contents

10.1 INTRODUCTION

In assessing medical-image quality, task-based approaches are desirable because both the meaning of the term, "medical-image quality," and the assessment method itself change depending on the task at hand. In addition, the imaging process consists of deterministic and stochastic components; hence, it is critical to understand the impact of both types of components on image quality. To take the stochastic components of the imaging process into account, statistical methods for considering various sources of randomness in the imaging chain are necessary. Therefore, statistical, task-based evaluation methods are advocated for the rigorous assessment of medical-image quality [1]. The task-based evaluation approach has the following requirements: (1) a task of interest, (2) an observer performing the task, and (3) a figure of merit for measuring observer performance. To rigorously perform image-quality assessment, each of the three requirements needs to be carefully chosen given the imaging system and its data sets. In this chapter, we discuss how to perform the statistical, task-based assessment and optimization of tomosynthesis systems with our focus on the task of binary signal detection and spatial-domain model observers. We will first give mathematical descriptions of the Bayesian ideal and Hotelling observers and their figures of merit in the two-dimensional (2D) case, then describe how the equations are modified to accommodate the three-dimensional

(3D) information in the model observer for the assessment of tomosynthesis systems. Since the high dimensionality of data is the bottleneck to making use of spatial-domain model observers in image-quality assessment, we will also describe the channelization of data and its utility in reducing large data dimensionality while preserving salient information for the model observer to adequately perform a given task of interest. We will also describe efficient and anthropomorphic channels for evaluating the system using projection and reconstruction data, respectively. Lastly, we will give a brief description of a few example studies that make use of spatial-domain model observers for evaluating digital breast tomosynthesis (DBT) systems.

10.2 OVERVIEW OF MODEL OBSERVERS

10.2.1 IMAGE FORMATION AND BINARY DETECTION

The planar image acquisition process can be mathematically written as

$$\mathbf{g} = \mathcal{H}\mathbf{f} + \mathbf{n}, \tag{10.1}$$

where $\mathbf{f}$ represents the object to be imaged, $\mathcal{H}$ represents a continuous-to-discrete projection imaging operator that maps

$\mathbf{f}(\equiv f(\mathbf{r}))$, a function of continuous variables in a Hilbert space of square-integrable functions, to an $M \times 1$ data vector $\mathbf{g}$, $\mathbf{r}$ is a 3D spatial coordinate, and the vector $\mathbf{n}$ is associated with noise in the imaging process and measurement. Note that $+$ does not necessarily mean the addition operation, for example, the noise can depend on the object $\mathbf{f}$, and that $\mathcal{H}$ accounts for the deterministic aspect of the imaging process.

Here, we focus on binary signal detection as our task of interest, which requires the following two hypotheses: signal-absent, H_0, and signal-present, H_1:

$$H_0 : \mathbf{g} = \mathcal{H}\mathbf{f}_b + \mathbf{n}, \tag{10.2}$$

$$H_1 : \mathbf{g} = \mathcal{H}(\mathbf{f}_b + \mathbf{f}_s) + \mathbf{n}, \tag{10.3}$$

where $\mathbf{f}_b$ represents either a known or random-background object and $\mathbf{f}_s$ represents either a known or random signal object. In the discussion on model observers for image-quality assessment, we focus on the signal-known-exactly (SKE) task paradigm in random backgrounds. For notational convenience, $\mathbf{b}$ and $\mathbf{s}$ henceforth denote the noiseless background image $\mathcal{H}\mathbf{f}_b$ and the noiseless signal image $\mathcal{H}\mathbf{f}_s$, respectively.

10.2.2 BAYESIAN IDEAL OBSERVER

For a binary detection task, the Bayesian ideal observer is optimal among all observers, either human or model [2]. In other words, the ideal observer uses all available statistical information from the data; it maximizes sensitivity for any specificity (i.e., it maximizes diagnostic accuracy); and it minimizes the average cost of making decisions. The ideal observer computes the likelihood ratio of data $\mathbf{g}$ as its test statistic, given by [2]

$$\Lambda(\mathbf{g}) = \frac{pr(\mathbf{g} \mid H_1)}{pr(\mathbf{g} \mid H_0)}, \tag{10.4}$$

where $pr(\cdot \mid H_j)$ are the probability density functions (PDFs) of image data $(\cdot)$ under the hypothesis H_j, $j = 0, 1$. To illustrate, if $\mathbf{g}$ is drawn from the hypothesis H_0, then $pr(\mathbf{g} \mid H_1)$ is more likely to be smaller than $pr(\mathbf{g} \mid H_0)$, and hence, $\Lambda(\mathbf{g})$ is more likely to be smaller than 1. Whereas, if $\mathbf{g}$ is drawn from the hypothesis H_1, then $pr(\mathbf{g} \mid H_1)$ is more likely to be bigger than $pr(\mathbf{g} \mid H_0)$, and hence $\Lambda(\mathbf{g})$ is more likely to be bigger than 1. Using data samples of $\mathbf{g}$ drawn from the two hypotheses, two sample distributions of $\Lambda(\mathbf{g})$ can be generated, which show the aforementioned trend with a sufficiently large sample.

The separability of these two distributions depends on the difficulty of the given task and it can be measured as follows. Each likelihood ratio is compared to a threshold to make a decision between signal-present and signal-absent hypotheses. By varying the threshold and plotting true-positive fraction (TPF) versus false-positive fraction (FPF), a receiver operating characteristic (ROC) curve can be generated for the task. Among figures of merit derived from ROC analysis, the area under the curve (AUC) is a common scalar figure of merit that is maximized by the ideal observer. The methods for estimating AUC via ROC analysis include parametric methods, such as LABROC and PROPROC by Metz et al. [3–5], and

nonparametric methods, such as the Wilcoxon–Mann–Whitney method [6,7]. The parametric methods transform the given test statistic estimates to fit binormal models and generate a parametric ROC curve, the area under which is used as an estimate of AUC for the given data. When the data are degenerate, or cause a hook on the ROC curve, PROPROC may be used to produce a properly shaped (i.e., convex) ROC curve. To generalize the AUC result to a population, multiple-reader multiple-case (MRMC) variance analysis is often used [8–13].

By definition, the ideal observer gives an absolute upper bound for the performance of any observer. It measures the amount of detectable information produced by an imaging system (or contained in images from the system) and provides the gold standard for improving the system or computer-aided methods for processing images to aid human perception. To compare with this gold standard, *efficiency* is defined as the ratio of *detectabilities* (which will be discussed in Section 10.2.3) of a given observer over the gold standard observer. The knowledge of the efficiency of either human or model observers relative to the ideal observer is useful because (1) it indicates how much improvement can be made to computer-aided methods for aiding humans in performing difficult tasks, (2) it provides relevant information for developing anthropomorphic model observers that can track human performance, and (3) it indicates any necessity for improving *suboptimal* model observers by accounting for more complicated data statistics (compared to simpler Gaussian statistics) to improve observer performance on difficult tasks.

When the data are Gaussian, the ideal observer is equivalent to the ideal linear observer, which makes use of the mean and covariance of data for making decisions. This can be readily seen by inserting the PDFs of Gaussian distributions under the two hypotheses into Equation 10.4. The ideal linear observer is also called the *Hotelling observer*, which will be discussed in depth in Section 10.2.3. When the data are not Gaussian, however, it is often infeasible to calculate the performance of the ideal observer due to the lack of data PDFs as well as high data dimensionality. In the past decades, a number of computational approaches using the Markov-chain Monte Carlo (MCMC) have been developed to help improve the calculation of ideal-observer performance [14–17]. Kupinski et al. [14] developed an MCMC algorithm for performing a detection task involving non-Gaussian backgrounds. He et al. [15] and Abbey et al. [16] adopted the Kupinski MCMC and modified it for investigating detection tasks involving data from single photon emission computed tomography (SPECT) and breast computer tomography (CT) systems, respectively. But these MCMC approaches have common limitations in that they use a parametric patient-object model and have too long a computational time to explore a large space of system parameters such as those present in tomosynthesis systems. Furthermore, in clinical applications, it is often infeasible to find a parametric model for representing complex, realistic background images, and even if a representative parametric model exists, using the aforementioned MCMC approaches is still time consuming. To improve upon the limitations of the Kupinski MCMC, Park et al. investigated methods of efficient feature selection for approximating ideal-observer performance [18,19] and MCMC approaches for rapid estimation of ideal-observer performance [17,20]. One of the MCMC approaches [17] is an extension of the Kupinski MCMC, which uses the same proposal density as that of the Kupinski MCMC but a different formula for estimating acceptance probability in the channelized data space. With this extended MCMC, a Markov chain of channelized data vectors can be generated to estimate the performance of the ideal observer constrained to the given channels, a *channelized-ideal observer* (CIO). This MCMC enables the comparison between the ideal observer and the CIO in detection tasks involving non-Gaussian lumpy backgrounds. We discuss the details on the other MCMC [20] and efficient feature-selection approaches [18,19] in Sections 10.2.4 and 10.2.5, respectively.

10.2.3 IDEAL LINEAR (OR HOTELLING) OBSERVER

As discussed in Section 10.2.2, when the data are non-Gaussian, it is often too difficult or impossible to estimate the performance of the fully informed, nonlinear Bayesian ideal observer. In that case, the ideal linear (Hotelling) observer is often a good alternative for use in a task-based assessment of image quality. The Hotelling observer is optimal among all linear observers and maximizes the signal-to-noise ratio (SNR) [1]. It linearly applies a *Hotelling template*, $\mathbf{w_g}$, to signal-present and signal-absent data, $\mathbf{g}|H_j$, to generate the corresponding signal-present and signal-absent test statistics:

$$t_j(\mathbf{g}) = \mathbf{w}_\mathbf{g}^t [\mathbf{g} \mid H_j]. \tag{10.5}$$

The Hotelling template $\mathbf{w_g}$ is calculated using the ensemble mean and covariance of the data via

$$\mathbf{w_g} = \mathbf{K_g}^{-1} \Delta \mathbf{s}, \tag{10.6}$$

where $\mathbf{w_g}$ is an $M \times 1$ vector and $\mathbf{K_g}$ is an $M \times M$ matrix that is the average of the covariance matrices of the signal-absent and signal-present image data, $\mathbf{K}_{\mathbf{g}|H_0}$ and $\mathbf{K}_{\mathbf{g}|H_1}$. The vector $\Delta \mathbf{s}$ is another $M \times 1$ vector that is the mean difference between the signal-present and signal-absent image data. More specifically, we may write the mean difference and the covariance as

$$\Delta \mathbf{s} = \bar{\mathbf{g}}_1 - \bar{\mathbf{g}}_0, \quad \bar{\mathbf{g}}_j = \langle \mathbf{g} \rangle_{\mathbf{g}|H_j}, \quad j = 0,1, \tag{10.7}$$

$$\mathbf{K_g} = \frac{1}{2}[\mathbf{K}_{\mathbf{g}|H_0} + \mathbf{K}_{\mathbf{g}|H_1}], \tag{10.8}$$

$$\mathbf{K}_{\mathbf{g}|H_j} = \langle (\mathbf{g} - \bar{\mathbf{g}}_j)(\mathbf{g} - \bar{\mathbf{g}}_j)^t \rangle_{\mathbf{g}|H_j}, \tag{10.9}$$

where $\mathbf{g}|H_j$ is the data drawn from the hypothesis H_j, $\langle (\cdots) \rangle_{(\cdot)}$ denotes the average of $(\cdots)$ over a vector of random variables $(\cdot)$, and t denotes the transpose operator.

For the case of an additive SKE signal where noise is independent of the signal, $\mathbf{K_g} = \mathbf{K}_{\mathbf{g}|H_0}$ and for the case of a low-contrast SKE signal where noise is not independent of the signal, $\mathbf{K_g}$ is approximately $\mathbf{K}_{\mathbf{g}|H_0}$. For the case of SKE binary detection, with noisy data from real imaging systems, such as repeated scan images of a phantom including a lesion to detect, an SKE task is not truly SKE (e.g., only the middle of the signal is known). In this case, the mean signal is estimated using the given noisy data

and also suffers from noise due to finite sample size effects. On the contrary, the true signal, $\mathbf{s}$, is known exactly in simulation and hence, there is no uncertainty in the signal. Thus, the true signal, $\mathbf{s}$, can be used as the mean difference signal, $\Delta \mathbf{s}$, without having to be calculated from a finite number of noisy signal-present image realizations, which will likely result in a blurry version of the signal $\mathbf{s}$. Generally, we want to use a model observer that is as optimal as possible to give an upper bound on system performance for hardware optimization; so, we recommend the use of the true signal $\mathbf{s}$ for $\Delta \mathbf{s}$ whenever it is available and reasonable.

For measuring observer performance given a finite sample of patient or phantom cases, the SNR can be computed using the test statistics, t_j, $j = 0,1$, given in Equation 10.5 [1]:

$$\mathrm{SNR}_t = \frac{\langle t_1 \rangle - \langle t_0 \rangle}{\sqrt{(1/2)(\sigma_0^2 + \sigma_1^2)}}, \tag{10.10}$$

where σ_j^2, $j = 0,1$, indicates the variance of t_j. By plugging Equation 10.5 into Equation 10.10, SNR_t becomes

$$\mathrm{SNR}_t = \frac{\mathbf{w}_\mathbf{g}^t \Delta \mathbf{s}}{\sqrt{\mathbf{w}_\mathbf{g}^t \mathbf{K_g} \mathbf{w_g}}}. \tag{10.11}$$

Using the template $\mathbf{w_g}$ given in Equation 10.6 into Equation 10.11, it yields another expression of SNR_t:

$$\mathrm{SNR}_t = \sqrt{\Delta \mathbf{s}_1^t \mathbf{K_g}^{-1} \Delta \mathbf{s}_2}, \tag{10.12}$$

where $\Delta \mathbf{s}_1$ and $\Delta \mathbf{s}_2$ are estimated from two different (independently sampled) data sets: one for training the observer (i.e., estimating the observer template $\mathbf{w_g}$) and the other for testing the observer (i.e., estimating the test statistics, $\{t_j\}$), respectively. Another method of computing the SNR can then be written as

$$\mathrm{SNR}_{\mathrm{resub}} = \sqrt{\Delta \mathbf{s}^t \mathbf{K_g}^{-1} \Delta \mathbf{s}}, \tag{10.13}$$

by approximating the formula of SNR_t given in Equation 10.12 with the use of the same data set for estimating both the $\Delta \mathbf{s}_1$ and $\Delta \mathbf{s}_2$. Here, the subscript "resub" indicates the resubstitution of data sets.

$\mathrm{SNR}_{\mathrm{resub}}$ may yield biased performance estimates when it is not justified to use the same data for the estimation of $\Delta \mathbf{s}_1$ and $\Delta \mathbf{s}_2$. But owing to its simplicity, $\mathrm{SNR}_{\mathrm{resub}}$ is often used instead of SNR_t without careful consideration of bias in SNR estimates. Therefore, the importance of the appropriate use of this formula cannot be overemphasized. To accurately estimate the SNR, the first $\Delta \mathbf{s}$ should be calculated using a data set, which is distinct from the one used for calculating the second $\Delta \mathbf{s}$ to avoid biasing SNR trends. But in practice, when the sampling of patient cases for estimating SNR represents the patient population well, or when a sufficient sample size is used, the $\mathrm{SNR}_{\mathrm{resub}}$ formula works well without having to use two different independent data sets for calculating the left and right $\Delta \mathbf{s}$'s. Thus, it is important to obtain as much prior knowledge on the system's noise and data statistics as possible to decide which metric is appropriate to use for image-quality assessment. In realistic situations where the sampling of patients representative of the patient population is difficult and/or there are only limited data sets, the users of this approach should carefully make sure that SNR estimates using $\text{SNR}_{\text{resub}}$ in Equation 10.13 are either unbiased or updated with bias correction. Since channelized Hotelling observers (CHOs), which we will discuss in detail in the next section, are more often used than the Hotelling observer, the bias and variance of SNR estimates of a CHO is perhaps a more practical concern. One method for estimating the variance and bias of CHO SNR estimates is proposed by Wunderlich and Noo [21–24].

As an alternative to the aforementioned SNR formulas, the AUC (area under the ROC curve) can be estimated using the signal-absent and signal-present test statistics, $t_j, j = 0,1$, as discussed in Section 10.2.2. Then the SNR can be computed via [1]

$$\text{SNR}_{\text{AUC}} \equiv 2\,\text{erf}^{-1}(2\,\text{AUC} - 1), \qquad (10.14)$$

where erf^{-1} is the inverse of the error function. This quantity, SNR_{AUC}, is also called the *detectability index, d_A*. The SNR_t and SNR_{AUC} are equivalent when each of the t statistics, t_1 and t_0, follows a normal distribution. With increasing signal intensity and size, low noise, and so on, AUC approaches one, which yields infinite SNR_{AUC}. In that case, either SNR_t or $\text{SNR}_{\text{resub}}$ may be used judiciously to avoid infinite values.

One of the advantages of the Hotelling observer or CHO is that the overall data covariance $\mathbf{K_g}$ can be decomposed into a number of components representing various sources of variability in the data $\mathbf{g}$. For example, in digital radiography, $\mathbf{K_g}$ can be decomposed into the following expression without assuming that different noise sources are independent of each other [1]:

$$\mathbf{K_g} = \mathbf{K_g}^{(\text{elec})} + \mathbf{K_g}^{(\text{x-ray})} + \mathbf{K_g}^{(\text{object})}, \qquad (10.15)$$

where the terms represent, respectively, the electronic noise, the noise associated with x-ray interactions with the detector material, and the effect of object variability. The second term in Equation 10.15 can be further decomposed into

$$\mathbf{K_g}^{(\text{x-ray})} = \mathbf{K_g}^{(x)} + \mathbf{K_g}^{(\text{gain})} + \mathbf{K_g}^{(Kx)}, \qquad (10.16)$$

where the terms represent the Poisson statistics of the x-rays as reflected through the gain mechanism, the excess noise of the gain mechanism, and the effect of reabsorbed Compton-scattered and K x-rays, respectively. The decompositions given above are the result of averaging conditional integrals over appropriate sources of variability including the object itself. Readers are referred to the work by Barrett and his collaborators [1,25] for further discussion of these decompositions. Decomposing the covariance in this way facilitates a more practical and efficient use of the Hotelling observer or CHO in objective image-quality assessment studies. That is, by including the terms for the sources of variability of interest and neglecting the other terms, the impact of the sources of variability can be studied individually or collectively in a systematic fashion. Then, the sources of variability that most affect image quality can be isolated and further used for investigating methods to improve image quality.

For incorporating a deterministic system response through an indirect conversion process, one can employ a linear operator $\mathcal{B}$ that consists of system-response functions at the detector face. Then the covariance of object variability incorporating the detector's system response can be written as $\mathcal{B}\mathbf{K_g}^{(\text{obj})}\mathcal{B}^t$. For the evaluation of tomosynthesis systems, it is useful to consider angle-dependent system-response functions to form a more relevant $\mathcal{B}$ because the image quality of tomosynthesis systems is affected by the choice of projection angles. Angle-dependent system-response functions can be simulated using Monte Carlo, measured in the laboratory, or modeled analytically using a combination of Monte Carlo and laboratory measurements [26–29].

10.2.4 CHANNELIZED MODEL OBSERVERS

Medical imaging applications that use digital detectors, such as tomosynthesis and CT, produce high-dimensional data. This high dimensionality of data is one of the major bottlenecks in the use of both nonlinear and linear ideal observers in image-quality assessment. To estimate the nonlinear Bayesian ideal observer (i.e., when data complexity is higher than that of a Gaussian distribution), the likelihood ratio is calculated by evaluating one integral for each dimension of the data, which is often infeasible even with known PDFs. Compounding the difficulty is the tendency that the PDFs for data arising from realistic applications are unknown. Accurately estimating data covariance for the ideal linear (Hotelling) observer is difficult when the dimensionality of the data $\mathbf{g}$ is large due to the inevitable small sample size relative to data dimensionality. In addition, inversion of the covariance matrix is difficult due to high data dimensionality. Furthermore, for many realistic applications, data covariance can be nearly singular, making the estimation and inversion of covariance even more difficult. These problems lead to unstable, unreliable performance (either SNR or AUC) estimates. Therefore, in most cases, where data complexity is higher than that of an uncorrelated Gaussian distribution, it is essential to reduce data dimensionality while preserving salient information for either of the model observers to perform the given task as optimally as possible.

To reduce the dimension of the data, a linear transformation $\mathbf{T}$, which consists of N_c rows of channels for the given data $\mathbf{g}$, can be applied to $\mathbf{g}$:

$$\mathbf{v} = \mathbf{Tg}, \qquad (10.17)$$

where $\mathbf{T}$ is an $N_c \times M$ matrix, $\mathbf{g}$ is an $M \times 1$ data vector, and $\mathbf{v}$ is the resulting $N_c \times 1$ channelized vector. Channels that make up the matrix $\mathbf{T}$ need to be appropriately chosen so that the channelized model observer approximates the performance (or performance trend) of either the ideal observer or the human observer, in which cases, channels are referred to as *efficient* and *anthropomorphic*, respectively. For system design and optimization, we advocate the use of efficient channels for approximating the Hotelling (or nonlinear Bayesian ideal) observer whenever possible because ideal-observer performance gives upper bounds for system performance as discussed in Section 10.2.2. However, efficient channels, or surrogate channels for giving a similar trend, have not yet been fully identified for data arising from tomosynthesis applications. This is an active area of investigation. In the following section, we describe the existing 2D channels for the task-based evaluation of planar imaging systems. In Section 10.3.3, we discuss more details on ways for utilizing the existing channels and on directions for finding more efficient channels for the task-based evaluation and optimization of tomosynthesis systems.

A linear observer that uses the mean and the covariance of **v** given in Equation 10.17 is called the CHO. Then, all the formulas given in Section 10.2.3 can be used by replacing the data **g** with this channelized data **v**. In particular, the template and test statistic of the CHO are given by

$$\mathbf{w_v} = \mathbf{K_v}^{-1}\Delta\mathbf{v}_s, \tag{10.18}$$

$$t_\mathbf{v} = \mathbf{w}_\mathbf{v}^t\mathbf{v}, \tag{10.19}$$

where $\Delta\mathbf{v}_s$ is the mean difference between the signal-present and signal-absent channelized data and the matrix $\mathbf{K_v}$ is the covariance of the channelized data, which is in fact $\mathbf{TK_gT}^t$. A nonlinear model observer that uses the full statistical information of **v**, that is, PDFs of **v**, is called the CIO. The test statistic of the CIO is given by

$$\Lambda(\mathbf{v}) = \frac{pr(\mathbf{v} \mid H_1)}{pr(\mathbf{v} \mid H_0)}, \tag{10.20}$$

where $pr(\mathbf{v}|H_j)$ are the PDFs of channelized data **v** under the hypotheses H_j, $j = 0,1$. If the channels in the matrix **T** are efficient, the performance of the CIO (or CHO) approximates that of the unconstrained nonlinear (or linear) ideal observer using Equation 10.4 (or Equation 10.5) as its test statistic. It is more difficult to find efficient channels for the CIO to approximate the nonlinear observer since the estimation of the nonlinear Bayesian ideal observer itself is very difficult or impossible. To estimate the CIO as a surrogate observer for rapidly calculating ideal-observer performance, Park and Clarkson [20] developed a computational approach using another MCMC, which they called a *channelized-ideal observer Markov-chain Monte Carlo* (CIO-MCMC). By sampling from a much-lower dimensional space of channelized data, the CIO-MCMC quickly forms a Markov chain and estimates the CIO likelihood ratio efficiently provided that there exist efficient channels for the data used for the given detection task. The CIO-MCMC method is general in that it can be used for estimating ideal-observer performance in binary classification tasks involving many types of backgrounds and signals [20]. This MCMC method, however, requires knowledge of the noiseless data (**b** and **s**) so that it is more suitable for computer-simulation studies for system optimization, where the noiseless data are available. In the following section, we discuss the selection of efficient as well as anthropomorphic channels in the 2D domain.

10.2.5 CHANNEL CHOICES FOR 2D DATA

In the case of 2D image data, various choices of efficient and anthropomorphic channels have been studied and validated with psychophysical data for a number of different detection tasks involving many background and signal types [18,19,30–36]. With regard to efficient channels, 2D channels that have been investigated for use in task-based image-quality assessment include Laguerre–Gauss (LG) functions [19,31,32], singular vectors of a linear imaging system using a singular value decomposition (SVD) [18], and a partial least-square (PLS) algorithm [19].

To be efficient for a CHO (i.e., for a CHO to approximate the Hotelling observer), LG channels require the stationarity of background data and the rotational symmetry of the signal. In other words, LG channels are efficient for the CHO [19,31,32] and are close to efficient for the CIO [19] when a detection task involves stationary random backgrounds and rotationally symmetric signals. But SVD or PLS channels do not have such requirements. Singular vectors for a given linear imaging system can be calculated via the SVD of the imaging system matrix that consists of a set of system-response functions. Singular vectors strongly associated with the signal-only image were found to be the most efficient for both CHO and CIO for a detection task with randomly varying lumpy backgrounds and symmetric and asymmetric signals [18]. In addition, singular vectors associated with the background-only image were also highly efficient but only with a larger number of singular vectors. The limitation of the SVD approach is that it requires the system to be linear and the system response to be known. In addition, by the way the singular vectors are generated, the singular vectors themselves do not incorporate the difference between the signal-present and signal-absent data generated from the imaging system so that a large number of SVD channels, in comparison with LG and PLS channels, are required for a channelized model observer (either CHO or CIO) to approximate the original unconstrained observer (either Hotelling or ideal observer, respectively) [18].

The PLS method has been used in regression models where new latent variables are formed as a weighted linear combination of the data to reduce data dimensionality. For channelized model observers, the PLS weights needed to form the latent variables (i.e., a channelized vector **v**) are used in the channel matrix **T** [19]. The PLS channels are estimated by maximizing the covariance between the data and the truth (signal-present and signal-absent) so that PLS channels effectively incorporate the difference between signal-present and signal-absent data and require fewer channels for approximating ideal-observer performance. PLS channels are highly efficient (superior to LG and SVD channels) for both CIO and CHO for detection tasks involving the same types of backgrounds and signals mentioned above. PLS channels are flexible in that they do not require the linearity of the imaging system or the system response to be known, which the SVD channels require. In addition, PLS channels are not limited to the types of background and signal statistics, for which LG channels respectively require the stationarity and the rotational symmetry. Therefore, with sufficient training on sample size, PLS channels can be generated for many types of background and signal. Unfortunately,

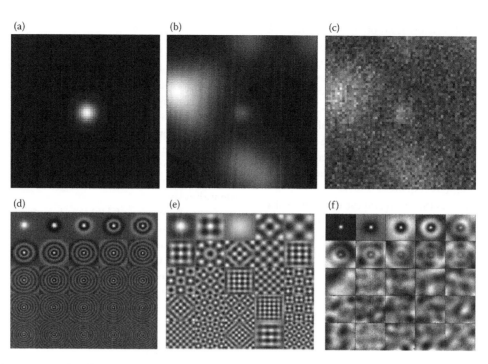

Figure 10.1 (a–c) A Gaussian signal at the center of the FOV, a noise-free lumpy background including the signal, and a resulting noisy image using the lumpy background and signal. (d–f) The first 25 LG, SVD, and PLS channels for efficiently detecting the given signal in randomly varying noisy lumpy backgrounds. Note that each of the 25 channels of each type has the same dimensions as those of the signal and background images given on top. (All images are reproduced from J. Witten, S. Park, and K. J. Myers, Partial least squares: A method to compute efficient channels for the Bayesian ideal observers, *IEEE Trans. Med. Imag.* 29(4), 1050–1058, 2010. With permission from the IEEE Transactions on Medical Imaging.)

using a typically small sample from realistic imaging systems leads to noisy PLS channels, which cannot be directly used for calculating the performance of the ideal observer. This is because using noisy PLS channels degrades the performance of a CHO compared to the performance of another CHO constrained to high-resolution PLS channels generated with a large sample or other types of efficient channels if available. But even with a small sample, the PLS method can still be useful for developers of model-observer approaches to learn the salient features of efficient channels and model high-resolution efficient channels. Figure 10.1 shows an example of LG, SVD, and PLS channels for detecting a Gaussian signal in randomly varying, noisy lumpy-background images.

For a model observer to track the trend of human performance, anthropomorphic channels can be employed. Anthropomorphic channels, which were used in a CHO performing detection tasks involving various images and tasks, include spatial-frequency selective [30], difference-of-Gaussians (DOGs) [33], Gabor [34,35], and square channels [36]. A main feature of these channels is that they have no response to zero frequency (i.e., the DC term). The features of these channels are supported by the following findings in vision science: (1) spatial-frequency channels are thought to exist in the human visual system as evidenced by the Nobel prize-winning work by Hubel and Wiesel [37], and subsequent work by others [38,39] and (2) humans are known to have poor sensitivity to low-frequency information [40]. Using these types of channels in a CHO does not necessarily guarantee that the CHO model closely predicts human performance. For an anthropomorphic model observer to predict human performance, additional factors that affect

human visual perception, such as internal noise and spatial and/or temporal contrast sensitivity, need to be incorporated in the design of observer models. The discussion of these factors is beyond the scope of this chapter; hence, readers who are interested in learning more on anthropomorphic model observers are referred to the aforementioned articles and other references [41–45].

10.3 APPLICATION TO TOMOSYNTHESIS

In this section, we discuss ways to modify the aforementioned Hotelling observer and CHO approaches to incorporate 3D spatial information in the evaluation of tomosynthesis systems. Since the dimension of both projection and reconstruction data from tomosynthesis imaging systems can be prohibitively large for rapidly and accurately estimating model-observer performance, it is often impractical to make use of the full-dimensional data in task-based system performance calculations. It is hence desirable to reduce data dimensionality by linearly transforming the data and calculating model-observer performance using the resulting lower-dimensional vector. Therefore, in this section, we also discuss the use of channels for incorporating 3D information in the modified model observer.

More specifically, in Section 10.3.1, we discuss how to modify the data format given in Equation 10.1 to accommodate projection and reconstruction data from tomosynthesis systems. In Section 10.3.2, we describe how to utilize the Hotelling observer for incorporating 3D spatial information using the modified data format. Then, in Section 10.3.3, we describe how to adapt the CHO, including some ideas on how to choose efficient channels as well as anthropomorphic channels for different CHO models. Lastly, in Section 10.3.4, we present a summary of a few studies that used spatial-domain model observers for the evaluation of DBT systems.

10.3.1 IMAGE FORMATION AND BINARY SIGNAL DETECTION

For assessing a 3D system such as tomosynthesis, we consider observer-performance analysis in both projection and reconstruction data spaces. Optimizing an imaging system using its "raw data" (i.e., projection data) first is often desirable when (1) reconstructing the data is regarded as postprocessing without needing to be optimized or (2) separating the effect of the reconstruction algorithm from the image-quality assessment simplifies the optimization problem, allowing for a rapid computation of *candidate* sets of optimal parameters. Then, to incorporate the impact of reconstruction effects on system performance and hence optimize the system accordingly, human observer studies are needed unless there are already anthropomorphic model observers validated for the given task and type of data. In any case, due to a large parameter space over which to optimize the system, it is more practical if human studies are performed for a small set of system and reconstruction parameters selected by appropriate model-observer studies *in silico*. Therefore, it may be of interest to some readers to assess the image quality of tomosynthesis systems in the following order: (1) perform the initial system optimization using projection data before incorporating the impact of reconstruction into choosing *candidate* sets of optimal parameters, (2) run another optimization study using reconstruction data to see the impact of different reconstruction algorithms or parameter values of a particular reconstruction algorithm on image quality, and (3) run human observer studies on reconstruction data produced from a tomosynthesis system with the candidate optimal parameters estimated from (1) and (2).

Equation 10.1 may be modified to accommodate projection data by concatenating different angular projections to form a single data vector $\mathbf{g}$ for use in observer-performance analysis:

$$\mathbf{g} = \{\mathbf{g}_k\}_{k=1}^{N_p}, \tag{10.21}$$

$$\mathbf{g}_k = \{\mathbf{g}_{k,i}\}_{i=1}^{M}, \tag{10.22}$$

where N_p is the number of angular projections and M is the number of pixels in each projection $\mathbf{g}_k$. The dimension of the resulting data vector $\mathbf{g}$ is $MN_p \times 1$. In this case, the imaging operator $\mathcal{H}$ also needs to be modified and it is now a collection of different angular projection imaging operators, that is

$$\mathcal{H} = \{\mathcal{H}_{\theta_k}\}_{k=1}^{N_p}, \tag{10.23}$$

where each $\mathcal{H}_{\theta_k}$ indicates a planar imaging operator with a projection angle θ_k, $k = 1,\ldots,N_p$. Then each angular projection $\mathbf{g}_k$ can be written as

$$\mathbf{g}_k = \mathcal{H}_{\theta_k}\mathbf{f} + \mathbf{n}_k, \tag{10.24}$$

where $\mathbf{n}_k$ represents the noise associated with the angular projection. Since $\mathbf{f}$ is either $\mathbf{f}_b + \mathbf{f}_s$ or $\mathbf{f}_b$, respectively, for the signal-present and signal-absent hypotheses, it follows that $\mathbf{g}_k = \mathbf{b}_k + j\mathbf{s}_k + \mathbf{n}_k$ for each hypothesis H_j. Similarly, we have

$$\mathbf{b} = \{\mathbf{b}_k\}_{k=1}^{N_p}, \quad \mathbf{b}_k = \{b_{k,i}\}_{i=1}^{M}, \tag{10.25}$$

$$\mathbf{s} = \{\mathbf{s}_k\}_{k=1}^{N_p}, \quad \mathbf{s}_k = \{s_{k,i}\}_{i=1}^{M}, \tag{10.26}$$

$$\mathbf{n} = \{\mathbf{n}_k\}_{k=1}^{N_p}, \quad \mathbf{n}_k = \{n_{k,i}\}_{i=1}^{M}. \tag{10.27}$$

With the aforementioned data descriptions, the modified signal-present and signal-absent hypotheses for the projection data are given by

$$H_1 : \mathbf{g} = \mathbf{b} + \mathbf{s} + \mathbf{n}, \tag{10.28}$$

$$H_0 : \mathbf{g} = \mathbf{b} + \mathbf{n}, \tag{10.29}$$

where the vectors $\mathbf{b}$ and $\mathbf{s}$ represent the noiseless projected background and signal images, respectively, $\mathbf{n}$ represents noise in tomosynthesis imaging, and $\mathbf{g}$ represents the resulting data vector.

Now, to assess reconstruction algorithms for considering human perception of displayed data, reconstruction data (i.e., estimate of the object $\mathbf{f}$) can be written as

$$\hat{\mathbf{f}} = \mathcal{O}\mathbf{g}, \tag{10.30}$$

where $\mathcal{O}$ is a reconstruction operator that maps $\mathbf{g}$ to a $P \times N_s$ matrix $\hat{\mathbf{f}}$, where $P(\geq M)$ is the number of pixels in each slice of the reconstruction data and N_s is the number of reconstruction slices. In performing model and human observer studies for investigating reconstruction effects on system evaluation and optimization, the reconstruction data $\hat{\mathbf{f}}$ can be used instead of the projection data $\mathbf{g}$. For the reconstruction data, the hypotheses can be modified to

$$H_1 : \hat{\mathbf{f}} = \mathcal{O}(\mathbf{b} + \mathbf{s} + \mathbf{n}), \tag{10.31}$$

$$H_0 : \hat{\mathbf{f}} = \mathcal{O}(\mathbf{b} + \mathbf{n}). \tag{10.32}$$

We note that contrary to the conventional use of the notation +, this notation here does not necessarily imply that the signal $\mathbf{s}$ and noise $\mathbf{n}$ are additive. However, for simplifying the problem, the noise $\mathbf{n}$ is often treated as additive and the signal $\mathbf{s}$ is often treated as additive and independent of the background $\mathbf{b}$. For analyzing data from real systems, these assumptions need to be made with careful consideration of the principles of imaging physics as well as signal and background data statistics. It is important to understand under which conditions these assumptions are valid.

10.3.2 3D PROJECTION OR 3D HOTELLING OBSERVER

The formulas for the test statistic, observer template, and figures of merit for the Bayesian ideal and Hotelling observers, which are given in Sections 10.2.2 and 10.2.3, can be used for analyzing both projection and reconstruction data types given in Equations 10.21 and 10.30. But to differentiate Hotelling observer models using these two different data types, we call the Hotelling observer using the concatenated vector given in Equation 10.21 *the 3D projection* (3Dp) *Hotelling observer* and the Hotelling observer using the reconstruction data given in Equation 10.30 *the 3D Hotelling observer*. When appropriately used, the 3Dp (or 3D) Hotelling observer approach allows for the incorporation of spatial correlation information between pixels in different angular projections as well as pixels within each projection (or between voxels in different slices as well as voxels within each slice). However, as discussed earlier, when data dimensionality is large, estimating data covariance can be unreliable or even computationally infeasible. Therefore, the use of these observer models is only desirable in cases (1) where background data statistics are known or simple enough (e.g., Gaussian) that a large sample is not required to account for the background statistics and (2) where the use of a small-enough ROI taken from any location in the full field of view (FOV) of each projection (or each slice) is justified for increasing the number of sample images (e.g., stationary background statistics).

When background statistics are more complex than those discussed above, we should consider the use of either a large or local ROI. For example, when it is important to include long-range correlation information in detecting the given signal, it is most appropriate to use a sufficiently large ROI to cover the correlation range. In another example, if local background statistics in the neighborhood of the signal to be detected are vastly different from those of the other areas of the background image, then the local ROI should be used to quantify detectability accurately. In these cases, given a small sample size and/or computational limitations for either approximating the performance of the unconstrained observer or predicting human performance, it is likely to be infeasible to reliably estimate the performance of the 3Dp and 3D Hotelling observers. In that case, a feature selection followed by a model observer constrained to a small set of chosen features can be used to reduce data dimension and produce reliable observer estimates. More discussion on this subject is presented in the following section.

10.3.3 3D PROJECTION OR 3D CHO

Tomosynthesis systems produce much more data than planar imaging systems (e.g., N_p times more data in the projections, where N_p is the number of angular projections). Then, the dimension of data covariance $\mathbf{K_g}$ becomes much larger (e.g., N_p^2 times larger than that of a single-projection covariance), making it inevitable to reduce data dimension using either efficient or anthropomorphic channels for a rigorous evaluation and optimization of tomosynthesis systems. For efficient channels, we focus our discussion on projection space because it is desirable to maximize signal information content for calculating optimal observer performance before reconstructing the data and processing them to display to the human observer. For assessing reconstruction data, it is natural to consider anthropomorphic channels since reconstruction is ultimately used for displaying the data to human observers. In this case, a model observer using anthropomorphic channels can be developed for predicting

human performance. But one can still consider efficient channels for system optimization in reconstruction space when the goal is to optimize the parameters of a reconstruction algorithm.

10.3.3.1 Evaluation in projection space

Here, we discuss the channelized version of the 3Dp Hotelling observer, which we call the 3Dp CHO, for system evaluation in projection space with the following two approaches. The first way to implement the 3Dp CHO is to apply channels, each of which is efficient for each angular projection. For notational convenience, we call this method I. Symbolically, N_p channel matrices $\{\mathbf{T}_k\}_{k=1}^{N_p}$, each of which efficiently extracts spatial correlation within the kth projection, can be applied to the data vector $\mathbf{g}(= \{\mathbf{g}_k\}_{k=1}^{N_p})$ to produce an $L_c \times 1$ channelized data vector $\mathbf{v}$:

$$\mathbf{v} = \{\mathbf{v}_k\}_{k=1}^{N_p}, \tag{10.33}$$

$$\mathbf{v}_k = \mathbf{T}_k\,\mathbf{g}_k, \tag{10.34}$$

where $\mathbf{T}_k$ is an $N_{c,k} \times M$ matrix $\{\mathbf{t}_{j,k}^t\}_{j=1}^{N_{c,k}}$, $\mathbf{t}_{j,k}$ is an $M \times 1$ channel vector, $N_{c,k}$ is the number of needed channels for each projection $\mathbf{g}_k$, and $L_c = \sum_{k=1}^{N_p} N_{c,k}$. Note that if each $\mathbf{T}_k$ has a set of tunable parameters to be efficient for the given data, $N_{c,k}$ also depends on the choice of parameter values. Given the choice of channels, the values of channel parameters and the number of channels $N_{c,k}$ also depend on the relationship between background and signal statistics.

Another way to implement the 3Dp CHO, which we call method II, is to apply "3D" channels in multiangle projection space, that is, channels that extract statistical information from both within each projection and between different angular projections. In this case, the $M \times N_p$ data vector $\mathbf{g}$ is regarded as "3D"-like data in the multiangle projection space. Then, to reduce data dimensionality, $\mathbf{T}_n$, $n = 1,\ldots, N_c$, each of which is an $M \times N_p$ matrix of a 3D channel, can be applied to the vector $\mathbf{g}$ via the dot product, yielding an $N_c \times 1$ channelized data vector $\mathbf{v}$:

$$\mathbf{v} = \{v_n\}_{n=1}^{N_c}, \tag{10.35}$$

$$v_n = \mathbf{T}_n \cdot \mathbf{g} = \sum_{k=1}^{N_p} \mathbf{t}_{k,n}^t\,\mathbf{g}_k, \tag{10.36}$$

where $\mathbf{T}_n = \{\mathbf{t}_{k,n}\}_{k=1}^{N_p}$ and each $\mathbf{t}_{k,n}$ is an $M \times 1$ vector. If these "3D" channels $\{\mathbf{T}_n\}_{n=1}^{N_c}$ are efficient, they allow for the 3Dp CHO to approximate the 3Dp Hotelling observer. For both methods I and II, the formulas using $\mathbf{v}$ given in Section 10.2.4 can be used to calculate the template $\mathbf{w}_\mathbf{v}$, the test statistic $t_\mathbf{v}$, and the figures of merit of the 3Dp CHO.

When efficient channels in multiangle projection space are not available, it is reasonable to consider making use of the existing 2D efficient channels, including those discussed in Section 10.2.5. In this situation, method I can be more useful because it incorporates spatial correlation within and between angular projections with the use of 2D channels so that 3D information from tomosynthesis is still incorporated in observer-performance analysis. But when the complexity of background (and/or signal) data statistics is high, such as in DBT applications, one potential bottleneck to making use of method I is that without a large sample, the dimension of channelized data **v** can still be too large to accurately estimate data covariance, invert the covariance, and hence produce stable SNR trends. Trade-offs between the number of training images and the accuracy and stability of covariance and inverse covariance estimation must be understood before deciding which observer model to use in image-quality assessment. The methods for inverting a large matrix include the matrix-inversion lemma, Neumann's series, iterative computation of the observer template, and pseudoinverse via SVD [1,46].

Another drawback of method I is that spatial correlation between different angular projections may still not be efficiently incorporated into CHO calculations, which can lead to a performance gap between the 3Dp Hotelling and CHO and/or an increase in the number of channels $N_{c,k}$ for each channel matrix $\mathbf{T}_k$. With respect to the performance gap, as long as the gap remains constant over a parameter set of interest, the use of this approach is justified as a surrogate for the *optimal* 3Dp CHO. However, because it is often infeasible to calculate the unconstrained observer, it is thus infeasible to verify such a trend between the unconstrained and constrained observers. To help improve the situation, it is useful to (1) utilize simulation for generating data with similar statistics and validating the existing 2D channels using the data or (2) justify the use of these channels by comparing the 3Dp CHO to other suboptimal observers and ensuring that each of the 2D channels is highly efficient for each projection and sufficiently incorporates the 2D signal information with respect to the statistics of the whole *volume* of signal and background projections.

In the case of method II, the dimension of the channelized data vector is only N_c, which is much smaller than the dimension MN_p of the original data vector **g**, or L_c of the channelized data vector **v** from method I. To utilize this method, it is most important to find efficient or appropriate surrogate channels for making up the channel matrix $\{\mathbf{T}_n\}_{n=1}^{N_c}$ to extract spatial correlation between the angular projections as well as within the projection. Unfortunately, 3D efficient channels in multiangle projection space are not yet identified and validated for most tomosynthesis systems. When the assessment of medical-image quality necessarily calls for real or realistic enough anthropomorphic images such as in DBT evaluation, it is difficult to define efficient channels especially when (1) there are only background data but no information on either the signal or relative difference between signal-present and signal-absent data, and (2) the complexity of background statistics is high and the number of available images is too small to capture efficient features for the given data and task even if signal-present and signal-absent data are available. Therefore, whenever possible, users of model-observer approaches are encouraged to obtain efficient channels via either simulation or analytical modeling for the types of random backgrounds and signals most relevant to their task. Among the methods for investigating feature selection to find efficient channels, the PLS algorithm [19] has the potential to provide efficient channels by either directly applying the PLS

algorithm using a large sample or by modeling salient features using PLS channels of poor quality generated by applying the algorithm with a small sample. We note that how big the sample size has to be for calculating statistically significant observer-performance trends depends on the complexity of data statistics and the number of system parameters of interest. The author and her collaborators [47] are currently extending the PLS method for choosing efficient channels in the 2D case, as investigated by Witten et al. [19], to the 3D case.

10.3.3.2 Evaluation in reconstruction space

To display 3D data to human eyes, reconstruction is an essential part of the imaging chain. Therefore, for assessing tomosynthesis systems based on reconstruction data, it is most appropriate to incorporate some aspects of the human visual system into the model-observer design. The most common method is to constrain the Hotelling observer to a set of features that represent the human visual system. That is,

$$\mathbf{v} = \mathbf{T}\mathcal{V}(\hat{\mathbf{f}}), \qquad (10.37)$$

where each row of **T** is a collection of channels that can extract information in a similar fashion to that of the human observer using the given viewing mode $\mathcal{V}$ for the reconstructed object $\hat{\mathbf{f}}$. Depending on the viewing mode, either of the methods discussed in Section 10.3.3.1 can be used to produce the channelized data **v** from the reconstruction data. While many different anthropomorphic channels for the 2D case have been identified, it is not yet clear what should be used as anthropomorphic channels for the 3D case. This is another area of active research [48–50]. What is currently clear is that since the stack mode is one common way of viewing reconstruction slices, anthropomorphic channels should somehow incorporate time-dependent human response to reconstruction slices, correlations, and feedback from the previous slice. As other ways of viewing and displaying the reconstruction data for tomosynthesis systems emerge in the future, other characteristics will be incorporated in **T**. In any case, to validate anthropomorphic model observers, human observer studies in well-defined settings are inevitable. The references for conducting a psychophysical study using planar and volumetric images include Refs. [51–53].

10.3.4 MODEL-OBSERVER STUDIES FOR DBT

The following list indicates some examples of spatial-domain model-observer studies for the assessment of DBT systems using a number of different designs of the Hotelling observer and CHO models. Here, we only give a brief description of the design of observer models and data generation. For details on study results, readers are referred to the original articles cited here in chronological order.

- Chawla et al. [54] implemented a 3Dp CHO using the same set of 10 2D LG channels for each $\mathbf{T}_k$ and generated AUC plots as a function of the number of angular projections and the range of span angle with the use of tomosynthesis projection data of 82 patients. The width parameter of the 2D LG channels was chosen slightly bigger than the diameter (3 mm) of the projected signal at the normal projection angle. By design, their version of the 3Dp CHO did not include spatial correlation between different angular projections.

- Park et al. [55] implemented the 3Dp Hotelling observer with the use of small nonoverlapping ROIs from all possible locations of the image FOV for comparing laboratory mammography and three-angle DBT systems using a physical phantom. The phantom consisted of spheres of different sizes and densities for generating variable-background structures. Two hundred different random structures (i.e., 200 "patients") were created using this physical phantom. The use of the ROIs from all locations of the FOV was justified owing to the stationary nature of background image statistics generated by the phantom.

- Young et al. [56] implemented one version of the 3Dp CHO using the same set of 2D LG channels for each $\mathbf{T}_k$ and generated SNR maps as a function of the number of angular projections and the range of span angle for an *in silico* DBT optimization. The projection data of 1000 anthropomorphic breast phantoms generated using the UPenn phantom software [57] were produced via ray tracing and were used for the system optimization after the sources of noise were "added" to the projection data. For realizing scatter through the object, the scatter of a uniform phantom of the same shape and size as well as the same glandular density as those of the anthropomorphic phantoms was used. The number and the width parameter of the 2D LG channels were optimized to achieve maximum observer performance. As a result, the number and width parameter of the 2D LG channels were respectively six and the same as the diameter (3 mm) of the projected signal at the normal projection angle. In their subsequent work [58], a similar observer model was used for exploring dose-delivery schemes using different patient classes defined by the average glandular density. In this subsequent study, the projection data were generated using a full Monte Carlo simulation to include more realistic scatter effects on observer performance.

- Platisa et al. [48,49] investigated the use of both 3Dp and 3D CHO models with 2D and 3D LG channels as 2D and 3D efficient channels for 3D data sets generated using different types of synthetic random-background models, such as white and colored noise, lumpy, and clustered–lumpy backgrounds. They also compared the 3Dp CHO model to two other variants. The two variants differ in how they use the 2D CHO template and incorporate correlation between different angular projections. Given a patient (or phantom), one variant calculates many 2D CHO templates corresponding to different angular projections, and hence, as many test statistics as the number of different angular projections. Then it computes the Hotelling template using the test statistics and its covariance to estimate the performance of the Hotelling observer as the final observer performance. The other variant uses only one 2D CHO template using the projection that contains the most signal information and applies it to all other projections. Then it calculates the final observer performance as described for the first variant.

- Zeng et al. [59] implemented a 2D CHO model with 12 2D LG channels for investigating the impact of

different reconstruction algorithms on the optimization of DBT systems. In their work, the projection data of 60 anthropomorphic phantoms using the UPenn phantom software [57] were generated via ray tracing and reconstructed using various reconstruction algorithms after applying a Poisson noise model. The 2D CHO was applied to the central slice of the reconstruction data that contained the maximum signal information. The width parameter of the 2D LG channels was not optimized and it was chosen to be half the diameter (8 mm) of the signal in the central slice.

- Zhang et al. [47] studied the use of PLS in generating "efficient" channels for the 3Dp CHO models that are discussed in Section 10.3.3.1. They used the DBT projection data of the thousand phantoms generated using UPenn breast phantom software and ray tracing [56,57], in which the signal embedded in each phantom was a sphere of diameter 3 mm. Their preliminary results indicate that the PLS method has the potential for producing fairly efficient channels even with limited sample size in comparison to the case of LG channels. They are currently expanding their study to the case of an asymmetric signal.

10.4 CONCLUDING REMARKS

While there has been an increasing number of model-observer studies for the objective assessment of tomosynthesis systems, there has not yet been a systematic comparison between model-observer approaches (e.g., whether suboptimal or optimal, Fourier domain or spatial domain) for understanding how much 3D information is needed to accurately estimate SNR trends as a function of system and object parameters. It is not yet clear when a suboptimal observer that uses only partial information from the data is sufficient and when the optimal observer that uses the full statistical information in the data is absolutely necessary to not bias the result of system evaluation and optimization.

Different data statistics require different levels of complexity in the design of observer models. To appropriately utilize the existing observer models or develop new observer models for investigating the impact of additional information from the data on system performance, it is critical to understand data statistics comprehensively: the statistics of each source of variability in the data, including tissue variations seen in real patients even within the same patient class, and how one source of variability is related to another. When it is not straightforward to achieve this, it is useful to compare the performance between different observer models to identify optimal or anthropomorphic observer models suitable for the given data and task. With enough effort and researchers, there can be a consensus in the near future on observer models for the rigorous assessment of tomosynthesis systems.

To avoid exposing real patients to unnecessary radiation, various types of physical and virtual phantoms are used as surrogate patients for system calibration and optimization (e.g., see Chapter 6). These phantoms are useful for evaluating the system for various purposes and likely generate different types of data statistics, which could require different observer models. The comprehensive understanding of data statistics as described above is also useful in relating study results using the different phantoms to the performance of a real system dealing with real patients.

ACKNOWLEDGMENTS

The author thanks Drs. Kyle Myers and Joshua Soneson for reviewing the multiple drafts of the book chapter and sharing their thoughts on how to improve them. She would also like to thank Drs. Nicholas Petrick and Kyle Myers for allowing her to work on this book chapter during her official work hours.

REFERENCES

1. H. H. Barrett and K. J. Myers, *Foundations of Image Science*, John Wiley and Sons, New York, 2004.

2. H. L. Van Trees, *Detection, Estimation, and Modulation Theory (Part I)*, Academic Press, New York, 1968.

3. C. E. Metz, B. A. Hermann, and J.-H. Shen, Maximum likelihood estimation of receiver operating characteristic (ROC) curves from continuously-distributed data, *Med. Decis. Mak.* 17, 1033–1053, 1998.

4. X. Pan and C. E. Metz, The "proper" binormal model: Parametric receiver operating characteristic curve estimation with degenerate data, *Acad. Radiol.* 4, 380–389, 1997.

5. C. E. Metz and X. Pan, Proper binormal ROC curves: Theory and maximum-likelihood estimation, *J. Math. Psychol.* 43, 1–33, 1999.

6. H. B. Mann and D. R. Whitney, On a test of whether one of two random variables is stochastically larger than the other, *Ann. Math. Stat.* 18(1), 50–60, 1947.

7. E. R. Delong, D. M. Delong, and D. L. Clarke-Pearson, Comparing the areas under two or more correlated receiver operating characteristic curves: A nonparametric approach, *Biometrics,* 44, 837–845, 1988.

8. D. D. Dorfman, K. S. Berbaum, and C. E. Metz, Receiver operating characteristic rating analysis: Generalization to the population of readers and patients with the jackknife method, *Invest. Radiol.* 27(9), 723–731, 1992.

9. N. A. Obuchowski and H. E. Rockette, Hypothesis testing of diagnostic accuracy for multiple readers and multiple tests: An ANOVA approach with dependent observations, *Commun. Statist. Simul.* 24(2), 285–308, 1995.

10. B. D. Gallas, One-shot estimate of MRMC variance: AUC, *Acad. Radiol.* 13(3), 353–362, 2006.

11. E. Clarkson, M. A. Kupinski, and H. H. Barrett, A probabilistic model for the MRMC method. Part I: Theoretical development, *Acad. Radiol.* 13(11), 1410–1421, 2006.

12. M. A. Kupinski, E. Clarkson, and H. H. Barrett, A probabilistic model for the MRMC method. Part II: Validation and applications, *Acad. Radiol.* 13(11), 1422–1430, 2006.

13. B. D. Gallas, A. Bandos, F. W. Samuleson, and R. F. Wagner, A framework for random-effects ROC analysis: Bias with the bootstrap and other variance estimators, *Commun. Stat. Theory Methods,* 38, 2586–2603, 2009.

14. M. A. Kupinski, J. W. Hoppin, E. Clarkson, and H. H. Barrett, Ideal observer computation using Markov-chain Monte Carlo, *J. Opt. Soc. Am. A* 20, 430–438, 2003.

15. X. He, B. S. Caffo, and E. Frey, Toward realistic and practical ideal observer estimation for the optimization of medical imaging systems, *IEEE Trans. Med. Imag.* 27(10), 1535–1543, 2008.

16. C. K. Abbey and J. M. Boone, *An Ideal Observer for a Model of X-Ray Imaging in Breast Parenchymal Tissue*, IWDM 2008, Tucson, AZ, E. A. Krupinski, ed. LNCS 5116, pp. 393–400, 2008.

17. S. Park, H. H. Barrett, E. Clarkson, M. A. Kupinski, and K. J. Myers, A channelized-ideal observer using Laguerre–Gauss channels in detection tasks involving non-Gaussian distributed lumpy backgrounds and a Gaussian signal, *J. Opt. Soc. Am. A* 24, B136–B150, 2007.

18. S. Park, J. M. Witten, and K. J. Myers, Singular vectors of a linear imaging system as efficient channels for the Bayesian ideal observer, *IEEE Trans. Med. Imag.* 28(5), 657–667, 2009.

19. J. Witten, S. Park, and K. J. Myers, Partial least squares: A method to compute efficient channels for the Bayesian ideal observers, *IEEE Trans. Med. Imag.* 29(4), 1050–1058, 2010.

20. S. Park and E. Clarkson, Efficient estimation of ideal-observer performance in classification tasks involving high dimensional complex backgrounds, *J. Opt. Soc. Am. A* 26, B59–B71, 2009.

21. A. Wunderlich and F. Noo, Estimation of channelized Hotelling observer performance with known class means or known difference of class means, *IEEE Trans. Med. Imag.* 28(9), 1198–1207, 2009.

22. A. Wunderlich and F. Noo, Confidence intervals for performance assessment of linear observers, *Med. Phys.* 38(S1), S57–S68, 2011.

23. A. Wunderlich and F. Noo, On efficient assessment of image quality metrics based on linear model observers, *IEEE Trans. Nucl. Sci.* 59(3), 568–578, 2012.

24. A. Wunderlich and F. Noo, New theoretical results on channelized Hotelling observer performance estimation with known difference of class means, *IEEE Trans. Nucl. Sci.,* 60(1), 182,193, 2013.

25. H. H. Barrett, K. J. Myers, N. Devaney, and C. Dainty, Objective assessment of image quality. IV. Application to adaptive optics, *J. Opt. Soc. Am. A* 23(12), 3080–3105, 2006.

26. B. D. Gallas, J. S. Boswell, A. Badano, R. M. Gagne, and K. J. Myers, An energy- and depth-dependent model for x-ray imaging, *Med. Phys.* 31(11), 3132–3149, 2004.

27. A. Badano, I. S. Kyprianou, R. J. Jennings, and J. Sempau, Anisotropic imaging performance in breast tomosynthesis, *Med. Phys.* 34(11), 4076–4091, 2007.

28. I. S. Reiser, R. M. Nishikawa, and B. A. Lau, Effect of non-isotropic detector blur on micro-calcification detectability in tomosynthesis, *Proc. SPIE* 7258, 7258Z–7258Z-10, 2009.

29. M. Freed, S. Park, and A. Badano, A fast, angle-dependent, analytical model of CsI detector response for optimization of 3D x-ray breast imaging systems, *Med. Phys.* 37, 2593–2605, 2010.

30. K. J. Myers and H. H. Barrett, Addition of a channel mechanism to the ideal observer model, *J. Opt. Soc. Am. A* 4, 2447–2457, 1987.

31. H. H. Barrett, C. K. Abbey, B. D. Gallas, and M. Eckstein, Stabilized estimates of Hotelling-observer detection performance in patient-structured noise, in *Medical Imaging 1998: Image Perception*, H. L. Kundel, ed., *Proc. SPIE* 3340, 27–43, 1998.

32. B. D. Gallas and H. H. Barrett, Validating the use of channels to estimate the ideal linear observer, *J. Opt. Soc. Am. A* 20, 1725–1738, 2003.

33. C. K. Abbey and H. H. Barrett, Human- and model-observer performance in ramp-spectrum noise: Effects of regularization and object variability, *J. Opt. Soc. Am. A* 18(3), 473–488, 2001.

34. Y. Zhang, B. T. Pham, and M. P. Eckstein, The effect of nonlinear human visual system components on performance of a channelized Hotelling observer in structured backgrounds, *IEEE Trans. Med. Imag.* 25(10), 1348–1362, 2006.

35. Y. Zhang, B. T. Pham, and M. P. Eckstein, Evaluation of internal noise methods for Hotelling observer model, *Med. Phys.* 34(8), 3312–3322, 2007.

36. H. H. Barrett, J. Yao, J. P. Rolland, and K. J. Myers, Model observers for assessment of image quality, *Proc. Natl. Acad. Sci. USA* 90, 9758–9765, 1993.

37. D. H. Hubel and T. N. Wiesel, Receptive fields, binocular interaction and functional architecture in the cat's visual cortex, *J. Physiol.* 160, 106–154, 1962.

38. M. B. Sachs, J. Nachmias, and J.G. Robson, Spatial-frequency channels in human vision, *J. Opt. Soc. Am.* 61, 1176–1186, 1971.

39. N. Graham and J. Nachmias, Detection of grating patterns containing two spatial frequencies—Comparison of single-channel and multiple-channel models, *Vis. Res.* 11, 251–259, 1971.

40. J. G. Robson, Spatial and temporal contrast sensitivity functions of the visual system, *J. Opt. Soc. Am.* 56(8), 1141–1142, 1966.

41. M. P. Eckstein, C. K. Abbey, and F. O. Bochud, A practical guide to model observers for visual detection in synthetic and natural noisy images, *Handbook of Medical Imaging*, Vol. 1, Chapter 10, eds. J. Beutel, H. L. Kundel, and R. L. Van Metter, SPIE Press, 2000.

42. C. K. Abbey and F. O. Bochud, Modeling visual detection tasks in correlated image noise with linear model observers, *Handbook of Medical Imaging*, Vol. 1, Chapter 11, eds. J. Beutel, H. L. Kundel, and R. L. Van Metter, SPIE Press, 2000.

43. C. K. Abbey and M. P. Eckstein, Observer models as a surrogate to perception experiments, *Handbook of Medical Image Perception and Techniques*, eds. E. Samei and E. Krupinski, Cambridge University Press, 2010.

44. S. Park, A. Badano, B. D. Gallas, and K. J. Myers, Incorporating human contrast sensitivity in model observers for detection tasks, *IEEE Trans. Med. Imag.* 28(3), 339–347, 2009.

45. A. Avanaki, K. Espig, C. Marchessoux, E. Krupinski, and T. Kimpe, Integration of spatio-temporal contrast sensitivity with a multi-slice channelized Hotelling observer, to appear in *Proc. SPIE* 8673, 86730H 2013.

46. H. H. Barrett, K. J. Myers, B. D. Gallas, E. Clarkson, and H. Zhang, Megalopinakophobia: Its symptoms and cures, *Proc. SPIE* 4320, 299–307, 2001.

47. G. Z. Zhang, K. J. Myers, and S. Park, Investigating the feasibility of using partial least squares as a method of extracting salient information for the evaluation of digital breast tomosynthesis. *Proc. SPIE 8673*, Medical Imaging 2013: Image Perception, Observer Performance, and Technology Assessment, 867311 (March 28, 2013).

48. L. Platisa, B. Goossens, E. Vansteenkiste, A. Badano, and W. Philips, Channelized Hotelling observers for the detection of 2D signals in 3D simulated images, *Proceedings of the IEEE Conference on Imaging Processing*, Kairo, Egypt, pp. 1781–1784, 2009.

49. L. Platisa, B. Goossens, E. Vansteenkiste, S. Park, B. D. Gallas, A. Badano, and W. Philips, Channelized Hotelling observers for the assessment of volumetric imaging data sets, *J. Opt. Soc. Am. A* 28(6), 1145–1161, 2011.

50. B. Goossens, L. Platisa, and W. Philips, Theoretical performance analysis of multislice channelized Hotelling observers, *Proc. SPIE* 8318, 83180U, 2012.

51. L. Platisa, A. Kumcu, M. Platisa, E. Vansteenkiste, K. Deblaere, A. Badano, and W. Philips, Volumetric detection tasks with varying complexity: Human observer performance, *Proc. SPIE* 8318, 83180S, 2012.

52. A. Kumcu, L. Platisa, M. Platisa, E. Vansteenkiste, K. Deblaere, A. Badano, and W. Philips, Reader behavior in a detection task using single- and multislice image data sets, *Proc. SPIE* 8318, 831803, 2012.

53. I. Reiser and R. M. Nishikawa, Signal-known exactly detection performance in tomosynthesis: Does volume visualization help human observers? *Proc. SPIE* 8318, 83180 K, 2012.

54. A. S. Chawla, E. Samei, R. S. Saunders, J. Y. Lo, and J. A. Baker, A mathematical model platform for optimizing a multiprojection breast imaging system, *Med. Phys.* 35(4), 1337–1345, 2008.

55. S. Park, R. Jennings, H. Liu, A. Badano, and K. J. Myers, A statistical, task-based evaluation method for three-dimensional x-ray breast imaging systems using variable-background phantoms, *Med. Phys.* 37(12), 6253–6270, 2010.

56. S. Young, P. Bakic, R. Jennings, K. J. Myers, and S. Park, A virtual trial framework for quantifying the detectability of masses in breast tomosynthesis projection data, *Medical Physics*, 40 (5), 051914, 2013.

57. P. R. Bakic, C. Zhang, and A. D. A. Maidment, Development of an anthropomorphic breast software phantoms based upon region-growing algorithm, *Med. Phys.* 38(6), 3165–3176, 2011.

58. S. Young, A. Badal, K. Myers, and S. Park, A task-specific argument for variable-exposure breast tomosynthesis, *Proceedings of the 11th International Workshop on Digital Mammography*, Philadelphia, PA, pp. 72–79, 2012.

59. R. Zeng, S. Park, P. Bakic, and K. J. Myers, Is the outcome of optimizing the system acquisition parameters sensitive to the reconstruction algorithm in digital breast tomosynthesis? *Proceedings of the 11th International Workshop on Digital Mammography*, Philadelphia, PA, pp. 346–353, 2012.

11 Observer experiments with tomosynthesis

Tony Martin Svahn and Anders Tingberg

Contents

11.1 INTRODUCTION

When a new medical imaging device has been developed, it is necessary to evaluate it thoroughly before it can be implemented into the clinical routine. In the early testing stage, physical measurements such as modulation transfer function, noise power spectrum, and detector quantum efficiency are useful to quantify the fundamental characteristics of the detector system. The next step is to determine the ability of the system to visualize various test patterns by imaging physical phantoms. This is often included in acceptance tests, commissioning, and during the periodic quality control of the system, and incorporates the imaging and display system as well as the human observer. However, the task is not clinical and does not describe the ability to identify abnormalities. Observer performance measurements, on the other hand, describe the entire imaging chain, including the task and the interpretation by the observer. These measurements characterize and compare the accuracy of diagnostic tests. The common purpose underlying a diagnostic test is to provide reliable information of the patient's condition to positively influence the healthcare provider's plan for managing the patient (Sox et al., 1988) or to understand disease mechanisms and natural history through research (McNeil and Adelstein, 1976). Diagnostic accuracy describes the property of discriminating among alternative states of health. It might involve tasks such as identifying disease, distinguishing between benign and malignant disease, assessing treatment response, or a prognostic task such as assessing an individual's risk of developing a disease.

There are three components to objective assessment of image quality: (1) the task, (2) the images, and (3) the observer(s). The task is connected with the type of observer performance method used and can include signal detection, parameter estimation, or combined variants of the two. The images are obtained in a way that enables the investigation of specific tasks. The ability of an observer to perform a relevant task is usually limited by the noise sources in the imaging chain, which includes the anatomical noise present in most images. The image quality concept is thus inherently statistical and based on variations in the image formation and on observer-related variations (inter and intra). As a result, there is a strong relationship between these three components and the subsequent observer performance. The observer can be a human reader, such as a radiologist, or a mathematical model, also called an "observer model." Mathematical observer models for tomosynthesis are described in Chapter 10. This chapter deals with the objective assessment of medical images with human observers. Such studies are often called "observer studies" or "observer experiments."

11.2 TYPES OF OBSERVER EXPERIMENTS

An observer experiment is conducted to measure diagnostic or clinical image quality, that is, how well the observer can use the image to solve the diagnostic question. There are various types of observer experiments. These can be based on direct comparison of visibility of relevant pathological findings, or they can measure the performance of a radiologist in a well-defined diagnostic task based on a set of clinical or simulated images. Observer performance studies can thereby be divided into methods with regard to (1) subjective assessment (visibility, conspicuity, characteristics, size, etc.) of defined anatomical or pathological structures and (2) objective assessment (i.e., detection studies). The first group involves methods of "nonblinded" (subjective) interpretation, for example, the presence of an anatomical structure is known to the observer, such as image criteria (European Commission, 1996), and visual grading analysis (VGA) (Båth and Månsson, 2007). In contrast to these, the methods in the second group are objective because the subjectivity of the interpretation is controlled for by including normal (disease-free) cases, and the experiment measures the ability to distinguish between abnormal (diseased) and normal cases. This type of observer performance experiments includes study designs such as alternative forced choice (AFC) (Burgess, 1995) and methods for evaluating diagnostic accuracy such as receiver operating characteristics (ROC) (Metz, 2000), free-response ROC (FROC) (Chakraborty, 2000), and the localization ROC (LROC) (Swensson, 1996) paradigms. In general, observer performance methods involve acquiring study cases with known disease status (i.e., known gold standard), roughly half disease-free. These relate to the use of phantoms (physical and simulated), hybrid images (simulated lesions inserted onto normal backgrounds), or patient images. When using simulated structures, it is easier to obtain cases with pathology of threshold (borderline) detectability, which may be preferred when examining images stemming from different imaging techniques with subtle radiographic differences or, for instance, when comparing images generated with different reconstruction algorithms. Collecting patient images with abnormalities of borderline detection generally require more effort, especially when studying subtle differences, because of the more demanding process of identifying and verifying appropriate cases. When simulated pathology is used, the truth (gold standard) is exactly known to the investigator; however, a caveat may be lack in clinical realism and generalizing the results to a clinical population may be difficult or not possible.

11.2.1 VISUAL GRADING ANALYSIS

In VGA (Månsson, 1994, Båth and Månsson, 2007), the level of visibility of a defined anatomical or pathological structure is judged and scored on a scale. The score could be a numerical value or simply words describing the visibility (e.g., "inferior," "equal," and "superior"). VGA can be performed in two ways: by scoring relative to one or more reference images (displayed side by side) or by scoring on an absolute scale (i.e., each image is displayed individually without a reference image). The scale steps used belong to an ordinal scale and the numerical representations are merely arbitrarily chosen names of the scale steps and do not represent numbers on an interval scale. Statistical analysis of visual grading data can be performed using the visual grading characteristics methodology (Båth and Månsson, 2007). Observer (inter and intra) and case variability can also be considerable in subjective assessment of image quality and should be taken into account. Recent work by Smedby et al. (Smedby and Fredrikson, 2010, Smedby et al., 2012, 2013) has shown a way to extend the analysis to populations of readers and cases, which would be a useful tool and further development of this is underway (Båth, private communication, 2012). VGA is a method characterized by simplicity and sensitive discriminating properties and resembles how radiologists assess image quality (Andersson et al., 2008). Radiologists are regularly faced with the task of comparing image quality obtained using different imaging systems or settings. The underlying assumption is that the impression of visibility (or image quality) of anatomical structures correlates with diagnostic performance. Hence, if radiographic details have superior visibility in one modality, it should indicate that the modality also would be able to show subtler features better and thus be more effective clinically. VGA has been shown to be useful in numerous studies (Sund et al., 2000, Tingberg et al., 2004), though there are studies in which the assumption has shown not to be valid (Tingberg et al., 2005). Sometimes, VGA is considered to be a "beauty contest," that is, the "prettiest picture" gets the highest scores (Båth, 2010). Although VGA has its limitations with respect to resembling the clinical task of the radiologists (i.e., to detect, localize, and classify pathology, and to differentiate between abnormal and normal/benign), it is a straightforward test, which may be preferred when optimizing clinical examination protocols. In addition, VGA is easy to understand and carry out and in some cases, such as when the true state of an image (normal or abnormal) has not been established, it is the only test that is possible to carry out.

11.2.2 BINARY DECISION TASKS

A binary decision task is a task that can have two outcomes. This could be the presence or absence of a signal in a background, or the presence or absence of disease in an image. The number of true positives (TPs) is defined as the number of correctly identified diseased (abnormal or signal-present) cases, while the number of true negatives (TNs) is the number of correctly identified healthy (normal or signal-absent) cases. The sensitivity is then the true-positive fraction (TPF), that is, the TPs divided by the number of abnormal cases in the population. Specificity is the true-negative fraction (TNF), that is, the TNs divided by the number of normal cases in the population. The terms so far describe the correct diagnosis associated with the abnormal and normal cases in the test population, in contrast to the false negatives (FNs) and the false positives (FPs), which are the numbers of undetected abnormal cases and incorrectly classified normal cases (i.e., classifying a normal case as abnormal), respectively. False-negative results in the clinic cause harm by delaying treatment and providing false reassurance. False positives lead to unnecessary, perhaps risky tests (psychological emergences/anxiety). The metrics of sensitivity and specificity/recall rate are anticorrelated and their results depend on the reader's individual threshold, which can make comparisons of two or more modalities difficult. Methods with combined performance measures are therefore commonly used such as ROC, LROC, or FROC.

The ability to detect a signal in a background depends on both the signal and the background characteristics. When the signal shape and location are known, and the background statistics is specified, the task of detecting the signal is called a "signal-known-exactly, background-known-exactly" (SKE/BKE) detection task. The Rose model (Rose, 1948) gives an equation

for computing detection performance of a flat-topped disk added to a white-noise background. Burgess et al. (1981) have shown that this model can predict the performance of human observers. Mathematical observer models (see Chapter 10) have been developed to predict human performance in more complicated backgrounds, such as medical images.

Two experimental designs have been used to measure detection task performance of human observers, namely, alternative-forced choice (AFC) experiments and ROC experiments.

11.2.2.1 Alternative forced choice experiments

In the AFC study (Burgess, 1995, Ruschin, 2006), the reader is simultaneously presented with two or more images, usually small regions of interest. One of these images contains the lesion, while the other(s) is lesion-absent. The lesion is typically centered in the image and/or the location of the lesion and a dummy location on the other image(s) are indicated. The reader is forced to choose the image that is most likely to contain the signal. Each presentation of the multiple images results in a decision and is referred to as a "trial." The performance measure in an AFC study is the percent correct responses (PC): the number of correctly identified images with lesions divided by the total number of trials. The higher the PC value, the better the system for that particular task. AFC studies provide effective estimations of subtle differences between various modalities, performed under highly controlled manners but differ from clinical readings when it comes to the display and interpretation mode. For example, in the clinic, the radiologist is rarely faced with a situation where it is required to decide which of the two images contains a lesion and the use of small regions of interest of images taken out of the clinical context.

11.2.2.2 Receiver operating characteristics

ROC methodology is widely used for evaluating diagnostic accuracy (i.e., the ability to discriminate between abnormal and normal cases) of one or more modalities (Metz, 1978, 1986, 2000). Based on statistical decision theory, the ROC method models the decision-making process used by radiologists in the clinic. The truth is unknown to the reader(s) participating in the study but known to the investigator. The reader is shown one single image (or an ROI of an image), and the task of the observer is to provide a rating based on his or her confidence level that the image contains a signal (i.e., is abnormal). The lesion location may be indicated, in which case the detection task would be an SKE task (Barrett et al., 1998). The ratings are used to conduct several binary studies simultaneously, leading to a more efficient design of the study. One of the most common study types is where several readers interpret all images for each modality under investigation, the multiple-reader multiple-case (MRMC) study design (Dorfman et al., 1992). This design is used to maximize statistical power. Continuous rating scales (e.g., 0–100), as well as discrete scales (e.g., 1–5), can be used. The ROC curve is a graphical plot of the sensitivity of a test or TPF versus its FPF or "recall rate" 1-specificity or 1-TNF, and begins at the origin, associated with the most stringent decision threshold in which all test results are negative for disease, and ends at unity, associated with the most liberal decision threshold for which all test results are positive for disease. Any number of levels in the confidence scale (responses)

Table 11.1 **An example of a discrete rating scale (confidence level responses) for ROC studies**

RATING	INTERPRETATION OF PRESENCE OF DISEASE
1	Definitely absent
2	Probably absent
3	Possibly present
4	Probably present
5	Definitely present

can be used, but commonly, a set of five confidence level responses is used (Table 11.1), with 1 indicating absolute certainty that a finding is disease-absent (normal) and 5 indicating absolute certainty that a finding is disease-present (abnormal).

The ratings are pooled together to form probability distributions of the two classes of cases (e.g., abnormal and normal). Figure 11.1 illustrates the two probability density distributions that represent the test results of the normal and the abnormal cases in a study population. A larger separation between the distributions means that it is easier for the reader to distinguish between normal and abnormal cases. The operating points are obtained by varying the decision threshold at different cut-offs (χ_t) and by computing the TPF and FPF at each threshold. The ROC curve has $(c-1)$ threshold cut-offs (t), χ_t, where c is the number of confidence levels. Based on the discrete rating scale in Table 11.1, this yields four cut-offs and thus four operating points.

For the first operating point (A) (Figure 11.2), only the TPs and FPs with rating 5 are counted and divided by the numbers of abnormal and normal images in the study population, respectively, which yields the TPF and FPF, that is, the y- and x-coordinates. For the second operating point, one considers the TPs and FPs with rating $5 + 4$ (B) in the same manner and for the third operating point, TPs and FPs with rating $5 + 4 + 3$ (C), and so on.

At this point, some curve fitting method needs to be applied to obtain the area under the curve (AUC), which is the primary performance measure in ROC. Other measures are

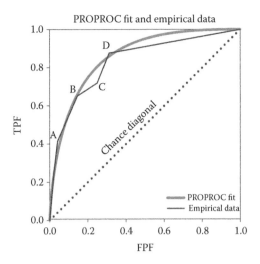

Figure 11.2 The smooth and empirical ROC curve with AUCs: 0.858 and 0.831, respectively. The data points A–D refer to the four operating points. The curve fit was done using the proper ROC model (PROPROC).

the sensitivity, specificity, or recall rate. The AUC rewards good decisions (TPs and TNs) and penalizes bad decisions (FNs and FPs). It ranges from 0.0 (perfectly inaccurate) to 1.0 (perfectly accurate). The empirical ROC curve (nonparametric method) that does not involve any curve fitting is obtained by connecting the operating points, including the "trivial points" (0, 0) and (1, 1) using straight-line segments. The empirical curve often results in a systematic underestimate of the AUC (Bamber, 1975, Hanley and McNeil, 1982) and it can be useful to fit a statistical model (parametric method) to the test results yielding a smooth curve. Both methods are illustrated in Figure 11.2. The empirical curve has potential advantages compared to the smooth curve in that it uses all data, without any assumptions and provides a plot that goes through all the points (Zweig and Campbell, 1993). One of the most common parametric methods is the binormal model approach, which has been shown to be biased in some cases (Goddard and Hinberg, 1990); an evident example of this is when it predicts an unphysical "hook" near the

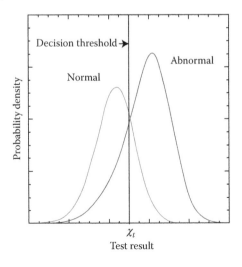

Figure 11.1 The two probability distributions of rating values representing the test result of a reader's interpretations on the normal and abnormal cases in a study set. The operating points are obtained by varying the decision threshold at different cut-offs (χ_t).

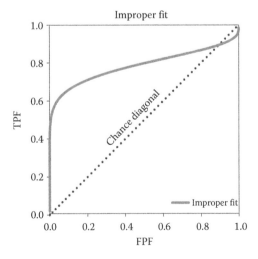

Figure 11.3 An example of an "exaggerated" improper fit using the binormal approach with the "hook" near (1, 1).

upper right corner for degenerate ROC datasets (Figure 11.3), which tend to go below the chance diagonal that is associated with pure chance (AUC = 0.5) and often considered the lowest level of diagnostic accuracy. Recently, parametric methods have been further developed such as the proper ROC (PROPROC) model (Pan and Metz, 1997) and the search-model (SM) (Chakraborty, 2006) that avoid the problem with the hook. The SM was originally developed for FROC data, but is also applicable for ROC data (Chakraborty and Svahn, 2011). It has been shown by Hajian-Tilaki et al. (1997) that parametric and nonparametric methods overall produce very similar results for the ROC curve estimates and its variance. Notwithstanding which method is used, the resulting AUC is defined as the probability that an abnormal image is rated higher than a normal image, or as the average value of specificity for all possible values of sensitivity (Hanley and McNeil, 1982, Metz, 1986, 1989) or vice versa.

11.2.2.3 Comparison of AFC and ROC

Because of the different study setup in AFC compared to ROC, as described earlier, AFC studies are faster to conduct than ROC studies but consistent with the faster reading times they also yield less information per interpretation. ROC does a better job by asking for ratings and the empirical AUC is calculated by comparing the ratings of abnormal and normal images. It has been demonstrated that the percent correct responses (PC) in 2AFC is equivalent to that of the area under the ROC curve under the assumption that the underlying probability density functions are normal (Burgess, 1995), which means that a PC of 0.5 in a 2AFC study is equal to pure guesswork (cf. the chance-diagonal in the ROC plot). This enables a direct comparison of the two methods. The ROC sampling statistics is superior to that of the 2AFC experiment. Say that an ROC study set contains N normal and A abnormal images, then the number of comparisons are NA/2. If N = A = 100, there are effectively 5000 comparisons, although the observer only interpreted 200 images. In 2AFC, the observer would have to interpret all 5000 comparisons to yield a similar statistical power. However, if all 5000 comparisons were done, then the estimated performance measure would likely be better by the 2AFC method, since direct comparisons are being used and the resulting coefficients of variation are lower (Burgess, 1995). About 200–500 comparisons are commonly used in AFC experiments. The ROC method requires about half as many image pairs as the 2AFC method for a given variance (Burgess, 1995). Also, in ROC, an abnormal and a normal image might both receive an equivalent rating, while in 2AFC, the observer is forced to choose, although the images involve the same confidence. The 2AFC method cannot show operating curves, or where the clinical operating point is, which is an important consideration, if one wants to know if the new modality gets higher AUC at the upper end of the ROC curve. At the actual operating point, the sensitivity could be the same but the operating point could shift to the right because of more recalls. In summary, ROC offers improved clinical relevance and statistical power compared to that of 2AFC, while advantages of 2AFC are in the faster reading times and lower variation due to the direct comparison. Free-response ROC, which is described in Section 11.2.3.1, provides further increase

in statistical power and clinical relevance (Chakraborty and Berbaum, 2004).

11.2.3 CONSIDERING THE LOCATIONS OF THE LESION

Most clinical tasks involve more information than can be handled in the binary ROC approach where the patient's state is classified as normal or abnormal. Limitations of the ROC method have been long recognized (Egan et al., 1961, Miller, 1969, Bunch et al., 1978) and concern the inability to handle the lesion locations and multiple lesions occurring on the same case (Chakraborty and Berbaum, 2004). In the clinic, the lesion location is highly relevant, for example, for guiding the subsequent surgical treatment. Figure 11.4 gives a simplified illustration of situations that yield loss in statistical power (i.e., sensitivity at detecting differences between the performances of modalities), when ROC analysis is used. In Figure 11.4a, the observer misses the lesion and marks another location. In Figure 11.4b, there are two lesions, but the observer can only state that the image is positive even if both lesions are detected.

Neglecting lesion location results in a loss of statistical power (Chakraborty and Berbaum, 2004). In an experimental study, there are no clinical consequences when ignoring the lesion location, but including lesion location resembles the clinical situation more. In the clinical practice, not using the information of all the lesions could result in a lumpectomy that only contains one of the lesions. A recall based on a disease-free region could result in the detection of the lesion in the diagnostic work up or nondetection, resulting in worse prognosis.

11.2.3.1 Free-response ROC

In the FROC paradigm, the investigator knows the number of lesions and their locations but the reader is blinded to this information. The reader's task is to mark the lesions and rate them using a confidence scale that reflects how certain the reader feels in the decision. Each mark and rating is location-specific and no limit is imposed on the number of mark-rating pairs that the reader can place on an image. Prior to analysis, the mark-rating pairs representing the FROC data are scored, that is, each mark is classified as a lesion localization (LL) if it falls within the boundary of the lesion, or else it is classified as a nonlesion localization (NL, indicating a disease-free region). The ability of the observer to localize the lesion is explicitly accommodated in the FROC scoring step. The lesion-localization fraction (LLF) is defined as the

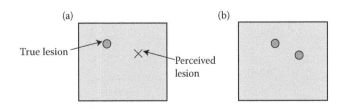

Figure 11.4 (a) The reader has marked an incorrect localization (perceived lesion), while the true lesion has a different location—the result is an ROC scored true positive (TP) yielding a correct decision for the wrong reason. (b) The right image shows two lesions; if both lesions are detected, the reader will only get rewarded for detecting one of them. If both the lesions were visible, it would increase the rating compared to if only one lesion was visible, and the ROC area under the curve (AUC) would increase.

number of correctly identified lesions divided by the total number of lesions present in the study, whereas the nonlesion-localization fraction (NLF) is the number of false localizations divided by the total number of abnormal images in the study.

11.2.3.2 Figure-of-merit for FROC

A common method for analyzing FROC data is the jackknife alternative free-response receiver operating characteristic (JAFROC) method (Chakraborty and Berbaum, 2004, Chakraborty, 2011). The JAFROC figure-of-merit (θ) is the probability that lesions, on abnormal images, are rated higher than NLs on normal images. It is analogous to the area under the ROC curve (AUC) used in ROC studies in the sense that it is a whole area measure. If the reader marks all lesions in the study population and does not mark any normal images, the figure-of-merit (FOM) would be unity and if the reader marked every normal image and did not mark any lesion, it would be zero. The JAFROC FOM is the trapezoidal (nonparametric, empirical) estimate of θ, which is defined by

$$\theta = \frac{1}{N_N N_L} \sum_{i=1}^{N_N} \sum_{j=1}^{N_L} \psi(X_i, Y_j)$$

$$\psi(X_i, Y_j) = \begin{bmatrix} 1.0 & \text{if} & Y_j > X_i \\ 0.5 & \text{if} & Y_j = X_i \\ 0.0 & \text{if} & Y_j < X_i \end{bmatrix}$$

where N_N is the number of normal/benign (disease-free) cases, N_L is the total number of lesions, X_i is the rating of the highest rated mark on the ith disease-free case, and Y_j is the rating of the jth lesion. The FOM is equivalent with the nonparametric area under the alternative FROC curve (AFROC; plot of LLF vs. FPF). An example of an AFROC curve is shown in Figure 11.5. The SM fitted curve (Chakraborty, 2006) does not extend continuously to (1, 1) as illustrated in Figure 11.5, which is a consequence of

a finite number of images with no decision sites. Like the ROC curve, the AFROC curve is contained within the unit square.

11.2.3.3 Inferred ROC data

By using inferred ROC quantities, one can define TPF, FPF, and an inferred ROC curve from the FROC data (Chakraborty, 2006). The rating of the highest rated mark on the image is usually used to infer the ROC rating for that image. This means that a single NL done on an abnormal image becomes a TP, while an NL on normal images becomes an FP, in the inferred ROC. If they occur in multiple events, the highest rated mark is used. Inferred ROC data can be analyzed by the Dorfman, Berbaum, and Metz (DBM) method (Dorfman et al., 1992).

11.2.3.4 Localization ROC

LROC analysis (Swensson, 1996, DeLuca et al., 2008) applies to detection tasks in which each image contains either no lesion or one lesion at most. In an LROC experiment, the task of the observer is to specify and rate the single location, if any, at which an abnormality is deemed to be present. The observers are informed that there can be at most one lesion per image. To score the location data, the lesion-containing images are usually divided into quadrants. Another possibility is to use coordinates with a specified acceptance criterion. A given mark is scored as a TP if it matches an actual lesion location, or as an FP if it occurs in a normal image. Based on the confidence level ratings, a curve is plotted in a similar way to that of the ROC method, except that a mark is scored as a TP only when the reported location is correct. The LROC curve measures the ability to detect and correctly localize the actual targets in the images. There are both empirical and parametric approaches available for the graphical LROC plot. These are described in more detail in Chakraborty (2011).

11.2.4 STATISTICAL CONSIDERATIONS

A study that fails to show an anticipated difference, for example, comparison of two modalities, is a nightmare for every researcher. If a study does not show a difference between the two modalities, then one cannot automatically draw the conclusion that there is no difference. The explanation could simply be that the study involved too few cases or observers. Before setting up the observer experiment, a sample-size estimation should be carried out to determine the number of cases and the number of observers needed to achieve a certain level of uncertainty.

Important factors that could influence the results and the interpretation of a study need to be reported. These include case selection, the level of enrichment of pathologic lesions, and the difficulty of the cases. Further, methods and criteria used to determine the true state of the images need to be reported.

11.2.4.1 Sample-size estimation

Sample-size estimation is an important consideration when evaluating the effect of a new modality in relation to a conventional one, especially if the study population comprises patients. A predicted difference can support the need to acquire additional patient images. An adequate number of readers and cases (sample size) allow the investigator to conclude that a clinically important difference was found rather than having inconclusive results because of an insufficient sample size. The sample size needed depends on how large the performance

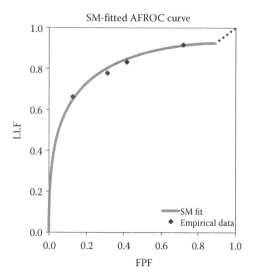

Figure 11.5 The search-model (SM) fitted AFROC curve (lesion-localization fraction as a function of false-positive fraction). The area under the nonparametric AFROC curve is used as a figure-of-merit in the JAFROC method.

difference (effect size) is between the two modalities; a large effect size yields a higher statistical power and therefore fewer readers and cases are needed to show a significant difference, and vice versa. Prior to conducting the study, this difference is unknown and thus it is valuable to conduct a pilot study to obtain such estimates. On the basis of the observed effect size of the pilot study, a sample-size estimation can be performed. Published methods for computing sample size to be applied on MRMC studies are the Hillis and Berbaum approach (2004) and the method of Obuchowski and Rockette (1995).

11.2.4.2 Standards for studies of diagnostic accuracy

In the planning stage of a study, it is important that the investigator ascertain which methods are used. The "standards for the reporting of diagnostic accuracy studies" (STARD) can bring valuable insights (available at http://www.stard-statement.org/). The objective of the STARD initiative is to improve the accuracy and completeness of reporting of studies of diagnostic accuracy. It allows readers of the published work to assess the potential for bias in the study (internal validity) and to evaluate its generalizability (external validity). Some of the itemized issues in the STARD checklist are a description of the reference standard, measures for diagnostic accuracy and statistical uncertainties (e.g., 95% confidence intervals), and the sampling method. The latter includes details such as description of the study population, inclusion and exclusion criteria, and setting and locations where the data was collected, if the recruitment was based on presenting symptoms, if the study population was a consecutive series of participants defined by the selection criteria, and so on.

11.2.4.3 Reference standard

The procedure that establishes the patient's condition status is referred to as the reference standard (truth or gold standard), and specifies whether a case is abnormal or normal/benign. The process of obtaining such a reference standard for clinical cases can be time-consuming, in particular when the full histopathology information and all available imaging techniques need to be evaluated.

In observer experiments on chest tomosynthesis and chest radiography, computed tomography (CT) has commonly been used as the reference standard (Dobbins et al., 2008, Vikgren et al., 2008, Zachrisson et al., 2009, Quaia et al., 2010), assuming that the lesions detected on CT represent the superset of lesions that can be found by other modalities. In mammography, abnormal findings are verified by fine needle or core biopsy followed by surgery and histopathologic examination of the specimens. In observer experiments on the accuracy of breast cancer size measurements in breast tomosynthesis (BT), the lesion size from the histopathologic examination has been used as a reference standard (Förnvik et al., 2010). Normal and benign cases are usually confirmed by following up patients for a certain time; 1 year has been used in several studies (Pisano et al., 2008, Gennaro et al., 2010, Wallis et al., 2012) to establish the absence of cancer.

To this point, the verification needed to conduct ROC studies has been completed; however, for FROC studies, the locations of the cancers need to be determined as well. Optimally, a truth panel, consisting of radiologists who are nonparticipants in the study and experienced in the modalities under investigation,

can outline the malignant regions in consensus (e.g., by using an electronic marker), while having access to all relevant data. For breast imaging, this means that information from BT, digital mammography (DM), ultrasonography, needle biopsy, and pathology may be used to define the outline of the lesions. For modalities that consist of several views, such as DM, the lesion should be outlined in both the craniocaudal (CC) and the mediolateral oblique (MLO) views. A way to use the 3D information in tomosynthesis is to perform the outlining in three slices: the initial, focus (central), and final slice. In cases where a breast cancer is not visible on the investigated imaging modalities, but seen and localized to a quadrant on ultrasonography images and the histopathologic examination, the lesion-containing quadrant can be outlined.

11.2.4.4 Patient images and sampling issues

Observer experiments that compare medical imaging systems based on readers' image-based decisions are inherently statistical and estimate performance measures from the interpretations, usually from a limited number of images. A concern is the extent to which the sample of patients and images included in an observer experiment adequately reflects the target population and how generalizable the results are. Therefore, before collecting patient images for a study, one needs to decide on a sampling method and be aware of potential limitations in the representation of the target population. Two completely different sampling methods are (1) the use of some type of selection criteria (aims to rank the performances of the imaging systems on a relative scale) and (2) random sampling from the population (aims to measure the absolute quantities of the imaging systems) (Metz, 1989). In random sampling, attempts are made to represent the target population as closely as possible. The use of a patient selection criteria, such as including only small-sized nodules (e.g., below a specified size) in chest x-ray imaging or the use of highly dense breasts in mammographic x-ray imaging, would result in cases that are more difficult for the reader, sometimes referred to as "challenge studies" (Zhou et al., 2011). Challenge studies usually involve larger performance differences between modalities than studies based on random sampling and are more cost-effective, but might exclude relevant cases. The underlying assumption in challenge studies is that if an imaging modality is better in performance for difficult cases, it would also be better, or at least equivalent, for easier cases. Conversely, commonsense suggests that if the cases are extremely easy, then even a poor imaging modality would perform well. Using easier cases, which are likely to be properly diagnosed in both modalities, tends to decrease statistical power, and also brings up ethical issues since the patients may be subject to unnecessary imaging procedures for a study of questionable statistical strength. In screening mammography where the disease prevalence is very low, about 4–7 cancers per 1000 women of ages 40–70 (Skaane, 2009), random sampling implies a large sample size.

11.2.4.5 Description of difficulty of the cases and the degree of enrichment

Since there are various ways to sample cases, it is important to describe how well the sampled cases mirror the real population. Such a description can also help to explain why results from different studies diverge. The STARD initiative covers a description of the patient sampling, but it does not, however, describe the difficulty of the cases that were collected (e.g., subtlety of abnormal findings and breast density). To increase the understanding of the results and to attempt to enable comparisons of similar studies, an independent classification of the abnormal findings could be performed according to cancer visibility (not visible, subtle, and visible), and in breast imaging studies, the cases should be classified according to BIRADS breast density (American College of Radiology, 1998): (1) fatty or <25% dense; (2) scattered fibroglandular densities or 25–50% dense; (3) heterogeneously dense or 50–75% dense; and (4) dense or >75% dense. This would be done in a nonblinded manner and might help relate proportions of subtlety of cancers that eventually becomes undetected in the blinded study.

11.2.4.6 Experimental studies

An experimental study (or "laboratory study") refers to a study performed in a controlled environment, which to some extent may differ from the clinical situation. There is a general assumption that experimental clinical studies correlate well with the diagnostic accuracy in the clinic, although it has been shown for radiologists that their performance can depend on the environment (Gur and Rockette, 2008). Possible reasons for this might be that the radiologists participating in an experimental study are aware of the fact that their results will not have any effect on patient care, or that they do not feel any pressure to reduce their false-positive rate; in the clinic, the radiologist must weigh the balance between higher sensitivity and higher false-positive rate (higher recall rate). However, most often, the goal of a study is to compare two or more modalities in which the discussed effects are likely to be of similar magnitude for all modalities (Beam et al., 1996).

11.2.4.7 Statistical analysis of ROC and FROC data

Both the ROC and JAFROC methods involve (1) defining a type of FOM, the AUC for ROC and the θ for JAFROC, and (2) estimating the FOM and its variance from the data. Both the multireader multicase (MRMC) ROC and MRMC FROC dataset are analyzed in two steps involving

1. Generation of pseudovalues
2. Analysis of pseudovalues using an ANOVA (analysis of variance) model

The calculation of the pseudovalues by use of the jackknife method (Efron, 1982) involves removal of each image, one at a time, recalculation of the FOM and determination of the effect of the removal of the image on the FOM. While NLs can occur on abnormal images, in the present analysis, they are ignored, as current methods of analyzing the data do not accommodate asymmetry between pseudovalues originating on normal and abnormal cases (Chakraborty and Berbaum, 2004), which could be due to satisfaction-of-search effects (Berbaum et al., 2010). Omitting the NLs from the abnormal images corrects this problem but at the expense of some loss in statistical power. Because the multiple responses on each image interpretation are reduced to a single pseudovalue, JAFROC analysis avoids having to make assumptions of independence between events on the same image, thereby addressing the concerns (Swensson, 1996)

with earlier FROC analyses (Chakraborty, 1989, Chakraborty and Winter, 1990). An *F*-test is used internal to the ANOVA, which yields a *p*-value for rejecting the null hypothesis (NH) of no difference between the modalities. If the observed *p*-value is smaller than 5%, the NH is rejected.

11.3 NATURE OF IMAGES FOR OBSERVER EXPERIMENT

When setting up an observer experiment for comparing for example a new modality to a conventional one, images are collected from the two modalities and presented to a group of observers. Ideally, the imaged "objects" are real patients sampled exactly as they present in the daily clinical routine. But for various reasons, other objects have to be used in the study. In some studies, it might be sufficient to use normal patient images with simulated lesions added to the images ("hybrid images"), or computerized models of humans. Anthropomorphic phantoms are easy to handle and can be exposed multiple times without radiation protection concerns, but they only represent one anatomical background. Also, test phantoms such as contrast-detail phantoms can be useful in some situations. Naturally, real patients represent the highest level of clinical realism of a study, and as the imaged object deviates more and more from real patients, the clinical realism of the study decreases. Studies using a CD-phantom might be useful in some cases but one has to consider how generalizable the results from phantom study are to humans. In this section, we discuss the pros and cons of using different types of objects for producing the images, and illustrate the discussion with a few studies.

11.3.1 ANTHROPOMORPHIC AND CONTRAST-DETAIL PHANTOMS

With regard to studies of different acquisition parameters or reconstruction algorithms, anthropomorphic phantoms can be useful. The major advantages are that they, in contrast to patients, can be exposed to radiation repeatedly without any risks of late effects, and that the images are highly reproducible. For breast imaging, anthropomorphic breast phantoms usually have limited resemblance to real breasts regarding anatomy and clinically occurring breast cancers, and varying the anatomical background, which is highly diverse in breasts, is usually not possible. These qualities tend to limit the clinical realism of a study. Currently, there are few commercially available anthropomorphic breast phantoms for tomosynthesis. An example of such a phantom is the model 020 phantom (Computerized Imaging Reference Systems Inc. (CIRS) (http://www.cirsinc.com)) (Figure 11.6), which is designed for both BT and breast-computed tomography.

The phantom contains simulated pathology (masses, fibers, and microcalcifications) and consists of interchangeable slabs simulating various glandular/adipose tissue proportions (50/50% and 25/75%), allowing for simulation of different breast densities to a certain degree. In order to appear as real lesions in 3D imaging, the objects are embedded into one of the phantom slabs and thus it is not possible to arbitrarily vary their locations or lesion characteristics. To the authors' knowledge, this type of phantom has only been used in one

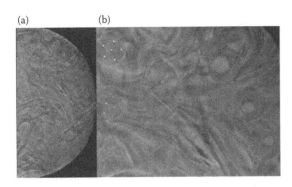

(a)   (b)

Figure 11.6 (a) A tomosynthesis image of an anthropomorphic breast phantom (Model 020, CIRS). (b) A closeup of structures within the phantom, simulating microcalcifications, fibers, and masses of various sizes. (Reproduced with permission from Vecchio S et al. 2011. *Eur Radiol*, 21, 1207–1213.)

publication to date (Vecchio et al., 2011). The corresponding phantom for two-dimensional (2D) mammography (RMI165, Radiation Measurement Inc., Middleton, WI, USA), in which it is possible to add various physical structures simulating pathology, has been used in several different published studies involving observer experiments (Kheddache et al., 1999, Svahn et al., 2007a, Yakabe et al., 2010). For body parts other than the breast, there is a wide variety of producers of anthropomorphic phantoms available (e.g., Kyoto Kagaku Co. Ltd, (http://www.kyotokagaku.com) and The Phantom Laboratory, (http://www.phantomlab.com)). These phantoms have the same limitations as described above. In both anthropomorphic and contrast-detail phantoms for tomosynthesis, spherical (or semispherical) objects are a natural choice in the resemblance of tumors, allowing precise measurements to be made. In mammography, the objects are preferably embedded in a bulk material of attenuation properties similar to that of polymethyl methacrylate (PMMA), for example, commonly used to simulate a "standard breast," defined as a 50-mm compressed breast containing about 50% glandular tissue (Zoetelief et al., 1996). A concern regarding the manufacturing process of tomosynthesis phantoms is the low reproducibility of phantoms of a specific type. For instance, the Model 020 phantom has anatomical-like patterns that are difficult to reproduce in a similar fashion in another phantom of this type. The manufacturing requirements are in general higher in three-dimensional (3D) imaging than in 2D. The objects in standard contrast-detail phantoms for 2D mammography (typically flat-topped disks) often consist of a single material, with contrasts being controlled by the thickness parallel to the detector plane, as opposed to tomosynthesis or CT where the contrast level primarily depends on the attenuation properties of the object. As a consequence, the object contrast level in 3D imaging is modulated by the use of materials with various attenuation properties (Figure 11.7). It is essential to realize that because there is higher variability in construction parameters and attenuation properties, the possibility to compare the results from different research groups based on the same type of phantom could be more limited in 3D than in 2D imaging. Recent research has shown that contrast-detail phantoms with a homogeneous background should be used with care because

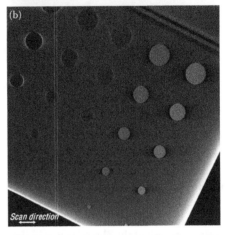

(a)

(b)

Scan direction

Figure 11.7 A photograph of a contrast-detail phantom (a) that can be used with tomosynthesis and computed tomography. The spheres are of various materials and sizes (4–20 mm) from left to right broadside: polyethene, acrylonitrilebutadienestyrene, nylon, polycarbonate, bakelite® (Etronax MF®), and polyoxymethylene. The object contrasts are material-dependent in 3D imaging as shown in the tomosynthesis (b). (Reproduced from Svahn et al. 2007b, *SPIE Medical Imaging*. San Diego, CA: SPIE Press.)

the absence of anatomical noise does not correlate with the real clinical situation (Månsson et al., 2005). However, for the appropriate study task (Johnsson et al., 2010) and for the use as a tool for quality control, these phantoms are well justified (Perry et al., 2006, Hemdal, 2009).

Johnsson et al. (2010) conducted an observer experiment using a contrast-detail phantom with embedded spheres of various sizes and densities (Figure 11.7). The phantom was scanned using both CT and chest tomosynthesis. In the experiment, six experienced thoracic radiologists, blinded to the true sphere diameters, independently performed size measurements in chest tomosynthesis and in CT. The results showed no significant difference in measurement accuracy between the two modalities and it was concluded that chest tomosynthesis might be used as an alternative to CT for size measurements of nodules.

11.3.2 SIMULATED 3D LESIONS ADDED TO COMPUTERIZED BREAST PHANTOMS

Zhou et al. (2006) used an anatomically realistic breast phantom to evaluate three reconstruction methods for BT: back projection (BP), simultaneous algebraic reconstruction technique (SART), and one statistical method; expectation maximization (EM) algorithm. A set of 3D phantoms were simulated with different

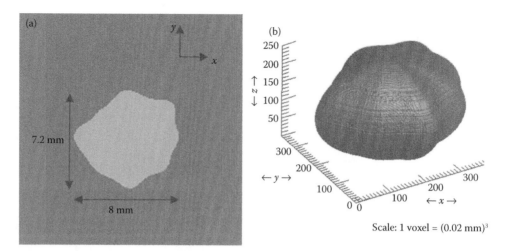

Figure 11.8 Tumor simulation example. (a) Central plane (*xy*) parallel to x-ray detector. (b) The same shape from (a) expanded to three dimensions in a voxel representation. The voxel side length of the simulation shown here is 0.02 mm. (Reproduced from Ruschin M et al. 2007. *SPIE Medical Imaging.* San Diego, CA, USA: SPIE Press.)

features with regard to the components of ductal structures, fibrous connective tissues, Cooper's ligaments, skin, and pectoralis muscle. In addition, breast structural noise (power law) was generated to account for variability of small-scale breast tissue. Random instances of 7–8 mm irregular masses were generated by a 3D random walk algorithm and placed in dense fibroglandular tissue. The task of the observers was to detect these masses in dense breasts. Five observers participated in an AFC experiment, in which 200 image pairs were shown and the potential lesion location was indicated for each reconstruction method. The EM algorithm yielded the highest performance (PC = 89%), followed by the SART algorithm (PC = 83%) and the BP algorithm (PC = 66%). The reader-averaged differences were statistically significant between all algorithm comparisons.

A recent study (Das et al., 2011) examined the iterative penalized maximum likelihood (PML) reconstruction method compared to filtered back projection (FBP), for the detection and localization of clusters of microcalcifications. Localization receiver operating characteristic (LROC) studies were conducted to evaluate the performance of the algorithms. BT projections were generated by computer simulations that accounted for noise and detector blur. Three different dose levels (0.7, 1.0, and 1.5 mGy) were examined in a 50-mm compressed breast. The compressed-breast phantoms were derived from CT images of mastectomy specimens and provided realistic background structures for the purpose of the detection task. Four observers interpreted the 98 test images individually for each combination of reconstruction method and dose level. All observers performed significantly better ($p < 0.05$) with the PML images than with the FBP images at all three dose levels, but there were no significant differences in observer performance for any of the dose levels.

11.3.3 SIMULATED 3D LESIONS ADDED TO CLINICAL IMAGES

Contrast threshold was examined by Ruschin et al. (2007) in a series of AFC studies for simulated 3D tumors added to patient breast images. Using a geometrical description of the tomosynthesis system, each simulated tumor was projected onto the 2D plane for each projection angle by mapping each

voxel within the simulated tumor to a detector element (i, j) (Figure 11.8). This yielded a set of tumor thickness images $x_k(i, j)$ corresponding to the integrated thickness of the tumor along the x-ray trajectories for each projection angle, k. If the projection images are described in log space as $\ln(\text{Im}_{0,k}(i, j))$, then the images with an added tumor are

$$\ln(\text{Im}_{1,k}(i, j)) = \ln(\text{Im}_{0,k}(i, j)) - x_k(i, j) \cdot \Delta\mu$$

where $\Delta\mu$ is the difference in attenuation coefficient between the existing breast tissue and the added tumor, and $x_k(i, j)\Delta\mu$ is the projected tumor signal intensity for the projection angle k ($k = 0$ is the central, zero-angle projection). The projection of the tumors was then added to the normal projection images. These were reconstructed using the FBP algorithm to form a 3D tomosynthesis volume (Figure 11.9). The tumor visibility was tested as a function of projected tumor signal intensity in BT and DM. An increased tumor visibility was found in favor of BT suggesting that the use of BT may lead to earlier detection of breast cancer. Timberg et al. (2010) studied this further using the FBP algorithm for lesions of various sizes and found that the detection of 1 mm and larger lesions was significantly better with BT, but for 0.2 mm lesions, DM outperformed BT.

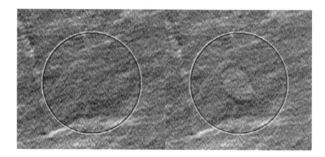

Figure 11.9 A set of BT images with the same anatomical background, lesion-absent (to the left), and lesion-present (to the right) reconstructed using the filtered back projection algorithm. (Reproduced from Ruschin M et al. 2007. *SPIE Medical Imaging.* San Diego, CA, USA: SPIE Press.)

11.3.4 BREAST SPECIMENS

Alternatives to phantoms are breast specimens obtained at mastectomy, which also are practical to use in the sense that they provide reproducible images and do not involve any patient dose. Breast specimens have been used in a study on image quality with regard to dose on various lesion types in BT by Timberg et al. (2008). The breast specimens came from patients who had abnormal findings detected by mammography and/or ultrasonography. There was a time window of about 40 min prior to pathological examination of the specimens in which multiple image acquisitions occurred. Breast specimens are generally thinner than *in vivo* breasts. To compensate for this, slabs of PMMA were placed onto the detector and the specimen was placed on top of the PMMA slabs. This allowed the researchers to simulate clinically relevant dose levels. The specimens were imaged at various different dose levels. During imaging, the compression paddle was placed above the specimens to resemble the clinical situation, but without any compression force applied on the specimens. The abnormalities were verified histologically and included various types of breast cancer morphology (Figure 11.10). This approach is more realistic compared to the use of an anthropomorphic phantom with regards to anatomy and breast cancer characteristics. Breast specimens differs compared to *in vivo* breasts, in that the skin line is not well defined and the anatomy is limited since only part of the breast is removed by surgery. As most mastectomy specimen contains breast lesions they are appropriate to use in visual grading studies. In the study, the collected BT images were subjected to observer studies in which various dose levels (approximately 2×, 1.5×, 1×, and 0.5× the total mAs at the same beam quality as used in a single digital mammography view under AEC conditions) were compared using paired data. Comparisons using paired data yield higher statistical power than random case comparison since various effects that can influence the results are avoided (e.g., differences in breast cancer contrasts, size, and type), which is a relevant point in the data analysis. There was a significant difference in the required exposure levels for achieving a

sufficient image quality for various lesion types. The study indicated that for low contrast lesions with diffuse borders, a higher exposure level (2×) was required, whereas for spiculated high-contrast lesions and lesions with well-defined borders, the exposure level was lower (<0.5×).

11.3.5 PATIENT IMAGES

A study using patients for producing the study cases represents the highest-level realism. Conclusions from such studies are often applicable in a real clinical situation. However, it is important to consider potential limitations of such studies and to be aware of how they might influence the results. In this way, it may be possible to avoid them in the study set-up. This type of study is more expensive to conduct compared to other types, but is essential in the validation process of a medical imaging device. Recent examples of observer experiments in tomosynthesis imaging using real patients can be found in Quaia et al. (2012) for chest, and Svahn et al. (2012b) for breast, and further examples are presented below.

11.4 PRACTICAL ISSUES WHEN SETTING UP AN OBSERVER EXPERIMENT

Before setting up an observer experiment, there are a number of practical issues that need to be considered. Does the observer experiment need to have radiologists as observers, or is it sufficient to have nonradiologists? A careful design of a study involves an efficient image viewing environment. Instructions to the readers and a training session prior to the study are vital for reducing intra- and interobserver variance. Feedback from the observers can be useful for interpreting the results from a study. This section describes these practical issues in detail to help the reader of this book to avoid potential pitfalls when conducting an observer experiment.

11.4.1 RADIOLOGISTS AS OBSERVERS

In the pilot stage of a study, attention should be on the study design and on factors that could negatively affect the result. In general, reader variability is the largest source of variability in ROC-based studies (Chakraborty, 2010). To reduce the effect of this problem, a sufficient number of readers need to be used to obtain a reliable average. The choice of readers is important. The intraobserver variation is generally smaller for experienced readers than for inexperienced readers (Kundel, 1988, Månsson, 1994), particularly for clinical studies. Using experienced readers in a study might require a smaller number of readers, compared to using inexperienced readers, to obtain an acceptable level of uncertainty in the results of the study. It should, however, be noted that this might not reflect the clinical situation where various experience levels are usually involved. When performing phantom or similar types of studies, on the other hand, the radiologists' experience might not be useful to the same degree. In an AFC experiment, it is often acceptable to use nonradiologists as observers, especially if the images contain simulated pathological structures, which are superimposed

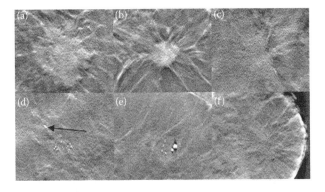

Figure 11.10 Example of BT images of breast specimens with various lesion characteristics. (a) A moderately spiculated mass, (b) a spiculated rounded mass, (c) less spiculated irregular mass with partly diffuse borders, (d) cluster of microcalcifications and/or ductal arrangement, (e) coarse calcifications, and (f) architectural distortion. (Reproduced from Timberg P et al. 2008. Impact of dose on observer performance in breast tomosynthesis using breast specimens. *SPIE Medical Imaging*. San Diego, CA, USA: SPIE Press.)

on anatomical backgrounds. However, for studies involving detection, localization, and classification (e.g., ROC, LROC, or FROC studies) where the study population comprises real patients (e.g., both the study task and the population are clinical-like), it is desirable to use radiologists in the observer experiment. In general, a naïve observer can be used when performing a perception task, whereas a radiologist is required when the task involves diagnostic expertise.

11.4.2 IMAGE DISPLAY

For increased accuracy and time efficiency in conducting observer experiments, various types of software platforms can be used. One example of such software is ViewDEX (viewer for digital evaluation of x-ray images) (Börjesson et al., 2005, Håkansson et al., 2010), which our group has used in several studies. The software presents the images on a viewing monitor and collects the evaluation data for further assessment. The study setup, image presentation, and image evaluation are designed as nonrestrictive as possible, and can be used for most of the study types described in this chapter. With regard to observer experiments in tomosynthesis, it is possible to display the slices using cine-loop mode at a user-controlled rate and to collect ratings data as well as mark-rating data (Figure 11.11). The software is available for download at http://sas.vgregion.se/sas/viewdex. Similar software has been developed by other research groups, such as the Sara platform (http://www.qaelum.com/products/sara.html) developed by Jacobs et al. (2011).

11.4.3 READER INSTRUCTIONS

It is important that all readers receive the same information about the study task and are prepared in the same way to avoid introducing this potential source of uncertainty. It is good to keep the instructions as simple as possible to avoid any misinterpretations (Chakraborty, 2009). The written instruction protocol should include the aim of the study, the study task,

and any specific issues that need to be brought to the readers' attention. An example of an instruction protocol for a free-response ROC study is given below:

Aim of the study
The purpose of the study is to compare digital mammography (DM) and breast tomosynthesis (BT) for the detection of malignant lesions.

The study task
You will interpret 30 DM cases (CC and MLO) and 30 BT volumes per session, in a total of eight sessions. You will interpret each case and indicate, with a mouse click, any and all regions suspicious for malignancy (BIRADS 3-5). Upon a mouse click a "cross" will be overlaid at the marked location. For each suspicious region, you will also indicate the probability of malignancy (Table 11.2) on a discrete BIRADS-like scale.

Notes
- You may find nothing to report on a case (equals no mark), one mark, two marks, and so on.
- If the same lesion is visible in both CC and MLO views, mark it in both views and give each mark a rating, for example, for instance 3 on CC and 4B on MLO.
- Try to mark the lesion accurately in the center of the suspicious region and additionally for BT images choose the slice (in the z-direction) where the lesion appears in best focus.
- There is a HIDE/SHOW toggle button, which does not erase the marks and scores, but only suppresses them for your convenience.
- Details of the user interface are in the "ViewDEX FROC Study User Guide" document, which we will discuss during the training sessions.

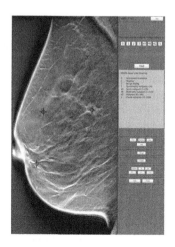

Figure 11.11 The image display for a breast tomosynthesis volume using the ViewDEX software platform. This example shows a free-response ROC study where the task is to mark and rate malignant findings. Various cine-loop rates are available for volume rendering on the task panel to the right. The observer can pan, zoom, and adjust the window and level settings. A toggle bar is available to hide/show a given mark.

Table 11.2 BIRADS-based malignancy rating scale

MEANING	MARK?	MEANING IN PROBABILITY OF MALIGNANCY (%)	DISCRETE BIRADS-LIKE RATING
Assessment incomplete OR Normal image	No	N/A	
Benign finding	Yes	0	2
Possibly suspicious for malignancy		<2	3
Probably suspicious for malignancy		3–33	4A
Suspicious for malignancy		34–64	4B
Quite suspicious for malignancy		65–94	4C
Highly suspicious for malignancy		≥95	5

11.4.4 TRAINING SESSION

The introduction of a new modality is often associated with learning curve effects for the radiologists, and it might require numerous cases before the readers reach a stable plateau in experience. By having a training session before the actual study, an attempt is made to reduce the most prominent part of such effects. In addition, the readers become accustomed to the graphical user interface of the image presentation and the data collection software, and the study procedure as a whole. The training session also yields an opportunity to test the output data; for instance, that all rating levels have been utilized in an ROC experiment to maximize the efficiency in the data representation. If this is not the case, it is still possible to extend the training further at this point. The efficiency of the training session may be increased if an expert radiologist familiarizes the readers after the training session with the appearance of normal tissue and various types of lesions. This would be done on the same set of training cases so the reader can get a brief feedback on potential interpretation errors. In general, radiographic features that are modality-specific and typically reflect noteworthy differences between the imaging modalities could be helpful to point out. The output data from the training sessions could be analyzed for test purposes but should not be used in the primary analysis, as it may be related to a lower performance due to more immediate learning curve effects. Prior to a training session, the investigator should test that everything works as expected concerning the image display, the study task, and the collection of reader data.

11.4.5 FEEDBACK FROM THE OBSERVERS

It is essential to ask for feedback from the participating readers after the training session. At this point, there is still an opportunity to make changes to the setup of the study, in case something is poorly executed. After the completion of a study, feedback can be helpful in explaining essential issues that the analysis of data cannot describe directly, or issues that can increase the understanding of the results. For instance, differences to what the readers are accustomed to in the clinic regarding comfort, blur, viewing settings, required effort in interpretation, and so on are worth noting. If most of the readers have the same opinion about a specific issue, it can be useful to consider this for future studies and even discuss it in the publication.

11.5 PUBLISHED STUDIES

Tomosynthesis has been commercially available for a few years and is being evaluated by several manufacturers and academic groups. In this section, we present how observer experiments have been used in selected published studies of tomosynthesis. In the early studies, the observer experiments served as "proof of concept" and in the more recent studies, the clinical potential of tomosynthesis have been evaluated in comparison to a conventional technique with regard to diagnostic accuracy, cancer visibility, BIRADS classification, view settings, accuracy of size measurements, reader experience, and more.

11.5.1 CLINICAL STUDIES OF CHEST TOMOSYNTHESIS

Vikgren et al. (2008) compared the diagnostic accuracy of lung nodules in chest tomosynthesis with conventional chest radiography in a FROC study. Eighty-nine patients were included in the study of which 42 patients were abnormal (containing 131 nodules) and 47 were normal. A multidetector CT served as the reference method. Four experienced chest radiologists searched images for pathology and scored their findings on a four-level scale. Only 16% of the nodules were detected using chest radiography, while 56% were detected using chest tomosynthesis. The study showed that tomosynthesis had a significantly higher diagnostic accuracy compared to conventional mammography. Other studies have shown similar results (Dobbins et al., 2008, Quaia et al., 2010). By repeating the study by Vikgren et al. one year later, Zachrisson et al. (2009) showed that the radiologists had reached a stable learning curve plateau already after 6 month.

Diagnostic performance and disease characterization of chest tomosynthesis were compared to that of conventional radiography in the detection of lung lesions in patients with pulmonary mycobacterial disease (Kim et al., 2010). The study population consisted of 100 patients (65 with mycobacterial disease and 35 control cases) imaged by both imaging modalities. Two chest radiologists independently interpreted the images and were instructed to indicate findings of mycobacterial disease in the images and to record the characteristics according to bronchiolitis, nodules, consolidation, cavities, and volume loss. CT was used as a reference standard and the observers matched the findings with information from CT. A higher detection of mycobacterial disease was found for chest tomosynthesis compared to chest radiography. Analysis of the cavities showed substantially more cavities detected using chest tomosynthesis. It was concluded that chest tomosynthesis is superior to chest radiography in detecting mycobacterial disease.

Most studies on chest tomosynthesis to date involve detection of lung nodules in comparison to those of chest radiography and chest tomosynthesis. With regard to subtler tasks such as examining dose reduction, various reconstruction algorithms, and acquisition parameters, it might be desirable to control the sizes, densities, and locations of lung nodules to obtain the necessary borderline cases. Tools for generating realistically looking nodules and adding them to disease-free patient images have been developed recently (Svalkvist and Båth, 2010, Svalkvist et al., 2010) and more observer experiments for optimization purposes are hence expected.

11.5.2 CLINICAL STUDIES COMPARING BT AND DM

In a pioneering work, Niklason et al. (1997) showed proof of concept of tomosynthesis. In the study, which involved four mastectomy specimens imaged with BT and screen-film mammography (SFM), the authors compared the visibility of a mass or a cluster of calcifications in each specimen. Three radiologists, experienced in mammography, scored lesion visibility, lesion margin visibility, and confidence in the classification of benign versus malignant lesions on a five-graded scale, in the slice image that best demonstrated the finding. The study showed that BT was superior to the conventional technique in three of the four cases. One of the first clinical studies involving patients that compared BT and conventional mammography was performed by Poplack et al. (2007), in which image quality and recall rates were compared in images of 98 patients recalled from screening mammography. Lesion conspicuity and visibility of mass margins and of calcifications were compared in BT and diagnostic SFM and recall rates were evaluated when BT was combined with DM. It was concluded that BT has similar or superior image quality compared to SFM in a diagnostic setting, and has the potential to reduce screening recall rates when used as an adjunct with DM.

Andersson et al. (2008) used visibility ratings and BIRADS classifications to compare 1-view BT with DM on a group of patients with abnormal findings. The BIRADS classifications were subject for VGA. The study showed substantially higher visibility ratings and upgrades in BIRADS classification (VGA: $p < 0.05$) for breast cancers imaged by BT. Förnvik et al. (2010) compared the accuracy of breast cancer size assessment of BT, DM, and US. An experienced radiologist performed the breast cancer size measurements on each modality. Pathology was used as a reference standard. The measured breast cancer sizes on BT and US correlated significantly better with pathology ($p < 0.05$) than those performed on DM. Tumor staging was also found to be significantly more accurate with BT than with DM.

11.5.2.1 Accuracy of breast cancer detection in BT versus DM

In general, clinical ROC-based studies and studies on sensitivity and specificity involve more variability than nonblinded studies, since they consist of blinded reading of a case-mix of abnormal and normal cases. At the time of writing, these study types have shown diverging results. Table 11.3 summarizes the main results of reported studies that compared 2-view digital mammography (2VDM) and variants of BT modalities in terms of diagnostic accuracy. In the table, 2VDM is the baseline modality to which the other modalities (column 2) were compared. Some studies employed multiple modalities and for these studies only the BT modality with the most image information is shown. For example, Gur et al. (2009) examined four modalities: 2VDM, 11 projection views, 2-view BT (2VBT) and 2VDM combined with 2VBT (denoted: 2VDM + 2VBT) but results are only listed for the 2VDM + 2VBT modality. Various observer data acquisition methods were used as listed in column 3. Good et al. (2008) and Svahn et al. (2010) used the free-response method while the rest used binary or ROC-based methods. In general, BIRADS or BIRADS-based ratings were used and one study used a binary rating (i.e., recall vs. no-recall). Some studies used a single measure of diagnostic accuracy, while some used more than one. Listed in column 4 are the numbers of abnormal and normal cases (which included breasts with benign lesions). Column 5 lists the numbers of readers in the study. The final column lists the difference in diagnostic accuracy measure (Δ) for the baseline modality (2VDM) minus the BT modality. If Δ is negative, it means that the BT modality has an improved performance compared to 2VDM, and if this improvement is significant at the 5% level, it is indicated with an asterisk.

Three of the studies presented in Table 11.3 found a statistically significant improved overall performance for BT, that is, when both the effects on sensitivity and specificity were considered. In the study by Gur et al. (2009), the specificity of 2VDM + 2VBT was found to be significantly higher than that of 2VDM when reader variability was ignored but the sensitivity

Table 11.3 List of published studies investigating breast tomosynthesis and digital mammography for breast cancer detection with regard to diagnostic accuracy and/or sensitivity and specificity

STUDY	BT MODALITY	DATA ACQUISITION FOM ANALYSIS	# ABN # NOR	# READERS	Δ(FOM), *p-VALUE < 0.05
Good et al. (2008)	2VBT	FROC θ NH: $\Delta\theta = 0$	25 5	9	$\Delta\theta = -0.020$
Gur et al. (2009)	2VDM + 2VBT	Binary SNS, SPC NH1: Δ (SNS) = 0 NH2: Δ (SPC) = 0	35 90	8	Δ (SNS) = −0.050 Δ (SPC) = −0.120* (reader variability ignored)
Teertstra et al. (2010)	2VBT	ACC, SNS, SPC NH: Δ (SNS) = 0, etc.	112 828	1	Δ (ACC) = 0.015 Δ (SNS) = 0.000 Δ (SPC) = 0.017
Gennaro et al. (2010)	1VBT	ROC AUC NH: Δ (AUC) = 0 &, noninferiority test	63 313	6	Δ (AUC) = −0.015 1VBT not inferior to 2VDM at 5% margin
Svahn et al. (2010)	1VDM + 1VBT	FROC θ NH: $\Delta\theta = 0$	25 25	5	$\Delta\theta = -0.114$*
Michell et al. (2010)	1VBT	ROC AUC NH: Δ (AUC) = 0	111 390	8	Δ (AUC) = −0.045*

Note: Only the primary analysis is presented. JAFROC, area under AFROC curve; AUC, area under the ROC curve; sensitivity, SNS; specificity, SPC; accuracy, ACC; normal, normal or benign.

difference was not significant and neither was the combined measure of these metrics. Table 11.3 shows no obvious trend between diagnostic accuracy of BT in relation to 2VDM and type of BT modality, data acquisition, and number of cases or readers.

11.5.2.2 BT in one or two views or used in adjunction with DM?

Rafferty et al. (2006) examined the use of one or two views (MLO or CC) in BT with regard to breast cancer visibility. Thirty-four patients were imaged with BT in the MLO and CC projection and subsequently underwent biopsy. Thirty-five percent of the lesions were more visible or only visible on either the MLO or the CC acquisition. In 33% of the cases, the MLO view was superior, while in 67% of the cases, the CC acquisition was superior. The study implies that it is important to acquire data in both views. It has also been demonstrated in ROC and FROC studies for enriched populations that performance is significantly improved when using 2-view BT (MLO and CC views) (Zanca et al., 2012) or BT MLO view combined with DM CC view (Svahn et al., 2010), in comparison to 2-view DM. In the same studies, the performance of 1-view BT was reported not to be significantly better than that of 2-view DM, but resulted in similar or higher performance. In a more recent study (Svahn et al., 2012b), containing a substantially larger population of cases than previously (Svahn et al., 2010), the performance of 1-view BT was also shown to be significantly better than that of 2-view DM. Some practical considerations that could favor a combined modality of BT MLO and DM CC views compared to 2-view BT might be (1) a faster reading time, (2) less effort to mentally fuse the information of the views, (3) a higher resolution with regard to the morphology of calcifications, and (4) possibly a gentler transition for the radiologists from the conventional 2-view DM in relation to 1-view BT or 2-view BT (e.g., lower learning curve effects). One disadvantage of the combined modality could be a lower performance than that of 2-view BT, which would be of interest to examine.

11.5.2.3 Reading time in BT and DM

A concern is the increased BT reading time compared to DM (Tingberg and Zackrisson, 2011) and as shown in ROC studies using patient images (Gur et al., 2009, Wallis et al., 2012, Zanca et al., 2012), it depends on views used for the BT modality (Table 11.4). Reading time is an important factor, especially in the screening context when a large number of cases have to be read

daily. Lång et al. (2011) investigated volume rendering at three different frame rates: 25 (fast), 14 (medium), 9 (slow) fps of BT volumes with added simulated lesions (calcifications and masses) viewed as cine loops in a FROC study but found that there were no substantial differences in detection accuracy between the different viewing procedures. Visualization tools must be developed for efficient reading and extensive work currently being done in this field to aid in the process of viewing, detection, and classification (Chen et al., 2007, Reiser et al., 2008).

11.5.2.4 Radiologist experience in breast tomosynthesis

It can be hypothesized that because BT has the potential to provide "clearer" images, for instance, by increased breast cancer visibility, the detection task should be less dependent on experience. As a result, the performance difference between BT and DM should be larger for less experienced radiologists, for example, the performance would go up on BT but remain lower on DM. Three studies have investigated this hypothesis so far: (1) Smith et al. (2008) performed an ROC study on BT (MLO, CC) combined with DM (MLO, CC) versus DM (MLO, CC), and did not find any such correlation but found that all readers improved their performance with the BT modality. (2) Wallis et al. (2012) examined the performances of BT (MLO, CC) and BT (MLO) versus that of DM (MLO, CC), also in an ROC study. No evidence of improved performance was found for experienced readers when using the BT modalities, while the less experienced readers performed significantly better on BT (MLO, CC) than on DM (MLO, CC), hence suggesting that the hypothesis is true. The difference was primarily due to a lower performance on DM for the less experienced radiologists, while the performance remained relatively stable on the BT modality for both experience categories. (3) In the third study (Svahn et al., 2012a), the performance of BT (MLO) was evaluated versus that of DM (MLO, CC) in relation to radiologist experience in a FROC study. The derived search-model parameter ν (Chakraborty, 2006), which denotes the probability that a breast cancer was considered for marking, increased in a linear fashion with increased experience. The search-model parameter μ (Chakraborty, 2006), which characterizes the ability of the reader to extract information from a signal site during cognitive evaluation and is influenced by external factors (e.g., complexity of the surround and lesion contrast) and observer-dependent factors (e.g., eyesight and expertise), was generally higher for BT with increased experience and the parameter difference between

Table 11.4 Average reading time in seconds and standard deviations per case for DM and various BT modalities reported from ROC Studies

STUDY	2VDM	1VBT	2VBT	2VBT + 2VDM
Gur et al. (2009)	73 ± 69.0		123 ± 87.6 (68)	143 ± 99.0 (96)
Zanca et al. (2012) Experienced radiologists	79 ± 10.4 74 ± 22.3	99 ± 18.7 (34)	134 ± 14.9 (70)	
Zanca et al. (2012) Less experienced radiologists	56 ± 8.9 74 ± 10.1	94 ± 23.6 (27)	115 ± 16.5 (105)	
Average	70.5 ± 24.1	96.5 ± 21.1 (37)	124 ± 39.7 (76)	143 ± 99.0 (96)

Note: Zance et al. (2012) examined reading times for both experienced radiologists (≥10 years of experience) and inexperienced radiologists (<10 years of experience). There were two groups of readers within each experience category depending on the modalities that were compared. The numbers in the parentheses refers to mean increase in reading time (%) in relation to standard 2-view digital mammography.

System performance

BT and DM (e.g., $\Delta\mu$) was larger for experienced readers. In contrast to the results of the two other studies, the performance difference was statistically significant for highly experienced and experienced radiologists but not for the less experienced readers. The less experienced radiologists had lower performance on BT, while the performance on DM was similar with regard to the various experience levels. In summary, the results of the studies show that while BT has the potential to provide "clearer" images to the radiologists than DM and may be particularly beneficial for less experienced radiologists, there might be detection and interpretation tasks in BT that are highly experience-dependent (Smith et al., 2008, Svahn et al., 2012a, Wallis et al. 2012).

11.6 CONCLUDING REMARKS

The papers described in this chapter generally report a benefit in the use of tomosynthesis, which due to suppression of tissue overlap often relates to an increased visibility of masses and in some cases also a greater clarity for microcalcifications (Kopans et al., 2011). Tomosynthesis has been evaluated in a variety of ways—from easier-to-conduct studies with regard to their setup and established truth (gold standard) to more advanced studies with results on benefits that are more valid for the general population. One way of assessing the clinical potential of tomosynthesis is by comparing it to a conventional method. This has been done extensively; for BT by comparing it to digital mammography, and for chest tomosynthesis by comparing it to chest radiography or CT. Below follows some points that partially summarize the contents in this chapter:

- There are various types of observer experiments available. Each of them has strengths and weaknesses. What type of observer experiment one should choose depends on the type of images available and the intention of the study. The closer the observer experiment resembles clinical interpretation of images, the higher the reliability and generalizability of the results of the study.
- The type of images to be chosen for a study also affects the reliability and generalizability of the results of the study. Test phantoms (e.g., contrast-detail phantom) are useful for quality control and similar tests, but should not be used for optimizing, for example, exposure parameters, since they do not consider the anatomic noise. Studies based on patients resemble the clinical situation best.
- The collection of images is critical for achieving a successful outcome of a study. A careful sampling strategy is needed to adequately represent a larger patient population.
- In most study types, the truth (gold standard) has to be known. The determination of truth can be time-consuming when patient images are used. Images with simulated pathology added to a normal image can be used with less effort; however, the introduced lesions have to be carefully validated with respect to their resemblance to real lesions, optimally, also regarding affected surrounding tissues such as derangement of anatomy.
- The clinical occurrence (frequency) and difficulty level of lesion (visibility, conspicuity, contrast, etc.) need to be considered in order to understand the clinical relevance, regardless of type of images used (clinical or partially/fully simulated images).

- In the planning stage of diagnostic accuracy studies, it is recommended to follow the STARD list to be aware of details on patient sampling, reference standard, and statistical methods.
- A training session before the actual study (with data collection) is absolutely essential to reduce inter- and intraobserver variance.

ACKNOWLEDGMENTS

The authors acknowledge Dr. Dev P. Chakraborty and Dr. Ingrid Reiser for contributions to this book chapter, and Dr. Ingvar Andersson, Professor Sören Mattsson, and Dr. Magnus Båth, for helpful comments. We would like to thank Dr. Federica Zanca who contributed with datasets. One of the authors (TS) was supported by the Franke and Margareta Bergqvist Foundation and by the Swedish Cancer Foundation, and the other (AT) was supported by Regionalt forskningsstöd from Södra Sjukvårdsregionen.

REFERENCES

American College of Radiology. 1998. *American College of Radiology. Breast Imaging Reporting and Data System (BI-RADS)*. 3rd ed. Reston, VA, USA.

Andersson, I., Ikeda, D. M., Zackrisson, S., Ruschin, M., Svahn, T., Timberg, P., and Tingberg, A. 2008. Breast tomosynthesis and digital mammography: A comparison of breast cancer visibility and BIRADS classification in a population of cancers with subtle mammographic findings. *Eur Radiol*, 18, 2817–2825.

Bamber, D. 1975. The area above the ordinal dominance graph and the area below the receiver operating graph. *J Math Psych*, 12, 387–415.

Barrett, H. H., Abbey, C. K., and Clarkson, E. 1998. Objective assessment of image quality. III. ROC metrics, ideal observers, and likelihood-generating functions. *J Opt Soc Am A Opt Image Sci Vis*, 15, 1520–1535.

Beam, C. A., Layde, P. M., and Sullivan, D. C. 1996. Variability in the interpretation of screening mammograms by US radiologists. Findings from a national sample. *Arch Intern Med*, 156, 209–213.

Berbaum, K. S., Franken, E. A. J., Caldwell, R. T., and Schartz, K. M. 2010. Satisfaction of search in traditional radiographic imaging. In: Samei, E., Krupinski, E. and Jacobson, F. (eds.) *The Handbook of Medical Image Perception and Techniques*. Cambridge, England: Cambridge University Press.

Bunch, P. C., Hamilton, J. F., Sanderson, G. K., and Simmons, A. H. 1978. A free-response approach to the measurement and characterization of radiographic observer performance. *J Appl Photogr Eng*, 4, 166–171.

Burgess, A. E. 1995. Comparison of receiver operating characteristic and forced choice observer performance measurement methods. *Med Phys*, 22, 643–655.

Burgess, A. E., Wagner, R. F., Jennings, R. J., and Barlow, H. B. 1981. Efficiency of human visual signal discrimination. *Science*, 214, 93–94.

Båth, M. 2010. Evaluating imaging systems: Practical applications. *Radiat Prot Dosimetry*, 139, 26–36.

Båth, M. and Månsson, L. G. 2007. Visual grading characteristics (VGC) analysis: A non-parametric rank-invariant statistical method for image quality evaluation. *Br J Radiol*, 80, 169–176.

Börjesson, S., Håkansson, M., Båth, M., Kheddache, S., Svensson, S., Tingberg, A., Grahn, A. et al. 2005. A software tool for increased efficiency in observer performance studies in radiology. *Radiat Prot Dosimetry*, 114, 45–52.

Chakraborty, D. P. 1989. Maximum likelihood analysis of free-response receiver operating characteristic (FROC) data. *Med Phys*, 16, 561–568.

Chakraborty, D. P. 2000. The FROC, AFROC and DROC variants of the ROC analysis. In: Beutel, J., Kundel, H. L. and Van Metter, R. L. (eds.) *Handbook of Medical Imaging. Volume 1. Physics and Psychophysics*. Bellingham, USA: SPIE Press.

Chakraborty, D. P. 2006. ROC curves predicted by a model of visual search. *Phys Med Biol*, 51, 3463–3482.

Chakraborty, D. P. (ed.) 2009. *Recent Development in FROC Methodology* Cambridge, UK: Cambridge University Press.

Chakraborty, D. P. 2010. Prediction accuracy of a sample-size estimation method for ROC studies. *Acad Radiol*, 17, 628–638.

Chakraborty, D. P. 2011. New developments in observer performance methodology in medical imaging. *Seminars Nucl Med*, 41, 401–418.

Chakraborty, D. P. and Berbaum, K. S. 2004. Observer studies involving detection and localization: Modeling, analysis, and validation. *Med Phys*, 31, 2313–2330.

Chakraborty, D. P. and Svahn, T. 2011. Estimating the parameters of a model of visual search from ROC data: An alternate method for fitting proper ROC curves. *SPIE Medical Imaging*. Orlando, FL: SPIE Press, 7966, 79660L-1-9.

Chakraborty, D. P. and Winter, L. H. 1990. Free-response methodology: Alternate analysis and a new observer-performance experiment. *Radiology*, 174, 873–881.

Chen, Y., Lo, J. Y., and Dobbins, J. T., 3RD 2007. Importance of point-by-point back projection correction for isocentric motion in digital breast tomosynthesis: Relevance to morphology of structures such as microcalcifications. *Med Phys*, 34, 3885–3892.

Das, M., Gifford, H. C., O'connor, J. M., and Glick, S. J. 2011. Penalized maximum likelihood reconstruction for improved microcalcification detection in breast tomosynthesis. *IEEE Trans Med Imaging*, 30, 904–914.

Deluca, P. M., Wambersie, A., and Whitmore, G. F. 2008. Extensions to conventional ROC methodology: LROC, FROC, and AFROC. *J ICRU*, 8, 31–35.

Dobbins, J. T., Mcadams, H. P., Song, J.-W., Li, C. M., Godfrey, D. J., Delong, D., Paik, S.-H., and Martinez-Jimenez, S. 2008. Digital tomosynthesis of the chest for lung nodule detection: Interim sensitivity results from an ongoing NIH-sponsored trial. *Med Phys*, 35, 2554–2557.

Dorfman, D. D., Berbaum, K. S., and Metz, C. E. 1992. Receiver operating characteristic rating analysis. Generalization to the population of readers and patients with the jackknife method. *Invest Radiol*, 27, 723–731.

Efron, B. 1982. *The Jackknife, the Bootstrap and Other Resampling Plans*, Montpelier: Capital City Press.

Egan, J. P., Greenberg, G. Z., and Schulman, A. I. 1961. Operating characteristics, signal detectability and the method of free response. *J Acoust Soc Am*, 33, 993–1007.

European Commission. 1996. *European Guidelines on Quality Criteria on Diagnostic Radiographic Images*, Brussels, EUR 16260.

Förnvik, D., Zackrisson, S., Ljungberg, O., Svahn, T., Timberg, P., Tingberg, A., and Andersson, I. 2010. Breast tomosynthesis: Accuracy of tumor measurement compared with digital mammography and ultrasonography. *Acta Radiol*, 51, 240–247.

Gennaro, G., Toledano, A., Di Maggio, C., Baldan, E., Bezzon, E., La Grassa, M., Pescarini, L. et al. 2010. Digital breast tomosynthesis versus digital mammography: A clinical performance study. *Eur Radiol*, 20, 1545–1553.

Goddard, M. J. and Hinberg, I. 1990. Receiver operator characteristic (ROC) curves and non-normal data: An empirical study. *Stat Med*, 9, 325–337.

Good, W. F., Abrams, G. S., Catullo, V. J., Chough, D. M., Ganott, M. A., Hakim, C. M., and Gur, D. 2008. Digital breast tomosynthesis: A pilot observer study. *AJR Am J Roentgenol*, 190, 865–869.

Gur, D., Abrams, G. S., Chough, D. M., Ganott, M. A., Hakim, C. M., Perrin, R. L., Rathfon, G. Y., Sumkin, J. H., Zuley, M. L., and Bandos, A. I. 2009. Digital breast tomosynthesis: Observer performance study. *AJR Am J Roentgenol*, 193, 586–591.

Gur, D. and Rockette, H. E. 2008. Performance assessments of diagnostic systems under the FROC paradigm experimental, analytical, and results interpretation issues. *Acad Radiol*, 15, 1312–1315.

Hajian-Tilaki, K. O., Hanley, J. A., Joseph, L., and Collet, J. P. 1997. A comparison of parametric and nonparametric approaches to ROC analysis of quantitative diagnostic tests. *Med Decision Making Int J Soc Med Decision Making*, 17, 94–102.

Hanley, J. A. and Mcneil, B. J. 1982. The meaning and use of the area under a receiver operating characteristic (ROC) curve. *Radiology*, 143, 29–36.

Hemdal, B. 2009. Evaluation of Absorbed Dose and Image Quality in Mammography. PhD thesis, Lund University.

Hillis, S. L. and Berbaum, K. S. 2004. Power estimation for the Dorfman-Berbaum-Metz method. *Acad Radiol*, 11, 1260–1273.

Håkansson, M., Svensson, S., Zachrisson, S., Svalkvist, A., Båth, M., and Månsson, L. G. 2010. VIEWDEX: An efficient and easy-to-use software for observer performance studies. *Radiat Prot Dosimetry*, 139, 42–51.

Jacobs, J., F. Zanca, and Bosmans, H. 2011. A novel platform to simplify human observer performance experiments in clinical reading Environments. *SPIE Medical Imaging*. Orlando, FL: SPIE Press. 7966, 79660B-1-9.

Johnsson, A. A., Svalkvist, A., Vikgren, J., Boijsen, M., Flinck, A., Kheddache, S., and Bath, M. 2010. A phantom study of nodule size evaluation with chest tomosynthesis and computed tomography. *Radiat Prot Dosimetry*, 139, 140–143.

Kheddache, S., Thilander-Klang, A., Lanhede, B., Mansson, L. G., Bjurstam, N., Ackerholm, P., and Bjorneld, L. 1999. Storage phosphor and film-screen mammography: Performance with different mammographic techniques. *Eur Radiol*, 9, 591–597.

Kim, E. Y., Chung, M. J., Lee, H. Y., Koh, W. J., Jung, H. N., and Lee, K. S. 2010. Pulmonary mycobacterial disease: Diagnostic performance of low-dose digital tomosynthesis as compared with chest radiography. *Radiology*, 257, 269–277.

Kopans, D., Gavenonis, S., Halpern, E., and Moore, R. 2011. Calcifications in the breast and digital breast tomosynthesis. *Breast J*, 17, 638–644.

Kundel, H. 1988. The evaluation of observer performance. In: Peppler, W. W. and Alter, A. A. (eds.) *Chest Imaging Conference*, Madison: Medical Physics Publishing Corp., 291–295.

Lång, K., Zackrisson, S., Holmqvist, K., Nyström, M., Andersson, I., Förnvik, D., Tingberg, A., and Timberg, P. 2011. Optimizing viewing procedures of breast tomosynthesis image volumes using eye tracking combined with a free response human observer study. *SPIE Medical Imaging*. Orlando, FL: SPIE Press. 7966, 796602-1-11.

Mcneil, B. J. and Adelstein, S. J. 1976. Determining the value of diagnostic and screening tests. *J Nucl Med Official Publication, Soc Nucl Med*, 17, 439–448.

Metz, C. E. 1978. Basic principles of ROC analysis. *Seminars Nucl Med*, 8, 283–298.

Metz, C. E. 1986. ROC methodology in radiologic imaging. *Invest Radiol*, 21, 720–733.

Metz, C. E. 1989. Some practical issues of experimental design and data analysis in radiological ROC studies. *Invest Radiol*, 24, 234–245.

Metz, C. E. 2000. Fundamental ROC analysis. In: Beutel, J., Kundel, H. L., and Van Metter, R. L. (eds.) *Handbook of Medical Imaging. Volume 1. Physics and Psychophysics*. Bellingham, USA: SPIE Press.

Michell, M., Wasan, R., Iqbal, A., Peacock, C., Evans, D., and Morel, J. 2010. Two-view 2D digital mammography versus one-view digital breast tomosynthesis. *Breast Cancer Res*, 12(Suppl 3), S3.

Miller, H. 1969. The FROC curve: A representation of the observer's performance for the method of free response. *J Acoustical Soc Am*, 46, 1473–1476.

Månsson, L. G. 1994. Evaluation of Radiographic Procedures. Investigations Related to Chest Imaging. Thesis. PhD, Göteborg University.

Månsson, L. G., Båth, M., and Mattsson, S. 2005. Priorities in optimisation of medical x-ray imaging—A contribution to the debate. *Radiat Prot Dosimetry*, 114, 298–302.

Niklason, L. T., Christian, B. T., Niklason, L. E., Kopans, D. B., Castleberry, D. E., Opsahl-Ong, B. H., Landberg, C.E et al. 1997. Digital tomosynthesis in breast imaging. *Radiology*, 205, 399–406.

Obuchowski, N. A. and Rockette, H. E. 1995. Hypothesis-testing of diagnostic-accuracy for multiple readers and multiple tests—An anova approach with dependent observations. *Commun Stat Simulat Comput*, 24, 285–308.

Pan, X. and Metz, C. E. 1997. The "proper" binormal model: Parametric receiver operating characteristic curve estimation with degenerate data. *Acad Radiol*, 4, 380–389.

Perry, N., Broeders, M., De Wolf, C., Törnberg, S., Holland, R. and Von Karsa, L. (eds.) 2006. *European Guidelines for Quality Assurance in Breast Cancer Screening and Diagnosis. Fourth Edition.* Luxembourg: European Commission.

Pisano, E. D., Hendrick, R. E., Yaffe, M. J., Baum, J. K., Acharyya, S., Cormack, J. B., Hanna, L. A. et al. 2008. Diagnostic accuracy of digital versus film mammography: Exploratory analysis of selected population subgroups in DMIST. *Radiology*, 246, 376–383.

Poplack, S. P., Tosteson, T. D., Kogel, C. A., and Nagy, H. M. 2007. Digital breast tomosynthesis: initial experience in 98 women with abnormal digital screening mammography. *AJR Am J Roentgenol*, 189, 616–623.

Quaia, E., Baratella, E., Cernic, S., Lorusso, A., Casagrande, F., Cioffi, V., and Cova, M. A. 2012. Analysis of the impact of digital tomosynthesis on the radiological investigation of patients with suspected pulmonary lesions on chest radiography. *Eur Radiol*, 22, 1912–1922.

Quaia, E., Baratella, E., Cioffi, V., Bregant, P., Cernic, S., Cuttin, R., and Cova, M. A. 2010. The value of digital tomosynthesis in the diagnosis of suspected pulmonary lesions on chest radiography: Analysis of diagnostic accuracy and confidence. *Acad Radiol*, 17, 1267–1274.

Rafferty, E. A., Jameson-Meehan, L., and Niklason, L. 2006. Breast Tomosynthesis: One View or Two? Scientific Assembly and Annual Meeting of the Radiological Society of North America, Chicago, IL, USA.

Reiser, I., Nishikawa, R. M., Edwards, A. V., Kopans, D. B., Schmidt, R. A., Papaioannou, J., and Moore, R. H. 2008. Automated detection of microcalcification clusters for digital breast tomosynthesis using projection data only: A preliminary study. *Med Phys*, 35, 1486–1493.

Rose, A. 1948. The sensitivity performance of the human eye on an absolute scale. *J Opt Soc Am*, 38, 196–200.

Ruschin, M. 2006. The Role of Projected Anatomy, Random Noise, and Spatial Resolution on Clinical Image Quality in Digital Mammography. PhD thesis, Lund University.

Ruschin, M., Timberg, P., Svahn, T., Andersson, I., Hemdal, B., Mattsson, S., Båth, M., and Tingberg, A. 2007. Improved in-plane visibility of tumors in patient breast tomosynthesis images compared to 2D digital mammograms. *SPIE Medical Imaging*. San Diego, CA: SPIE Press. 6510, 65101J-1-11.

Skaane, P. 2009. Studies comparing screen-film mammography and full-field digital mammography in breast cancer screening: Updated review. *Acta Radiol*, 50, 3–14.

Smedby, O. and Fredrikson, M. 2010. Visual grading regression: Analysing data from visual grading experiments with regression models. *Br J Radiol*, 83, 767–775.

Smedby, Ö., Fredrikson, M., De Geer, J., Borgen, L., and Sandborg, M. 2013. Quantifying the potential for dose reduction with visual grading regression. *Br J Radiol*, 86, 31197714.

Smedby, Ö., Fredrikson, M., De Geer, J., and Sandborg, M. 2012. Visual grading regression with random effects. *SPIE Medical Imaging*. San Diego, CA: SPIE Press. 8318, 831805-1-5.

Smith, A. P., Rafferty, E. A., and Niklason, L. 2008. Clinical performance of breast tomosynthesis as a function of radiologist experience level. *LNCS*, 5116, 61–66.

Sox, H., Blatt, M., Higgins, M., and Marton, K. 1988. *Medical Decision Making*, Boston, MA, USA: Butterworth-Heinemann.

Sund, P., Herrmann, C., Tingberg, A., Kheddache, S., Månsson, L. G., Almén, A., and Mattsson, S. 2000. Comparison of two methods for evaluating image quality of chest radiographs. SPIE Medical Imaging. San Diego, CA, USA. SPIE Press. 3981, 251–257.

Svahn, T., Andersson, I., Chakraborty, D., Svensson, S., Ikeda, D., Förnvik, D., Mattsson, S., Tingberg, A., and Zackrisson, S. 2010. The diagnostic accuracy of dual-view digital mammography, single-view breast tomosynthesis and a dual-view combination of breast tomosynthesis and digital mammography in a free-response observer performance study. *Radiat Prot Dosimetry*, 139, 113–117.

Svahn, T., Hemdal, B., Ruschin, M., Chakraborty, D. P., Andersson, I., Tingberg, A., and Mattsson, S. 2007a. Dose reduction and its influence on diagnostic accuracy and radiation risk in digital mammography: An observer performance study using an anthropomorphic breast phantom. *Br J Radiol*, 80, 557–562.

Svahn, T., Lång, K., Andersson, I., and Zackrisson, S. 2012a. Differences in radiologists' experiences and performances in breast tomosynthesis. *LNCS* 7361, 377–385.

Svahn, T., Ruschin, M., Hemdal, B., Nyhlén, L., Andersson, I., Timberg, P., Mattsson, S., and Tingberg, A. 2007b. In-plane artifacts in breast tomosynthesis quantified with a novel contrast-detail phantom. *SPIE Medical Imaging*, San Diego, CA, USA: SPIE Press, 65104R-1–65104R-12.

Svahn, T. M., Chakraborty, D. P., Ikeda, D., Zackrisson, S., Do, Y., Mattsson, S., and Andersson, I. 2012b. Breast tomosynthesis and digital mammography: A comparison of diagnostic accuracy. *Br J Radiol*, 85, e1074–82.

Svalkvist, A. and Båth, M. 2010. Simulation of dose reduction in tomosynthesis. *Med Phys* 37(1), 258–269.

Svalkvist, A., Hakansson, M., Ullman, G., and Bath, M. 2010. Simulation of lung nodules in chest tomosynthesis. *Radiat Prot Dosimetry*, 139, 130–139.

Swensson, R. G. 1996. Unified measurement of observer performance in detecting and localizing target objects on images. *Med Phys*, 23, 1709–1725.

Teertstra, H. J., Loo, C. E., Van den Bosch, M. A., Van Tinteren, H., Rutgers, E. J., Muller, S. H., and Gilhuijs, K. G. 2010. Breast tomosynthesis in clinical practice: Initial results. *Eur Radiol*, 20, 16–24.

Timberg, P., Bath, M., Andersson, I., Mattsson, S., Tingberg, A., and Ruschin, M. 2010. In-plane visibility of lesions using breast tomosynthesis and digital mammography. *Med Phys*, 37, 5618–5626.

Timberg, P., Båth, M., Andersson, I., Svahn, T., Ruschin, M., Hemdal, B., Mattsson, S., and Tingberg, A. 2008. Impact of dose on observer performance in breast tomosynthesis using breast specimens. *SPIE Medical Imaging*. San Diego, CA: SPIE Press. 6913, 69134J-1-10.

Tingberg, A., Båth, M., Håkansson, M., Medin, J., Besjakov, J., Sandborg, M., Alm-Carlsson, G., Mattsson, S., and Månsson, L. G. 2005. Evaluation of image quality of lumbar spine images: A comparison between FFE and VGA. *Radiat Prot Dosimetry*, 114, 53–61.

Tingberg, A., Herrmann, C., Lanhede, B., Almén, A., Sandborg, M., Mcvey, G., Mattsson, S. et al. 2004. The influence of the characteristic curve on the image quality of clinical radiographs. *Br J Radiol*, 77, 204–215.

Tingberg, A. and Zackrisson, S. 2011. Digital mammography and tomosynthesis for breast cancer diagnosis. *Expert Opin Med Diagn*, 5, 517–526.

Wallis, M. G., Moa, E., Zanca, F., Leifland, K., and Danielsson, M. 2012. Two-view and single-view tomosynthesis versus full-field digital mammography: High-resolution X-ray imaging observer study. *Radiology*, 262, 788–796.

Vecchio, S., Albanese, A., Vignoli, P., and Taibi, A. 2011. A novel approach to digital breast tomosynthesis for simultaneous acquisition of 2D and 3D images. *Eur Radiol*, 21, 1207–1213.

Vikgren, J., Zachrisson, S., Svalkvist, A., Johnsson, Å. A., Boijsen, M., Flinck, A., Kheddache, S., and Båth, M. 2008. Comparison of chest tomosynthesis and chest radiography for detection of pulmonary nodules: Human observer study of clinical cases. *Radiology*, 249, 1034–1041.

Yakabe, M., Sakai, S., Yabuuchi, H., Matsuo, Y., Kamitani, T., Setoguchi, T., Cho, M., Masuda, M., and Sasaki, M. 2010. Effect of dose reduction on the ability of digital mammography to detect simulated microcalcifications. *J Digit Imaging Official J Soc Comput Appl Radiol*, 23, 520–526.

Zachrisson, S., Vikgren, J., Svalkvist, A., Johnsson, A. A., Boijsen, M., Flinck, A., Mansson, L. G., Kheddache, S., and Bath, M. 2009. Effect of clinical experience of chest tomosynthesis on detection of pulmonary nodules. *Acta Radiol*, 50, 884–891.

Zanca, F., Wallis, M., Moad, E., Leifland, K., Danielsson, M., Oyen, R., and Bosmans, H. 2012. Diagnostic accuracy of digital mammography versus tomosynthesis: Effect of radiologists' experience. *SPIE Medical Imaging*. San Diego, CA, USA. SPIE Press. 8318: 83180W1–14.

Zhou, L., Oldan, J., Fisher, P., and Gindi, G. 2006. Low-contrast lesion detection in tomosynthetic breast imaging using a realistic breast phantom. *SPIE Medical Imaging*. San Diego: SPIE Press. 6142, 61425A-1-12.

Zhou, X.-H., Obuchowski, N. A., and Mcclish, D. K. 2011. *Statistical Methods in Diagnostic Medicine*, Hobroken, NJ, USA: Wiley.

Zoetelief, J., Fitzgerald, M., Leitz, W., and Säbel M. 1996. *European Protocol on Dosimetry in Mammography*, Brussels, Belgium: European Commission.

Zweig, M. H. and Campbell, G. 1993. Receiver-operating characteristic (ROC) plots: A fundamental evaluation tool in clinical medicine. *Clin Chem*, 39, 561–577.

Section V

Clinical applications

12 Clinical applications of breast tomosynthesis

Steven P. Poplack

Contents

12.1 CONVENTIONAL MAMMOGRAPHY

As evidenced by the most recent recommendations of the US Preventive Service Task Force, mammography has become the primary tool in breast cancer screening in women aged 50 y.o. to 74 y.o. (USPSTF 2009). Screening mammography has almost certainly contributed to the 2.2% annual reduction in death due to breast cancer in the United States over the 18-year period from 1990 to 2007 (American Cancer Society). Furthermore, there are probably additional benefits gleaned from detecting breast cancer with mammography above and beyond mortality reduction. In particular, it appears that breast cancer detected with screening mammography may require less toxic treatment regimens, with a decrease in the need for both mastectomy for local control and adjuvant chemotherapy for distal recurrence (Barth 2005).

Despite these accomplishments, screening mammography remains controversial in part due to notable limitations in both sensitivity, that is, the ability to detect cancer in women who have cancer, and specificity, that is, the ability to recognize normal or benign conditions in the absence of breast cancer. Mammography has often been touted as having a sensitivity approximating 90%. In reality, the sensitivity of mammography screening hovers around 75% and is greatly impacted by background breast composition (Poplack 2000). In fact in one large registry-based study, the sensitivity of mammography varied from a high of 88% in breasts composed predominantly of fat to a low of 62% in breasts that were composed mostly of fibroglandular tissue, that is, extremely dense breasts. Specificity is similarly dependent on background breast composition with a range of 97% in the fatty breast to 90% in the extremely dense breast (Carney 2003).

It had been hoped that digital mammography (DM) would mitigate the inaccuracies of film-screen mammography (FM) due to dense breast composition given its enhanced ability to penetrate dense breast tissue with higher energy x-rays, and still maintain high contrast resolution. The Digital Mammography Imaging Screening Trial (DMIST), which included nearly 50,000 women who underwent both FM and DM, was performed to evaluate the impact of digital technology on mammography outcomes. While the overall accuracy of FFDM was shown to be similar to FM, improvements were demonstrated in women with denser breast composition, age <50 y.o. and in pre- and perimenopausal women. In particular, the sensitivity of mammography in women with heterogeneously dense or extremely dense breasts improved by 14% (95%CI: 0.03–0.26) from 55% (FM) to 70% (FFDM), although there was a less striking 0.4% improvement from 90% to 91% in specificity (Pisano 2005).

Perhaps the biggest limitation to current mammography is its propensity for false-positive results and resulting low positive predictive value (PPV). PPV reflects the likelihood of cancer when the test (in this case mammography) is abnormal. One of the clinical benchmarks for screening mammography that reflects the PPV is the recall rate, that is, rate of screening exams that are interpreted as abnormal and recalled for additional testing, which, based on clinical best practice standards, has been set at or below 10% (Linver 1998). To illustrate the impact of a low PPV, one may look at a hypothetical high-quality mammography practice in which 10% of women undergoing screening mammography are recalled for additional imaging, 20% of them, based on additional imaging, are subsequently recommended for biopsy and 25% of those that undergo biopsy truly have cancer. In this example, the PPV of screening mammography (PPV[1]) is 5% (5/100 abnormal screens), while the PPV of biopsy (PPV[3]) is 25% (5/20). In one retrospective study of women aged 40–69 y.o, the false-positive rate of screening mammography over a 10-year period was 23.8% and the cumulative risk of a false positive after 10 screening mammogram exams was nearly 50%. The authors of this study estimated that approximately 25% of financial resources devoted to screening were spent on further evaluation of false-positive results, including additional imaging and biopsy (Elmore 1998). Of note, based on data from the DMIST trial, the advent of FFDM has offered a modest 1% improvement in PPV[1] from 3% to 4% (Pisano 2005).

12.2 BREAST TOMOSYNTHESIS: THEORETICAL ADVANTAGES

In theory, digital breast tomosynthesis (DBT) has the ability to address the mammographic limitations of diminished sensitivity and poor PPV by virtue of its tomographic nature. Unlike conventional mammography, which transmits x-rays through a volume of tissue and constructs an image of x-ray absorption of all the tissue in the path of the x-ray, DBT has the capacity to bring into focus a thin slice of tissue while blurring out all other tissues that are superficial and deep to the slice of interest because it takes a series of projection images from multiple different source positions. Many of the shortcomings of mammography are attributable to the composite projection of overlying fibroglandular tissue. In the case of cancer detection, dense fibroglandular tissue may obscure mammographic signs of malignancy resulting in hidden or "missed" cancers and decreased sensitivity. Even when a cancer is not completely hidden by dense tissue, a complex background may distract the interpreter's focus and obfuscate a subtle mammographic finding of malignancy, a "where's Waldo"-like phenomenon. (An example of this phenomenon is displayed in Figure 12.1.) On the other hand, fibroglandular tissue may superimpose to falsely mimic a mammographic abnormality leading to an unnecessary "recall" for additional imaging. Unlike many new breast cancer imaging technologies, which offer an improvement in one operating characteristic at the expense of another, for example, enhanced sensitivity with decreased specificity, DBT has the promise of improving accuracy on all fronts.

One of the other important theoretical benefits of DBT is its similarity to conventional mammography. While the interpretation of DBT involves review of a multitude of slices of breast tissue, each slice is viewed and interpreted similar to a conventional mammography projection image. In that regard, the technology does not require learning a new method of interpretation, and can thereby rely on interpretive practices and a knowledge base that have been gleaned over the 30+ year history of mammography. Furthermore, the routine of reviewing multiple contiguous tissue slices is commonplace in the overall radiology experience with cross-sectional imaging techniques such as computed tomography (CT) and magnetic resonance imaging (MRI).

12.3 CLINICAL APPLICATIONS

Mammography has two primary clinical applications, screening or the detection of early-stage breast cancer in asymptomatic women with standardized image projections, and diagnosis or the evaluation of a clinical (e.g., breast lump) or mammographic (e.g., screening recall) abnormality with standard and nonstandard image projections tailored to addressing the particular problem. DBT has been piloted in both of these areas with promising results.

12.3.1 SCREENING

To date, there have been no large-scale randomized trials of DBT so that screening mammography performance results of DBT relative to conventional technologies, such as DM, must be inferred from smaller less scientifically rigorous studies. One of the largest research screening mammography experiences with DBT comes from a multi-institutional multireader study conducted by Hologic Inc., a DBT manufacturer, for market approval. In this study, five academic institutions prospectively acquired routine screening exams (craniocaudal (CC) and mediolateral (MLO) views of each breast) using both DM (Selenia™, Hologic Inc.) and an investigational prototype DBT device (Hologic Inc.). In addition, a subset of patients with a recent (i.e., within 30 days) DM screening mammogram who were presenting for image-guided breast biopsy underwent routine screening views with DBT. From a cohort of 1192 subjects, two enriched reader studies were conducted, utilizing 312 cases with 48 (study 1) and 51 (study 2) malignant cases interpreted by 12 (study 1) and 15 (study 2) radiologist readers. The two studies differed based on the information provided by the reader and the nature of the reader training session that preceded the interpretive sessions. In study 1, readers noted the presence or the absence of an abnormality with breast imaging reporting and data system (BIRADS) and probability of malignancy scores, while in study 2, readers additionally noted the location and finding type of all abnormalities. Each reader viewed DM images alone, DM plus DBT MLO view, and DM plus DBT CC and MLO views and scored images using both BIRADS and probability of malignancy scoring systems. An aggregate receiver operator curve over all observers was generated to indicate detection accuracy. For all readers, the area under the curve (AUC) was largest for FFDM plus 2vDBT, then FFDM plus 1vDBT, and smallest for DM only. The mean AUC increased significantly by 7.2% (study 1) and 6.8% (study 2) when comparing DM alone to DM plus 2vDBT. Sensitivity increased by 10.7% (study 1) and 16.0% (study 2) with the largest increase in sensitivity due to invasive cancer detection. Recall rates for nonmalignant abnormalities decreased by a mean of 16.7% (study

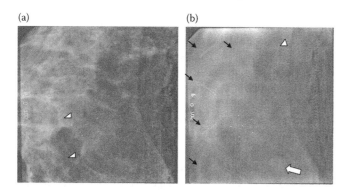

Figure 12.1 **(See insert.)** (a) DM and (b) DBT images of an American College of Radiology quality assurance phantom imbedded in a mastectomy specimen which has a heterogeneously dense breast composition. Two speck groups (arrowheads) are visible in the DM image (a), while the DBT image at the plane of interest reveals a third speck group (arrowhead), as well as five bar inclusions (arrows) and 1 mass inclusion (block arrow). (Courtesy of Andy Smith, PhD, Hologic, Inc.)

1) and 12.3% (study 2) and decreased significantly for all readers (Rafferty 2013). Representative images from a subject enrolled in this trial whose malignancy was more easily detected by DBT are displayed in Figure 12.2.

Other studies have demonstrated similar results. Gennaro and coworkers evaluated 376 breasts, including 63 malignancies, 177 biopsy-proven benign abnormalities, and 136 normal screening exams with six readers who underwent single-view (MLO) DBT using an investigational prototype (General Electric Inc.) in addition to 2-view DM (Senographe 2000D™, General Electric Inc.). The aggregate AUC was 0.851(DBT) versus 0.836(DM) and the authors concluded that single-view DBT was not inferior to 2-view DM (Gennaro 2010). One small study conducted by Andersson and coworkers provides insight into the cancer detection capability of DBT. In this study, single-view DBT using an investigational prototype (Siemens Inc., Erlangen, GE) was pitted against 2-view DM (Mammomat Novation DR™, Siemens Inc., Erlangen, Germany) in the evaluation of subtle malignancies. Subtle malignancies were gleaned from positive screening mammography of asymptomatic women or from symptomatic women with negative mammography and suspicious sonography. The DBT view was selected to correspond with the DM view in which the finding of malignancy was most subtle or the MLO view in mammographically occult cases. In this series, four cases were occult on both 2vDM and 1vDBT and 24 were clearly visible on both modalities. Two malignancies were visible on 1vDBT but not visible on 2vDM, while one case was visible on 2vDM but not visible on 1vDBT. Of note in the latter case that was detected by 2vDM but occult on DBT, the abnormality was not included on the DBT exam due to breast positioning (Andersson 2008).

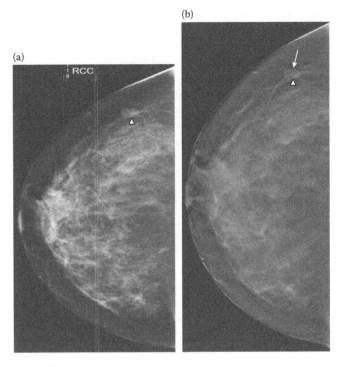

Figure 12.3 (**See insert.**) (a) Craniocaudal DM and (b) DBT images of a trial subject with a benign intra-mammary lymph node visualized as a nonspecific mass on DM (arrowhead) and a definitely benign intra-mammary lymph node (arrowhead) due to the presence of a fatty hilum (arrow) recognizable on DBT.

With regard to recall rate reduction, one study evaluated 99 women who were recalled from DM screening mammography (Selenia, Hologic Inc.) and had 1–3 adjunctive DBT views using an investigational prototype (Genesis™, Hologic Inc.) (Poplack 2007). When adjusted for recall threshold differences between the study and clinical mammographers, there was a 40% recall rate reduction when DBT was used adjunctively. Of the 45 exams that were not recalled because DBT was added, most (32/45) reflected superimposition-related abnormalities that were not seen at all on DBT, while a substantial minority (12/45) had recall averted because of better lesion characterization of a definitely benign entity (see Figure 12.3).

In summary, the available preliminary data suggests that DBT is a viable screening mammography technology whether it is used as a stand-alone modality or adjunctively with DM. When employed as a separate device, its performance appears to be at least equivalent to DM using receiver operator curve (ROC) analysis as the outcome measure. When used adjunctively along with DM, one may expect a significant reduction in recall rate and a modest increase in cancer detection.

12.3.2 DIAGNOSTIC IMAGING

The clinical application of DBT to diagnostic imaging evaluation has also been studied initially and appears to be a worthy alternative to DM. There have been a few studies that have compared DBT to diagnostic imaging with conventional mammography.

The largest of these studies was conducted by Teertstra and coworkers and involved 513 women with a diagnostic indication of screening recall or clinical abnormality (Teertstra 2010).

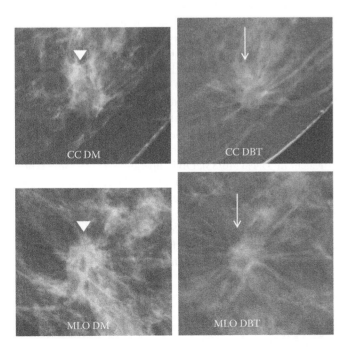

Figure 12.2 (**See insert.**) Cropped craniocaudal (CC) and medio-lateral oblique (MLO) digital mammography (DM) and digital breast tomosynthesis (DBT) images demonstrating an invasive ductal carcinoma seen as a vague asymmetry (arrowhead) with possible architectural distortion on DM and corresponding conspicuous mass with spiculated margins (arrow) on DBT.

Subjects were managed in the diagnostic clinic with standard practice, including comparison to prior imaging studies, and additional evaluation with mammography (Selenia, Hologic Inc.), sonography, or rarely MRI and BIRADS assessments were rendered by the clinical radiologist. Subjects underwent 2-view DBT with an investigational prototype (Hologic Inc.), though the DBT was evaluated and categorized with BIRADS at a later date (1–3months hence) by a single-study radiologist with clinical information but without the benefit of the presenting mammography exam, additional imaging, or comparison mammography. DBT and DM performance depended on the definition of a positive test result (BIRADS 0,3,4,5 vs. BIRADS 0,4,5). With BIRADS category 3 as a positive exam, DM outperformed DBT with a similar sensitivity of 92.9% and enhanced specificity 86.1% (DM) versus 84.4% (DBT). However, when BIRADS category 3 was considered to be a negative exam outcome, DBT was more accurate than DM, demonstrating a better sensitivity of 80% (DBT) versus 73% (DM) and a similar specificity of 96% (DBT) versus 97% (DM). The authors felt that the high rate of malignancy in the Probably Benign Category 3 subset was due to the high prevalence of cancer in this particular patient population. Of note, five of the eight cancers that were occult on DM were categorized as Suspicious Category 4 on DBT, and felt to be visible in retrospect on DM though less conspicuous. Two of the malignancies that were falsely negative on DBT were categorized as suspicious by DM although a note was made that in one there was insufficient positioning and in the other there was patient motion. While the combination of DBT and DM detected 109 of 112 cancers, the authors concluded that DBT was of little value in detecting additional malignancy given that these malignancies were readily diagnosed by other imaging technologies such as sonography and MRI.

It is difficult to interpret the results of this study in light of the vast differences in the conditions in which DBT and DM were interpreted. Nevertheless, it highlights some of the potential advantages and limitations of DBT. It is troubling that such a high proportion of Probably Benign (Category 3) assessments were malignant, which suggests the importance of adhering to the strict criteria of probably benign findings proposed by Sickles with both conventional mammography and DBT (see the discussion Section 12.3.3) (Sickles 1991). On the other hand, it emphasizes the ability of DBT to demonstrate abnormalities more conspicuously and thereby arrive at a diagnosis with a higher degree of confidence.

Some smaller studies of DBT in diagnostic evaluation with more traditional scientific study designs are maybe more illustrative of the capacity of DBT in this arena. Poplack and coworkers evaluated 99 women recalled from DM screening and matched DBT views (Genesis, Hologic Inc.) with the projections obtained at the time of film-screen (FS) (Lorad MIV, Hologic, Inc.) diagnostic evaluation. One study radiologist subjectively rated image quality based on finding type and modality, DBT versus DM. The authors found that DBT was equivalent (52%, 51/99) or superior (37%, 37/99) to FS diagnostic evaluation in 89% of the cases but that results were dependent on the nature of the finding. DBT was subjectively superior to FS diagnostic imaging for asymmetry, masses, and architectural distortion but

inferior to FS for calcifications. Indeed, calcifications accounted for 73% (8/11) of the instances in which DBT was considered inferior to FS. The authors felt that the long exposure time of 19 s associated with this DBT investigational prototype (Genesis, Hologic Inc.) likely contributed to this outcome (Poplack 2007). Hakim and coworkers demonstrated similar results when comparing DBT to DM diagnostic imaging of noncalcified mammographic abnormalities. In this study, four radiologists evaluated 25 women and subjectively rated the image quality of DM plus 2-view DBT (Gemini™, Hologic Inc.) versus DM plus DM diagnostic additional imaging. Of the 100 aggregate interpretations, DBT was deemed superior in 50% (50/100) and inferior in 19% (19/100), although of the 19 inferior ratings, 5 were considered inferior because the abnormality was not positioned within the field in one of the DBT views. Interestingly, the authors commented that in 12/100 aggregate ratings, sonography could have been avoided if DBT had been used rather than DM additional imaging (Hakim 2010).

Since the publication of the aforementioned studies, some additional data on the DBT evaluation of calcifications has become available. Spangler and coworkers reported on the evaluation of calcifications in 100 women who underwent both DM (Selenia, Hologic Inc.) and 2-view DBT utilizing an investigational prototype with 4.3 s acquisition time (Hologic Inc.) The cohort was composed of 40 normal screening exams, and 60 exams with abnormal calcifications including 40 biopsy-proven benign conditions and 20 malignancies. CC and MLO views from both modalities were rated by five study radiologists without access to additional diagnostic imaging or clinical information. Sensitivity for detecting calcifications was 84% for DM and 75% for DBT. More importantly, as it pertains to diagnostic imaging for the 198 calcification cases detected by readers across both modalities, and defining a positive result as a BIRADS assessment of 4 or 5, the sensitivity was 96.0% and 87.8%, and the specificity was 77.2% and 65.3% for DM and DBT, respectively. However, for the ROC analysis, there was no statistical difference in the AUC for FFDM, AUC = 0.76, and DBT, AUC = 0.72 (Spangler 2011). In contrast, Destounis and coworkers presented a comparative analysis of DM with DBT of 83 women with suspicious DM who underwent both DM and 2-view DBT. A single-study radiologist subjectively rated image quality as equivalent in 55.4% (46/83), DBT superior in 43.4% (36/83) or inferior in 1.2% (1/83) (Destounis 2011). The conflicting results of these two studies underline the current uncertain status of the utility of DBT evaluation for calcifications.

In summary, DBT appears to offer a possible advantage in the diagnostic evaluation of noncalcified imaging abnormalities when compared to conventional mammography. Published studies to date suggest that DBT is inferior to conventional mammography in the evaluation of calcifications, though newer data may call to question that assertion.

12.3.3 LIMITATIONS OF DBT

Despite its many possible advantages, DBT has limitations that should also be considered when adopting it to clinical practice. The potential disadvantages of this technology include challenges for work flow and implementation as well as pitfalls of DBT interpretation and the potential for additional radiation exposure.

While one of the great advantages to DBT is its similarity to conventional mammography, which allows for rapid learning, there are notable differences between the technologies that detract from the accuracy of DBT. Some limitations stem from the display of thin tomographic slices on which interpretation is based and may make it more difficult to detect and/or characterize certain types of mammographic abnormalities. For example, this may make it more difficult to characterize the distribution of calcifications. In the case of a cluster of calcifications, one or two particles noted on contiguous slices may be dismissed as benign and only appreciated as a cluster of calcifications when the multiple contiguous slices are summed. This same phenomenon may also be seen with calcifications in a segmental distribution (see Figure 12.4). Similarly, focal asymmetry may appear less dense and therefore be less apparent

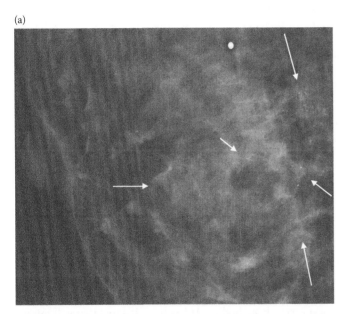

(a)

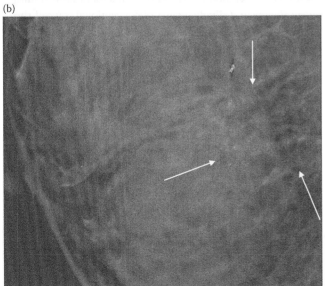

(b)

Figure 12.4 **(See insert.)** (a) DM image demonstrating malignant calcifications (arrows) in a segmental distribution versus (b) DBT thin slice reconstruction image of the same abnormality which shows clustered calcifications only.

when viewed on an individual thin slice reconstruction. This phenomenon can be easily addressed by reconstruction algorithms that create thicker slabs (e.g., ≥5 mm) of tissue rather than thin (e.g., 1 mm) slices or by having a corresponding projection image to cross-reference. Both these solutions are being pursued by current vendors of the DBT technology.

Another potential pitfall of interpretation relates to abandoning the usual mammographic interpretive practice. In one research study, a well-defined mass was misinterpreted as benign by some readers because of exquisitely circumscribed margin characteristics (Rafferty 2013). Some of the radiologists concluded that this mass was benign and did not require follow-up because it was so likely to be a cyst based on margin features in spite of a complex lobular shape. This was later shown to be a complex cystic mass by sonography, with subsequent pathology of an intracystic papillary carcinoma. While unusual, this highlights the importance of maintaining usual mammographic practice when interpreting DBT. At this point in time, there is no data to suggest that certain mammographic features are more predictive of benignity based on DBT evaluation than conventional mammography. However, this is not to say that additional information gleaned from DBT should not be factored into DBT evaluation. For example, experts have suggested that a screening exam with an aggregate of three or more masses with circumscribed margins can be considered benign due to the high negative predictive value of this finding (Leung 2000). DBT may provide the benefit of detecting more than two circumscribed masses, when a conventional projection mammogram might only visualize one or two circumscribed masses due to obscuration from superimposed fibroglandular tissue.

Like conventional mammography, DBT will also probably have some dependence on background breast composition. In the setting of the extremely dense breast, important findings may be obscured on DBT just as they are obscured on conventional mammography. Dense tissue within the plane of interest has the same limiting effect on DBT as it does with projection mammography.

Additional radiation exposure with the attendant risk of radiation-induced malignancy is also a concern for DBT. To date, two different approaches to mitigate this problem have been employed by DBT vendors. One approach is to reduce additional dose by limiting the exam to a single view. With this approach, the radiation dose from the DBT acquisition is typically twice that of DM exposure, as the beam quality of the DBT exam has been optimized for image clarity as opposed to dose reduction. Assuming the DBT dose is twice the usual radiation exposure as DM, by performing only a single DBT view, one arrives at the same cumulative dose, if DBT is employed as a stand-alone exam, that is, an alternative rather than an adjunct to DM. Most of the published literature, albeit limited to a few articles, suggests that single-view DBT is at least comparable if not more accurate than 2-view DM (Gennaro 2010). Another approach to lowering radiation exposure from DBT involves 2-view DBT with beam modification that minimizes radiation exposure through the selection of different x-ray target filter combinations, and results in an exposure that is comparable to the lower dose range of DM. One vendor has developed a DBT system with a tungsten/silver target/filter combination, which has less radiation exposure than a comparable DM unit that utilizes a molybdenum/molybdenum

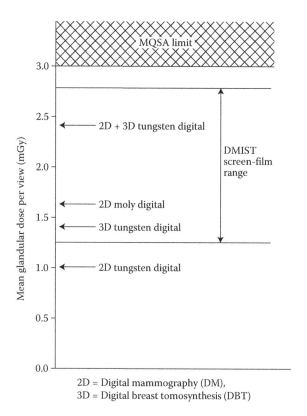

2D = Digital mammography (DM),
3D = Digital breast tomosynthesis (DBT)

Figure 12.5 Mean glandular dose in milliGray (mGy) of different anode types for DM and DBT with range of film–screen mammography exposure as recorded by the Digital Mammography Screening Trial (DMIST). (Courtesy of Jay Stein PhD, Hologic Inc.)

target/filter combination. This vendor's combined DM/DBT device has an aggregate radiation exposure that is greater than DM alone but well within the radiation dose limits legislated by the Mammography Quality Standards Act (Figure 12.5).

12.3.4 CLINICAL IMPLEMENTATION ISSUES

In addition to the limitiations noted above, there are challenges to the seamless integration of DBT into clinical practice. While technological advances in DBT have reduced acquisition and reconstruction times to the level of a DM exam, the time required for exam interpretation has almost certainly increased. Some of the increase relates to the learning curve and familiarity with a different method of image display. Conventional DM is interpreted at a digital workstation typically with two viewing monitors that display 1–4 images per monitor. Tailored "hanging" protocols can be programmed to sequence the flow of specified images, including comparison exams, and thereby minimize the time of interpretation. In contrast, DBT involves at least an order of magnitude greater number of images to be viewed. Most DBT vendors reconstruct the digital data set into 1-mm equivalent slice thickness, which means for an average breast that compresses to 4 cm, 40 1-mm slices must be viewed and interpreted. Dedicated DBT workstations allow for scrolling through images or automated cine display. Nevertheless, the sheer number of images requires additional interpretation time as well as concentration and effort. One pilot study evaluated the time for interpretation of 125 exams, including 30 exams with breast malignancy, of two view per breast DM versus DBT read by eight

experienced mammographers. The mean time of interpretation of DBT (of 2.39 min per exam) was more than twice that of DM (1.22 min). While this involved 2-view DBT rather than single-view DBT, all exams were viewed without comparisons, which would necessitate additional time (Gur 2009).

There are other workflow issues that represent significant challenges to incorporating DBT into clinical practice. DBT is a capital-intensive expensive technology, making it difficult for mammography facilities to convert all their mammography units from DM (or FS) to DBT. By financial necessity, many medium- to large-volume facilities will not be able to replace all their existing mammography units at once. Rather, DBT will be incorporated a unit or two at a time, which will require decisions on how and with whom to apply the new technology, that is, screening versus diagnostic imaging, first come first served or preferentially to specified subset populations like baseline exams or women with nonfatty breast composition. There are advantages and disadvantages to both these critical decisions. DBT offers great potential benefit in the screening population by reducing recalls and simultaneously improving cancer detection. However, it will require significantly more radiologist time to interpret the exam and initially may only be used in a subset of patients. In contrast, DBT will likely provide more accurate diagnostic imaging, and, due to the smaller overall numbers of diagnostic patients, may be applied to a larger proportion of diagnostic imaging patients. Furthermore, in many cases, diagnostic imaging that involves multiple views may well be replaced by one or two DBT views with resulting improved work flow.

In the current U.S. healthcare environment, DBT presents a number of challenges for successful integration into clinical practice. Until additional reimbursement is provided for the service, it will almost certainly "cost" practices additional time and financial outlay. Nevertheless, many mammographers and mammography facilities are anxious to begin using this technology for the improvement in breast imaging care that it almost certainly embodies. At last there is a tool that will mitigate some of the greatest disadvantages of conventional state-of-the-art mammography while at the same time improve detection of early-stage breast cancer and hopefully translate into improved breast cancer survivorship.

REFERENCES

American Cancer Society. Breast Cancer Facts & Figures 2011–2012. At: http://www.cancer.org/Research/CancerFactsFigures/BreastCancerFactsFigures/breast-cancer-facts-and-figures-2011-2012 [Accessed: May 25, 2012].

Andersson I, Ikeda DM, Zackrisson S, et al. Breast tomosynthesis and digital mammography: A comparison of breast cancer visibility and BIRADS classification in a population of cancers with subtle mammographic findings. *Eur Radiol*. 2008 Dec;18(12):2817–25.

Barth RJ Jr, Gibson GR, Carney PA, et al. Detection of breast cancer on screening mammography allows patients to be treated with less-toxic therapy. *AJR Am J Roentgenol*. 2005 Jan;184(1):324–9.

Carney PA, Miglioretti DL, Yankaskas BC, et al. Individual and combined effects of age, breast density, and hormone replacement therapy use on the accuracy of screening mammography. *Ann Intern Med*. 2003 Feb 4;138(3):168–75.

Destounis S, Murphy P, Seifert P, et al. Clinical experience with digital breast tomosynthesis in the characterization and visualization of breast microcalcifications. Abstract presented May 2, 2011. ARRS Annual Meeting, Scientific Session 1, no. 007.

Elmore JG, Barton MB, Moceri VM, et al. Ten-year risk of false positive screening mammograms and clinical breast examinations. *N Engl J Med*. 1998 Apr 16;338(16):1089–96.

Gennaro G, Toledano A, di Maggio C, et al. Digital breast tomosynthesis versus digital mammography: A clinical performance study. *Eur Radiol*. 2010 Jul;20(7):1545–53.

Gur D, Abrams GS, Chough DM, et al. Digital breast tomosynthesis: Observer performance study. *AJR Am J Roentgenol*. 2009 Aug;193(2):586–91.

Hakim CM, Chough DM, Ganott MA, et al. Digital breast tomosynthesis in the diagnostic environment: A subjective side-by-side review. *AJR Am J Roentgenol*. 2010 Aug;195(2):W172–6.

Leung JW, and Sickles EA. Multiple bilateral masses detected on screening mammography: Assessment of need for recall imaging. *AJR Am J Roentgenol*. 2000 Jul;175(1):23–9.

Linver MN, and Rosenberg RD. Callback rate after screening mammography. *AJR Am J Roentgenol*. 1998 Jul;171(1):262–3.

Pisano ED, Gatsonis C, Hendrick E, et al. Digital Mammographic Imaging Screening Trial (DMIST) Investigators Group. Diagnostic performance of digital versus film mammography for breast-cancer screening. *N Engl J Med*. 2005 Oct 27;353(17):1773–83.

Poplack SP, Tosteson AN, Grove MR, et al. Mammography in 53,803 women from the New Hampshire mammography network. *Radiology*. 2000 Dec;217(3):832–40.

Poplack SP, Tosteson TD, Kogel CA, Nagy HM. Digital breast tomosynthesis: Initial experience in 98 women with abnormal digital screening mammography. *AJR Am J Roentgenol*. 2007 Sep;189(3):616–23.

Rafferty EA, Park JM, Philpotts L, Poplack SP, Sumkins J, Halpern E, Niklason L. Assessing radiologist performance using combined digital mammography and breast tomosynthesis compared to digital mammography alone: Results of a multi-center, multi-reader trial. *Radiology*. 2013 Jan; 266(1):104–113.

Sickles EA. Periodic mammographic follow-up of probably benign lesions: Results in 3,184 consecutive cases. *Radiology* 1991 May;179(2):463–8.

Spangler ML, Zuley ML, Sumkin JH, et al. Detection and classification of calcifications on digital breast tomosynthesis and 2D digital mammography: A comparison. *AJR Am J Roentgenol*. 2011 Feb;196(2):320–4.

Teertstra HJ, Loo CE, van den Bosch MA, et al. Breast tomosynthesis in clinical practice: Initial results. *Eur Radiol*. 2010 Jan;20(1):16–24.

US Preventive Services Task Force. Screening for breast cancer: U.S. Preventive Services Task Force recommendation statement. *Ann Intern Med*. 2009 Nov 17;151(10):716–26, W-236.

Chest tomosynthesis

Magnus Båth and Åse Allansdotter Johnsson

Contents

13.1 SHORT INTRODUCTION TO CHEST RADIOLOGY AND CLINICAL USE OF DIFFERENT MODALITIES

This chapter will describe the present status of chest tomosynthesis and present an overview of potential applications of the technique. As a background to the clinical interest of chest tomosynthesis, already established modalities in chest radiology will first be briefly discussed. Chest radiology is a subspecialty of radiology, devoted to imaging and diagnosis of diseases within the respiratory and cardiovascular systems as well as to procedures such as lung biopsy, fluid drainage, and radiofrequency ablation of tumors. This chapter will focus on diagnostic imaging of the respiratory system.

13.1.1 CONVENTIONAL CHEST RADIOGRAPHY

After more than 100 years, conventional chest radiography is still a fundamental examination in chest radiology. It gives an instant overview of the patient's cardiopulmonary status. Heart size, pulmonary vascular pattern, radiodensity of the lungs, contours of the mediastinum, diaphragm, pleura, and chest wall can be reviewed within seconds after image presentation. This allows for the detection of a variety of different pathological changes in the chest, and for a number of patients, chest radiography is the only radiological examination performed prior to treatment. This includes patients with pneumothorax, pneumonia, and pulmonary edema. Advantages of chest radiography include short examination time, easy accessibility, low radiation dose, and low cost. A typical effective dose reported for chest radiography is 0.1 mSv, which is considered low in relation to the average natural background radiation of 2–3 mSv in many places throughout the world.

In chest radiography, the three-dimensional (3D) anatomy of the chest is presented superimposed on a two-dimensional (2D) image. This results in relatively low sensitivity, which is considered a major drawback for the modality. For instance, subtle pathology, such as a small pulmonary neoplasm, is often not detected on a chest radiograph. In the clinical setting, the radiologist often finds opacities overlying the lungs, which requires explanation. Such opacities should be regarded as pathological until proven otherwise and in this task, chest radiography might fail. More images such as oblique, apical, and lordotic views or the use of dual-energy imaging may be used in resolving the nature of the opacity, but the solution is far from guaranteed since the superimposition of surrounding anatomy is the main limiting factor for many detection tasks in chest radiography. In these cases, it is not uncommon that a follow-up examination or the use of another modality is recommended by the radiologist to solve the problem.

13.1.2 LINEAR TOMOGRAPHY

Linear tomography is the process of using combined motions of the x-ray tube and image receptor to generate an image where only one plane of the depicted anatomy remains in sharp focus whereas other planes are smeared out, partially resolving the problem with superimposed anatomy. If several planes are of interest, the acquisition has to be repeated. Thus, the radiation dose to the patient increases with the number of planes imaged. Prior to the era of computed tomography (CT), beginning in the late 1970s for chest imaging, linear tomography was a frequently used problem solver in the work-up of suspicious findings found on chest radiography images. Today, linear tomography of the chest is considered an historical method and has been replaced by CT.

13.1.3 COMPUTED TOMOGRAPHY

CT refers to the situation where the x-ray tube rotates around the patient, creating multiple x-ray projection data that are used to reconstruct cross-sectional images of the patient. Although the spatial resolution is lower compared to conventional radiography, the fact that a CT image contains virtually no superimposed anatomy and that the contrast resolution is far better compared to chest radiography, which enables the visualization of subtler abnormalities, has positioned CT as the most used reference method in chest radiology. Today, cross-sectional images of the patient can be reconstructed in any plane, and CT can be considered a 3D imaging technique. More and more indications for chest CT examinations are recognized among referring clinicians and today CT is used to diagnose and stage, for example, traumatic injuries, pulmonary embolism, aortic syndromes, coronary artery disease, and diffuse lung disease as well as pulmonary neoplasms.

A drawback when using CT to evaluate the cardiovascular system is the need for the use of contrast media. Nevertheless, the faster examination time, higher spatial resolution, better availability, and lower cost compared to magnetic resonance imaging (MRI) have made CT the preferred examination for many indications regarding the cardiovascular system. When CT

is performed primarily to evaluate the lung parenchyma, the use of contrast media is not necessary due to the inherent contrast of the air-filled lung parenchyma surrounding both normal anatomical and pathological structures. Thus, the most important limitation concerning CT of the respiratory system is the use of ionizing radiation and the fact that the image quality is strongly dependent on the radiation dose. An increased use of CT over the years has raised fear regarding radiation-induced cancer mortality in the population. In a study by Sodickson et al. (2009), comprising 31,462 patients, it was concluded that cumulative CT radiation exposure added incrementally to baseline cancer risk in the cohort.

Typical effective doses reported for chest CT are 4–8 mSv, although the radiation doses vary considerably depending on indication and scan protocol used. Furthermore, today, technical developments such as iterative image reconstruction point at considerable dose-saving while maintaining image quality (Silva et al. 2010). Previously, low-dose CT in chest radiology has mainly been used in lung cancer screening programs and an average effective dose of 1.5 mSv has been reported (Aberle et al. 2011a). Currently, it is possible to perform a chest CT at an effective dose of less than 1 mSv and preliminary reports point at dose levels lower than 0.5 mSv.

13.1.4 POSITRON EMISSION TOMOGRAPHY– COMPUTED TOMOGRAPHY

Positron emission tomography–computed tomography (PET–CT) images combine functional images of biochemical activity of an injected radionuclide (PET) with anatomical images (CT). PET–CT using a radionuclide-labeled glucose analog, fluorine-18-deoxyglucose (FDG), has revolutionized the radiological staging of lung cancer and provides a possibility to characterize a pulmonary nodule as probably benign or malignant. The technique is based on the fact that malignant tissue typically exhibits increased rates of glucose metabolism, enabling detection of tumor involvement not evident from anatomical images alone. Drawbacks include the high radiation doses resulting from combining x-rays and gamma rays in one examination and effective doses of 13–32 mSv have been reported (Huang et al. 2009). It is also important to remember that all pulmonary neoplasms do not demonstrate increased rates of glucose metabolism.

13.1.5 MAGNETIC RESONANCE IMAGING

MRI provides an improved contrast resolution of soft tissues compared to CT and has the benefit of not using ionizing radiation. In cardiovascular radiology, MRI is a reference method for assessing malformations, myocardial viability, aortic pathology, and so on. MRI is also a valuable problem solver in many cases, such as for the evaluation of mediastinal or chest wall invasion by tumor, characterization of masses, and for the assessment of diaphragmatic abnormalities. Difficulties in MRI of the lung comprise the consequences of cardiac and respiratory movement and the extremely low proton density of normal lung. With the numerous imaging sequences, gating techniques, and planes of sections available to the radiologist, no single protocol can be prescribed for a thoracic MRI examination. More than for any other imaging investigation of the chest, the protocol needs to be tailored to the clinical question to be answered

(Hansell et al. 2010). The longer examination time, higher cost, more limited availability, and the fact that not all patients can be examined (e.g., patients with pacemakers) restrict the use of MRI compared with the use of CT.

13.2 CHEST TOMOSYNTHESIS

13.2.1 BACKGROUND

As hinted at in Section 13.1, there is a fundamental difference between a 2D imaging technique, such as conventional chest radiography, and a 3D imaging technique, such as CT, in that the superimposition of anatomy in the 2D technique results in images where many types of pathology are much more difficult to detect—if they at all can be detected. Imaging a 3D object (the patient) in 3D solves this issue, although this is paid for, for example, by a substantial increase in radiation dose to the patient. From a health perspective, it would therefore be valuable if some of the advantages of suppressing the superimposed anatomy could be obtained at only a modest increase in radiation dose to the patient.

As previously described, a linear tomography acquisition results in an image where a plane of the patient is in focus whereas other planes are out of focus, leading to a blurring of the superimposed anatomy that reduces its negative impact on the detection of pathology. However, although being less visible due to the blurring, the anatomy surrounding the plane of interest is still present in the image, which is a disadvantage of the technique. Furthermore, the radiation dose to the patient increases proportionally to the number of planes imaged, as a new acquisition is needed for each new image. However, as described in Chapter 1, tomosynthesis solves both these problems. The increasing use of CT (leading to a constant increase in the radiation dose to the population from diagnostic imaging), the relatively recent introduction of large-area flat-panel detectors, and the availability of computer power necessary for efficient removal of surrounding anatomy with advanced reconstruction algorithms have led to a reawakened interest in this low-dose technique.

Chest tomosynthesis refers to the technique of acquiring a number of discrete projection radiographs of the chest at extremely low doses over a limited angular range and using these radiographs for reconstructing the chest in section images. These reconstructed slices contain much less of the superimposed anatomy than do the original radiographs, suggesting an improvement in detectability. Initial investigations have also shown that the detection of pathology is substantially increased compared with conventional chest radiography. As opposed to CT, the radiation dose from a chest tomosynthesis examination is comparable to that from a conventional chest radiography examination and—depending on the system used—effective doses in the order of 0.1–0.2 mSv have been reported (Sabol 2009, Båth et al. 2010, Yamada et al. 2011). Additionally, as the financial cost for a chest tomosynthesis examination usually is much lower than for the corresponding CT examination (existing chest tomosynthesis systems are modified digital chest radiography systems) and the patient throughput is higher due to, for example, shorter examination times, it may be beneficial for healthcare if chest tomosynthesis can be used for certain tasks for which CT is used today.

13.2.1.1 Theoretical requirements

As tomosynthesis is built on the principle of acquiring a large number of projection radiographs of the imaged object at slightly different angles and using these radiographs to reconstruct section images, successful implementation of chest tomosynthesis demands a large, dose-efficient, high-resolution detector that is able to acquire the desired number of radiographs within one breath hold. It was not until the relatively recent introduction of flat-panel detectors that such detectors became commercially available. Further, as the reconstruction algorithms are demanding and the total amount of data used is large, the hardware requirements are high. Even with modern computers, the reconstruction of a single examination may take several minutes. Nevertheless, the technical developments in this area in recent years have enabled the introduction of commercial chest tomosynthesis systems into healthcare.

13.2.1.2 Initial potential of chest tomosynthesis

Conventional chest radiography is today by far the most common imaging technique for the diagnosis and follow-up of pulmonary diseases. This is due to important advantages such as short examination time, low cost, and easy access. However, the low sensitivity of the technique for many clinical tasks is well known. As described above, this limitation is mainly due to the fact that chest radiography is a projection imaging technique and the detectability of many pathologic findings is therefore limited by the superimposed anatomy.

As chest tomosynthesis provides section images of the chest, the problem of the superimposed anatomy is reduced. Although each section image still contains more of the surrounding anatomy than does a CT image—which contains virtually no surrounding tissue—superimposition of tissue is a much smaller problem than in chest radiography. Consequently, chest tomosynthesis has the potential of leading to better diagnostics than conventional chest radiography for pathological conditions for which the detection is mainly hindered by the superimposed anatomy. As such conditions include the presence of a pulmonary neoplasm, it is easily understood that chest tomosynthesis may become an important modality for chest radiology.

CT is, due to its ability to depict anatomy in three dimensions, better contrast resolution, and short examination time, the gold standard in chest radiology today. However, the high sensitivity of chest CT unfortunately comes with a large proportion of false-positive findings regarding nodules that might be possible small lung cancers. In the clinical practice when CT is used to work up a suspicious finding on chest radiography, the detected opacity might or might not correspond to a pathologic finding. The CT examination could be completely normal, but it is not unlikely that a nodule that requires several follow-ups is detected. The cost-effectiveness of low-dose CT screening for lung cancer in a high-risk population is currently being analyzed (Aberle et al. 2011b), and the benefit from CT screening in a low-risk population is not known. Thus, in the clinical setting, chest tomosynthesis might prove a powerful tool in the characterization of incidental lesions detected on chest radiography and to select patients who should be referred to further work up with chest CT.

13.2.2 IMAGE ACQUISITION, RECONSTRUCTION, AND REVIEW

13.2.2.1 Available systems

At present (autumn 2013), at least three systems for chest tomosynthesis are commercially available. The manufacturers of these three systems are GE Healthcare, Shimadzu, and Fujifilm. All systems use large-area flat-panel detectors, although the GE and Fujifilm detectors are based on indirect conversion (CsI) whereas the Shimadzu detector is based on direct conversion (a:Se). The systems are similar in that the x-ray tube moves linearly in one direction during the acquisition of the projection radiographs. The GE and Fujifilm detectors are stationary whereas the Shimadzu detector performs a linear movement in the opposite direction of the x-ray tube. For all systems, the x-ray tube rotates during the motion so that the radiation field automatically is directed toward the detector. With the GE system, normally 60 low-dose projection images are acquired in 10 s, whereas the Shimadzu system acquires a slightly higher number of projections in 5 s. The default angular range is ±15° for the GE system and ±20° for the Shimadzu system. Regarding the settings of the Fujfilm system, at the source-to-image distance of 180 cm typically used for chest radiography, the angular range can be varied from ±5° to ±13.5°, the total exposure time from 6 to 12 s, and the number of projections from 30 to 60.

13.2.2.2 Patient setup and exposure

The patient setup for a chest tomosynthesis examination closely resembles that of a conventional chest radiography examination. The difference is that as the tomosynthesis projection images are collected during several seconds, the patient is required to stand still and hold his or her breath longer in order not to introduce motion artifacts. Normally, a chest tomosynthesis examination is performed with the patient standing in the postero-anterior (PA) or antero-posterior (AP) position, leading to coronal section images. However, the use of tomosynthesis to obtain sagittal section images (the patient being imaged in the lateral direction) has also been reported, although this setup results in a substantially higher radiation dose (the effective dose for a lateral examination is typically 3–5 times higher).

As automatic exposure control (AEC) cannot be used during the acquisition of the projection images, the exposure parameters need to be set prior to the examination. The tube load can be determined either manually or with the help of an initial scout image (a conventional chest radiograph). For example, on the GE system, the tube current–time product used for the AEC-determined scout image is multiplied by a constant factor (typically 10) and divided evenly over the 60 projection radiographs of the tomosynthesis examination.

13.2.2.3 Image reconstruction

After the exposure, the acquired projection images are used to produce tomosynthesis section images. The three clinically available chest tomosynthesis systems all use filtered back projection algorithms for the reconstruction of the section images. Other reconstruction techniques exist (see, e.g., Chapters 7 and 8), but have not yet been implemented in commercially available chest tomosynthesis systems. Normally, section images are reconstructed every 5 mm, but this can be altered by the user. Thus, in the typical case, about 60 coronal section images are obtained from a tomosynthesis examination, covering the entire chest of the patient.

13.2.2.4 Review

The review process of a chest tomosynthesis is very similar to reading a chest CT. It is preferably done in cine mode, identifying normal anatomy and searching for pathologic changes. It is more time consuming to read multiple chest tomosynthesis images compared to having all information presented in a single chest radiography image. On the other hand, the radiological confidence in judging what findings that represent normal anatomy in contrast to pathology is much greater. The increase in radiological confidence should, at least in theory, result in a less uncertain report to the referring clinician.

13.3 CLINICAL APPLICATIONS

Although initial clinical experiences of chest tomosynthesis have indicated that the modality has the potential to be useful for a number of clinical tasks, only a few scientific evaluations have been reported after these first few years of clinical use of the modality. Here, a short overview of recent such evaluations will be given.

13.3.1 DETECTION OF PATHOLOGY

Chest radiography is the preferred first examination for patients emanating from general practitioners (GPs). It is an important diagnostic tool for GPs and seems a cost-effective diagnostic test (Geitung et al. 1999, Speets et al. 2006). In a study from Australia (Simpson and Hartrick 1987) regarding requests for chest CT from GPs, 68% of the CT examinations were judged to be inappropriate and in some countries GPs have limited access to chest CT. However, the low sensitivity of chest radiography for detecting nodules, that is, possible lung cancers, is considered a significant clinical problem. This is especially troublesome since pulmonary lesions often are visible in retrospect when reviewing previous radiographic images of patients with a detected lung cancer. When tomosynthesis emerged as a possible solution to this problem, the use of the modality for detection of pulmonary nodules immediately earned a large interest.

Initial evaluations have concluded that the sensitivity of chest tomosynthesis may be three times higher than that of conventional chest radiography for nodules within the clinically important size range of 4–10 mm diameter, and even higher for smaller nodules (Dobbins et al. 2008, Vikgren et al. 2008, Jung et al. 2012). The nodule detection is affected by size, CT attenuation value, and location of the nodule in both chest tomosynthesis and radiography. Sensitivity increases with increasing nodule size and CT attenuation value and decreases for nodules with subpleural location (Yamada et al. 2011).

Another important field is mycobacterial disease, which remains a substantial cause of infection worldwide. Early identification and treatment of active cases are essential for tuberculosis control and detection of cavities is an important radiological task. There is currently one study concerning this topic, reporting that only 19% of cavities found on chest CT were detected on chest radiography and that 77% of the cavities were detected at chest tomosynthesis, indicating a fourfold increase in sensitivity for cavities in favor of chest tomosynthesis (Kim et al. 2010).

13.3.2 CLASSIFICATION OF PATHOLOGY

Only one clinical investigation has addressed the ability of chest tomosynthesis to characterize a suspected pulmonary lesion detected on chest radiography (Quaia et al. 2010). The accuracy in classifying a lesion as pulmonary or extra pulmonary was lower than 50% for chest radiography but as high as 90% for chest tomosynthesis, using CT as the reference method. Pulmonary lesions misinterpreted on tomosynthesis images were all located in the anterior part of the lung parenchyma close to the thoracic wall. The majority of the extrapulmonary lesions classified as pulmonary lesions on chest tomosynthesis were pleural plaques, a topic that is discussed further in Section 13.4.2.

13.4 LIMITATIONS

As the projection images in tomosynthesis are not acquired over 360°, the depth resolution in reconstructed section images is limited. Thus, the isotropic resolution possible with modern CT equipment cannot be achieved with tomosynthesis. This prevents the complete removal of superimposed tissue in tomosynthesis. Instead, anatomy surrounding a plane of interest will to some extent always be present in a given section image and a given object will show up in more than one section image.

Figure 13.1 exemplifies the low depth resolution by presenting a sagittal reconstruction from coronal tomosynthesis section

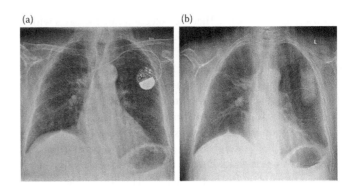

Figure 13.2 (**See insert.**) (a) A PA chest radiograph of a patient with a pacemaker and (b) a coronal tomosynthesis section image at the tracheal level. The low depth resolution of tomosynthesis results in the pacemaker showing up as an artifact in all section images, thus making the detection of pathology almost impossible.

images. The extremely low resolution in the z direction, in comparison with CT, is not obvious from the corresponding coronal section images. (In chest tomosynthesis, the z direction is orthogonal to the coronal plane, constituting the x–y plane.) The limited depth resolution makes it difficult to correctly localize a structure in the z direction, which may result in pitfalls for many clinical tasks. Also, the limited depth resolution may lead to artifacts, as objects will show up in section images they should not be present in. This is of special concern when high-density objects are superimposed on tissue with low density (Figure 13.2).

13.4.1 ARTIFACTS

The combination of the facts that the acquisition of the original projection images takes several seconds, that all projection images are used for the reconstruction of each section image, and that the chest contains organs with substantial movement (mainly the heart) leads to motion artifacts being relatively common in chest tomosynthesis examinations, even if the projection images are collected within one breath hold. Further, patient breathing during the image acquisition may result in severe artifacts in the section images, rendering the examination useless. Even subtler motion can significantly reduce the conspicuity of a tumor within the lung parenchyma. An example of such a lesion is given in Figure 13.3.

13.4.2 PITFALLS

The likelihood of detection and correct characterization of a pulmonary nodule in chest tomosynthesis depends on size,

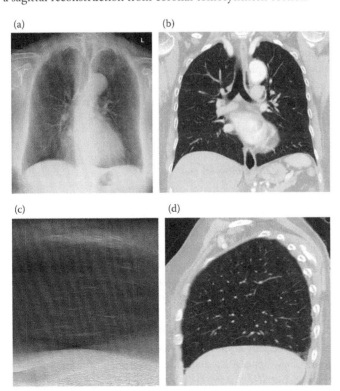

Figure 13.1 (**See insert.**) (a) A coronal tomosynthesis section image and (b) a coronal CT image at the tracheal level of the patient showing good agreement in the depicted anatomy. In (c), a sagittal reconstruction of the right lung from the coronal tomosynthesis section images is presented to illustrate the low depth resolution due to the limited angular range used. In (d), a sagittal CT image at a similar position is shown for comparison.

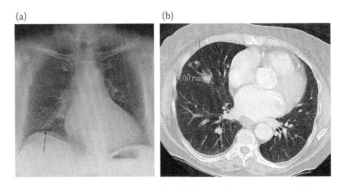

Figure 13.3 (**See insert.**) A coronal tomosynthesis section image (a) showing the presence of motion artifacts, reducing the visibility of the nodule clearly visible in the corresponding axial CT (b).

localization, and density. That the detection rate increases with nodule size is well known from chest radiography. Localization of pleural and subpleural nodular lesions has been identified as a major problem in analyzing chest tomosynthesis images, especially in the anterior and posterior parts of the lungs, where the direction of the radiation beam is not tangential to the pleural border. The distinction between pleural and subpleural lesions is an important clinical task, since a pleural nodule or plaque is less likely to represent a pulmonary neoplasm than a pulmonary nodule and follow-up of pleural nodules is thus not considered mandatory. The plausible explanation for the difficulty in distinguishing pleural from subpleural nodule location is the limited depth resolution of chest tomosynthesis.

High-density opacities such as skeletal changes can be problematic when reading a chest tomosynthesis examination. For instance a bone island, a benign focus of compact bone within cancellous bone, may be misinterpreted as a nodule when the bone island is reproduced in several of the chest tomosynthesis images. Costochondral calcifications may also be misinterpreted as nodules, but the experienced thoracic radiologist will probably find them less problematic than bone islands due to their often irregular appearance and the typical location. In addition, most pleural plaques are calcified, that is, exhibit high density. An illustration of the problem with high-density objects is given in Figure 13.4, where a calcified lymph node posterior to the trachea on the lateral chest radiograph is sharply reproduced as a possible foreign body within the trachea in the coronal chest tomosynthesis image.

Another topic is low-attenuation nodules, especially nonsolid nodules (ground-glass opacities) that even in retrospect may not be visible in chest tomosynthesis images due to the inferior contrast resolution compared with CT. Most nonsolid nodules are benign, but they have an overall higher malignancy risk than other solitary pulmonary nodules.

Part-solid nodules are the most likely of nodules less than 1.5 cm in diameter to contain lung cancer. These nodules might

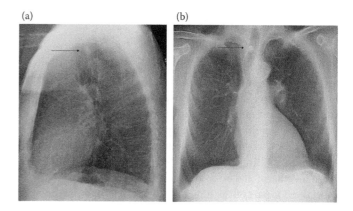

(a) (b)

Figure 13.4 (See insert.) (a) A lateral chest radiograph and (b) a coronal tomosynthesis section image at the tracheal level. The calcified lymph node posterior to the trachea on the lateral chest radiograph is depicted as a possible foreign body within the trachea in the tomosynthesis section image.

both be less conspicuous to the radiologist and be underestimated in size in chest tomosynthesis.

Furthermore, lymph nodes in hilar and mediastinal node stations may be perceived as pulmonary nodules and pulmonary nodules close to the hilum may be misinterpreted as lymph nodes. Small nodules located closely to vessels, especially at branching points, might be regarded as part of a somewhat tortuous vessel.

As an initial interest of chest tomosynthesis has been the detection of pulmonary nodules, the reporting of pitfalls has mainly been associated with this task. Table 13.1 presents a number of such pitfalls, related to both false-positives and false-negatives, as well as suggestions on how to avoid these (taken from Asplund et al. (2011)). Although mistakes cannot be avoided, it can be expected that as chest tomosynthesis becomes more common, more solutions on how to avoid certain pitfalls will develop.

Table 13.1 Suggestions on how to avoid potential pitfalls regarding nodules in chest tomosynthesis

FALSE-POSITIVES		FALSE-NEGATIVES	
PITFALL	SOLUTION/COMMENT	PITFALL	SOLUTION/COMMENT
Subpleural and pleural changes may often be misinterpreted as nodules because of their proximity to pleural borders, where skeletal structures overlap anatomy and pathology	This may possibly be prevented by relating the location where the ribs are in focus to the position of the suspicious finding	Nodules situated close to the pleural border may often be misinterpreted as pleural or subpleural changes, because skeletal structures may overlap nodules at such locations	This may possibly be prevented by relating the location where the ribs are in focus to the position of the suspicious finding
Lymph nodes may sometimes be misinterpreted as nodules close to hilar and mediastinal node stations	Even though the probability is high that the structures are lymph nodes, it is not possible to characterize them	Nodules located closely to vessels, especially at branching points, may appear as part of the vessel itself	These nodules are usually too small (<5 mm) to be properly distinguished from the vessel that they are close to
Skeletal changes, including costochondral calcifications, may be misinterpreted as nodules, especially those located posteriorly and anteriorly	This may possibly be prevented by relating the location where the skeletal structure is in focus to the position of the suspicious finding	Very small nodules (2–3 mm): sometimes discharged by radiologists as unspecific findings	It is important to bear in mind that small nodules may be very well depicted with tomosynthesis

Source: Taken from Asplund, S. et al. 2011. Acta Radiol. 52:503–512.

13.5 OUTLOOK

13.5.1 TRENDS IN THE USE OF CHEST TOMOSYNTHESIS

13.5.1.1 Problem solver

The radiologist analyzing a chest radiograph and detecting an unexplained opacity overlying the lung is facing a diagnostic problem. Many of these opacities turn out to be the result of superimposition of vascular and bone structures of the chest. A CT of the chest is likely to solve the problem, but the radiologist might hesitate to send the patient to CT, due to radiation dose concern, possible waiting time, and the risk of incidental findings, unless there is a certain degree of confidence that the opacity represents pathology. In this scenario, chest tomosynthesis provides an attractive alternative as a problem solver. One study has so far addressed the use of tomosynthesis as a problem solver and the results suggest that the majority of suspicious findings detected on chest radiography can be resolved by chest tomosynthesis, reducing the need for CT (Quaia et al. 2013). Figure 13.5 exemplifies the benefit of chest tomosynthesis as a problem solver. An opacity that was judged as tumor suspicious proved to be a healed rib fracture, and the patient required no further diagnostic imaging.

To be able to use chest tomosynthesis as an effective problem solver, the clinical situation requires reading of the chest radiograph while the patient is in the radiology department. Then, as the tomosynthesis examination is performed at a modified digital chest radiography system, a request of an additional chest tomosynthesis examination can be treated in the same way as a request for additional projections to the chest radiography examination. The radiologist is thus given an opportunity to explore the nature of the detected unexplained opacity. In many cases, the opacity will prove to be the result of superimposition of vascular and bone structures of the chest and a confident report of no pathological findings can be sent to the referring clinician. In some cases, real pathology will be clearly evident and the patient could be referred for further investigation without the diagnostic delay of a follow-up examination.

13.5.1.2 Increasing sensitivity

Many nodules detected on chest CT do not represent lung cancer or metastases. In the National Lung Screening Trial (Aberle et al.

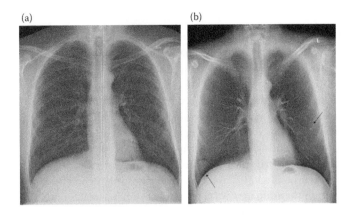

Figure 13.6 (See insert.) (a) A PA chest radiograph and (b) a coronal tomosynthesis section image at the hilum level. The tomosynthesis section image clearly shows two nodules, which are more difficult to discern on the chest radiograph.

2011b), 94.6% of the positive screening results from low-dose CT were false-positive results. The high incidence of indeterminate nodules on chest CT has resulted in clinicians advocating chest radiography as the method of choice in the search for pulmonary metastases for patients with colorectal cancer (Grossmann et al. 2010). One could argue that these patients could benefit from a chest tomosynthesis examination, enabling the detection of more patients with pulmonary metastases. However, introducing a more sensitive method will also render more false-positive findings and the effects of replacing chest radiography with chest tomosynthesis in the search of metastases have not been evaluated.

Figures 13.6 and 13.7 illustrate the increase in detection sensitivity when using chest tomosynthesis compared to chest radiography. The chest radiograph of the patient in Figure 13.6a was initially reported as normal, but the result of a double reading indicated one or possibly two unexplained opacities. The chest tomosynthesis examination revealed 15 lesions, two of which are shown in Figure 13.6b. The chest tomosynthesis examination in Figure 13.7 was ordered due to high clinical suspicion of mycobacterial disease and clearly depicted a cavity in the left upper lobe, which was not evident on the initial chest radiograph.

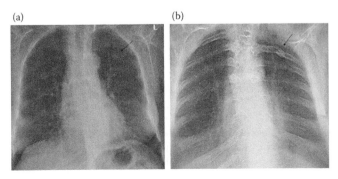

Figure 13.5 (See insert.) (a) A PA chest radiograph and (b) a coronal tomosynthesis section image at the level of the posterior parts of the ribs. The suspected opacity in the chest radiograph is clearly identified as a healed rib fracture on the tomosynthesis section image.

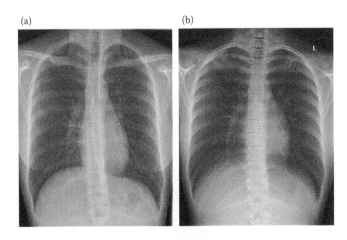

Figure 13.7 (See insert.) (a) A PA chest radiograph and (b) a coronal tomosynthesis section image at the spinal level. The tomosynthesis section image clearly shows a cavity in the left upper lobe, which is more difficult to discern on the chest radiograph.

Clinical applications

13.5.1.3 Follow-up of known pathology

As chest tomosynthesis has a larger amount of superimposed anatomical structures in the images than does CT, the sensitivity of chest tomosynthesis can never reach that of CT for structures for which detection is limited by the anatomical background. However, it has been reported that most nodules detected using chest CT are visible retrospectively on chest tomosynthesis images (Vikgren et al. 2008). This indicates the potential of using chest tomosynthesis for follow-up of known pathology. Initial studies on measurement agreement between tomosynthesis and CT indicate that size estimates on chest tomosynthesis images are comparable to size estimates on CT (Johnsson et al. 2010, 2012). Although scientific evidence are lacking, some clinicians today have already chosen to replace CT with chest tomosynthesis in the follow-up of pulmonary lesions visible on chest tomosynthesis but not on chest radiography. In order to choose a method for future follow-ups, the patient is referred to a chest tomosynthesis shortly after the first CT examination, or more often in conjunction to the first follow-up with CT, in order to evaluate if the lesion is adequately depicted. In cases where chest tomosynthesis accomplishes this task, it is likely to be chosen as the preferred follow-up examination, saving radiation dose and costs. Naturally, the decision for the follow-up method depends on a number of factors, including whether the patient is considered a high-risk or low-risk patient for a pulmonary neoplasm. An illustration of a nodule that according to chest tomosynthesis has remained stable for 2 years is given in Figure 13.8.

If the nodule cannot be seen on a follow-up chest tomosynthesis, should the patient undergo a chest CT? There is no answer to that question in the scientific literature today, although many radiologists would probably advocate the use of CT to confirm that the lesion has disappeared. Future research has to address this issue.

13.5.2 FUTURE OF CHEST TOMOSYNTHESIS

As chest tomosynthesis has only been available in the clinic for a few years, it is difficult to predict the future use of the method. However, Dobbins and McAdams (2009) have proposed four different potential strategies for incorporating chest tomosynthesis into the clinical arena regarding pulmonary nodules: (1)

additional or replacement test in *all* patients undergoing chest radiography; (2) additional or replacement test in *some* patients undergoing chest radiography; (3) evaluation of suspicious lesions seen by conventional chest radiography; and (4) follow-up of known nodules.

The scenario where chest radiography is replaced by tomosynthesis for all patients (Strategy 1) might not be appropriate, since the increased sensitivity also may render more false-positive findings that need to be worked up with CT (Vikgren et al. 2008). Patients with known or suspected malignancy, undergoing chest radiography in the search for pulmonary metastasis, would probably benefit from an approach where chest tomosynthesis is used as an additional or replacement test (Strategy 2). To prove this, prospective randomized-controlled studies should be conducted.

Chest tomosynthesis may also be used as a problem-solving tool trying to obviate unnecessary CT, thus reducing radiation exposure and cost (Strategy 3). As described above, phantom studies and preliminary work based on clinical patients also indicate that chest tomosynthesis could be an alternative for follow-up of known nodules (Strategy 4), as the accuracy of tomosynthesis regarding determination of nodule size seems adequate. Strategies 3 and 4 take advantage of the reduction of noise from overlying anatomy in chest tomosynthesis images but not the increased sensitivity for nodule detection (Dobbins et al. 2008, Vikgren et al. 2008).

In summary, chest tomosynthesis has during its short existence already introduced itself as a valuable imaging technique, being able to produce section images of the chest at a very low radiation dose. In comparison to chest radiography, it has been shown to improve detection of pathology such as pulmonary nodules, possibly representing small neoplasms, and cavities in mycobacterial disease. In the future, chest tomosynthesis may play an important role in chest radiology, despite its limitations related to the inferior depth and contrast resolution compared to CT.

REFERENCES

Aberle, D. R., C. D. Berg, W. C. Black et al; National Lung Screening Trial Research Team. 2011a. The National Lung Screening Trial: Overview and study design. *Radiology* 258:243–253.

Aberle, D. R., A. M. Adams, C. D. Berg et al; National Lung Screening Trial Research Team. 2011b. Reduced lung-cancer mortality with low-dose computed tomographic screening. *N. Engl. J. Med.* 365:395–409.

Asplund, S., Å. A. Johnsson, J. Vikgren et al. 2011. Learning aspects and potential pitfalls regarding detection of pulmonary nodules in chest tomosynthesis and proposed related quality criteria. *Acta Radiol.* 52:503–512.

Båth, M., A. Svalkvist, A. von Wrangel, H. Rismyhr-Olsson, and Å. Cederblad. 2010. Effective dose to patients from chest examinations with tomosynthesis. *Radiat. Prot. Dosimetry* 139:153–158.

Dobbins, J. T. 3rd, H. P. McAdams, J. W. Song et al. 2008. Digital tomosynthesis of the chest for lung nodule detection: Interim sensitivity results from an ongoing NIH-sponsored trial. *Med. Phys.* 35:2554–2557.

Dobbins. J. T. 3rd and H. P. McAdams. 2009. Chest tomosynthesis: Technical principles and clinical update. *Eur. J. Radiol.* 72:244–251.

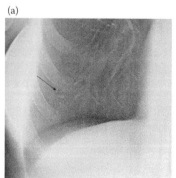 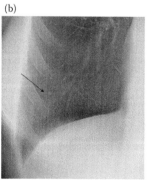

(a)   (b)

Figure 13.8 (**See insert.**) (a) A close-up of a chest tomosynthesis section image containing a 5-mm nodule in the middle lobe and (b) the corresponding image at the follow-up examination after 2 years. The images reveal no apparent nodule growth.

Geitung, J. T., L. M. Skjaerstad, and J. H. Göthlin. 1999. Clinical utility of chest roentgenograms. *Eur. Radiol.* 9:721–723.

Grossmann, I., J. K. Avenarius, W. J. Mastboom, and J. M. Klaase. 2010. Preoperative staging with chest CT in patients with colorectal carcinoma: Not as a routine procedure. *Ann. Surg. Oncol.* 17:2045–2050.

Hansell, D. M., D. A. Lynch, H. P. McAdams, and A. A. Bankier. 2010. *Imaging of Diseases of the Chest, 5th Edition.* Philadelphia, PA: Elsevier.

Huang, B., M. W. Law, and P. L. Khong. 2009. Whole-body PET/CT scanning: Estimation of radiation dose and cancer risk. *Radiology* 251:166–174.

Johnsson, Å. A., A. Svalkvist, J. Vikgren et al. 2010. A phantom study of nodule size evaluation with chest tomosynthesis and computed tomography. *Radiat. Prot. Dosimetry* 139:140–143.

Johnsson, Å. A., E. Fagman, J. Vikgren et al. 2012. Pulmonary nodule size evaluation with chest tomosynthesis. *Radiology* 265:273–282.

Jung, H. N., M. J. Chung, J. H. Koo, H. C. Kim, and K. S. Lee. 2012. Digital tomosynthesis of the chest: Utility for detection of lung metastasis in patients with colorectal cancer. *Clin. Radiol.* 67:232–238.

Kim, E. Y., M. J. Chung, H. Y. Lee, W.-J. Koh, H. N. Jung, and K. S. Lee. 2010. Pulmonary mycobacterial disease: Diagnostic performance of low-dose digital tomosynthesis as compared with chest radiography. *Radiology* 257:269–277.

Quaia, E., E. Baratella, V. Cioffi et al. 2010. The value of digital tomosynthesis in the diagnosis of suspected pulmonary lesions on chest radiography: Analysis of diagnostic accuracy and confidence. *Acad. Radiol.* 17:1267–1274.

Quaia, E., E. Baratella, G. Poillucci, S. Kus, V. Cioffi, and M. A. Cova. 2013. Digital tomosynthesis as a problem-solving imaging technique to confirm or exclude potential thoracic lesions based on chest x-ray radiography. *Acad. Radiol.* 20:546–553.

Sabol, J. M. 2009. A Monte Carlo estimation of effective dose in chest tomosynthesis. *Med. Phys.* 36:5480–5487.

Silva, A. C., H. J. Lawder, A. Hara, J. Kujak, and W. Pavlicek. 2010. Innovations in CT dose reduction strategy: Application of the adaptive statistical iterative reconstruction algorithm. *AJR Am. J. Roentgenol.* 194:191–199.

Simpson, G. and G. S. Hartrick. 2007. Use of thoracic computed tomography by general practitioners. *Med. J. Aust.* 187:43–46. Erratum in: *Med. J. Aust.* 187:256.

Sodickson, A., P. F. Baeyans, K. P. Andriole et al. 2009. Recurrent CT, cumulative radiation exposure, and associated radiation-induced cancer risks from CT of adults. *Radiology* 251:175–184.

Speets, A. M., Y. van der Graaf, A. W. Hoes et al. 2006. Chest radiography in general practice: Indications, diagnostic yield and consequences for patient management. *Br. J. Gen. Pract.* 56:574–578.

Vikgren, J., S. Zachrisson., A. Svalkvist et al. 2008. Comparison of chest tomosynthesis and chest radiography for detection of pulmonary nodules: Human observer study of clinical cases. *Radiology* 249:1034–1041.

Yamada, Y., M. Jinzaki, I. Hasegawa et al. 2011. Fast scanning tomosynthesis for the detection of pulmonary nodules: Diagnostic performance compared with chest radiography, using multidetector-row computed tomography as the reference. *Invest. Radiol.* 46:471–477.

Clinical applications

14

Tomosynthesis applications in radiation oncology

*Devon J. Godfrey, Lei Ren, Q. Jackie Wu,
and Fang-Fang Yin*

Contents

14.1 INTRODUCTION TO IMAGE-GUIDED RADIATION THERAPY

The most common form of radiation therapy, external beam radiation therapy (EBRT), irradiates the patient with high-energy electrons or photons generated with a linear accelerator or "linac." Conventional "fractionated" treatment regimens prescribe several weeks of daily EBRT, and thus, a patient often receives 25–35 treatments during their therapeutic course, for a typical total cumulative dose of 50–70 Gy. Historically, EBRT was delivered using parallel opposed fields (2D treatment plans) or very simplistic multiple-field geometries, which resulted in large areas of anatomy receiving a therapeutic radiation dose and only a gradual falloff of the dose away from the intended target. Recent

decades have witnessed the introduction of far more complex three-dimensional (3D) and intensity-modulated radiation therapy (IMRT) techniques, which allow the planner to create very conformal dose distributions that closely match the contour of the intended target and fall off quickly in the surrounding normal tissues. Improved conformality can be utilized to either reduce the normal tissue complication probability (NTCP) or for dose escalation to enhance the tumor control probability (TCP). Sample isodose distributions from 3D and IMRT plans of a subject undergoing salvage radiotherapy of the prostate bed are displayed in Figure 14.1, and illustrate the improved dose conformality and normal structure sparing provided by IMRT.

The recent drastic improvement in dose delivery precision has led to the need for better targeting accuracy, lest the tightly conformal dose distribution miss portions of the target and

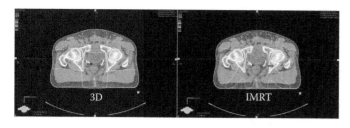

Figure 14.1 3D and IMRT plans for salvage radiotherapy of a prostate bed. IMRT provides better dose conformality around the target volume, reducing the dose to adjacent critical structures (the bladder in the rectum in this example).

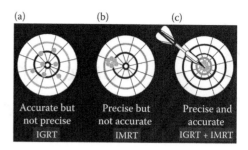

Figure 14.2 (a) Accurate but not precise treatment delivery, (b) precise but not accurate treatment delivery, and (c) precise and accurate delivery of treatment beams. (Reproduced from *Sem Rad Onc*, **16**, Yin FF, Das S, Kirkpatrick J, Oldham M, Wang Z, and Zhou SM, Physics and imaging for targeting of oligometastases, 85–101, Copyright 2006, with permission from Elsevier.)

instead irradiate critical normal tissues. The largest contributor to targeting inaccuracy is usually the patient setup uncertainty, which describes the variability in positioning the patient on the day of treatment exactly as they were positioned for their treatment planning imaging sessions, and can also include internal anatomic motion or deformation. Setup variability as well as mechanical and dosimetric uncertainties are accounted for by expanding the clinical target volume (CTV), which includes the target tumor and areas of potential microscopic disease spread, by several mm to create a planned target volume (PTV), which ideally accounts for all possible setup and delivery uncertainty. Treatment plans are designed to cover the entire PTV with the prescribed dose, which ideally ensures that the smaller CTV will always fall within the therapeutic dose region, even given the day-to-day variability in patient setup, and so on.

Employing image guidance to localize the target with the patient in their treatment position has proven to be a very effective method for reducing daily setup variation, and thus the magnitude of the PTV expansion. The frequency of image guidance is selected in accordance with the anatomic site, immobilization, and patient-specific characteristics, and varies from weekly imaging for relatively stable setups, to daily imaging for sites that are prone to motion or for treatment plans that require very small expansion margins due to the critical surrounding anatomy. Furthermore, the type of image guidance required depends on whether or not the rendering of bony anatomy is sufficient to localize the target, or if the target is likely to move or deform relative to the skeletal structure, thus requiring soft-tissue visualization. Finally, if the irradiated site involves tissues affected by respiratory motion, it is sometimes advantageous to irradiate under breath hold or respiratory gating conditions, necessitating four-dimensional (4D) image guidance.

Conformal dose delivery alone is not sufficient to ensure high-quality radiotherapy, and in fact, may be worse than a nonconformal technique if it is inaccurately aimed at the patient's internal anatomy. Optimal radiotherapy can only be achieved by combining the precise delivery of techniques such as IMRT with the localization accuracy of IGRT, as illustrated in Figure 14.2.

14.1.1 CONVENTIONAL RADIOGRAPHIC IGRT METHODS FOR EBRT

14.1.1.1 2D radiographic imaging

The most traditional form of EBRT image guidance employs the megavoltage (MV) treatment beam itself, coupled with a film/screen system or an electronic portal imaging detector (EPID),

to acquire orthogonal radiographs (e.g., antero-posterior and right lateral views) of the patient for general setup purposes, as well as "portal" images of the planned treatment field geometries and field shapes for verification of the actual treatment windows. MV imaging has the advantage of providing a beams-eye view of the patient, along with the superimposed shape of the treatment field portal, and can be implemented without the addition of an auxiliary imaging system. However, the MV beam produces almost exclusively Compton scattering interactions within the patient, which provides very little contrast between the bone and soft tissue, as this effect depends only on the electron density of the tissue, not on its composition. The high beam energy also results in low inherent detection efficiency for EPIDs, which require a thick scintillator layer to compensate. Thus, MV IGRT is characterized by high imaging dose, poor resolution, and only minimal anatomic contrast.

Improved imaging characteristics can be achieved with an on-board kilovoltage (kV) imaging (OBI) system, mounted on the accelerator, orthogonal to the treatment beam axis. Most clinical accelerator vendors now offer such auxiliary systems, including both a kV x-ray source and flat-panel detector, and their use is quickly becoming a standard practice within radiation oncology centers, worldwide. As kV imaging produces substantial photoelectric effect (PE) interactions, kV IGRT techniques yield images with excellent contrast of bony anatomy. Furthermore, kV imaging can be performed at very low exposure to the patient, enabling responsible daily IGRT protocols for patients (especially pediatric) who might otherwise be at high risk for secondary cancers due to accumulated imaging dose. In addition to acquiring orthogonal setup radiographs of bony anatomy, kV IGRT is often employed for the quick daily localization of the prostate and other mobile targets, using implanted metal fiducials as target surrogates.

14.1.1.2 3D cone-beam computed tomography

Visualization and localization of soft-tissue structures can be obtained from CBCT reconstructions acquired with the aforementioned MV or kV imaging systems. CBCT acquisition in this setting simply requires collecting projection radiographs while rotating the linac gantry and the attached imaging system around the patient. Both kV and MV CBCT yield excellent 3D soft-tissue information, but impart a substantial imaging dose (~1–10 cGy, depending on the anatomic site), require long acquisition times of 30–60 s due to gantry rotation speed regulations, and present mechanical collision concerns for certain patient setup configurations. MV CBCT generally results in higher imaging dose and lower anatomic contrast than kV CBCT, but has the advantage of substantially suppressing metal artifacts from implanted devices and/or prosthetics and of providing more accurate information for treatment dose modeling, which can be useful if the treatment plan needs to be adapted due to changes in patient anatomy.

14.1.1.3 Breath hold and 4D IGRT methods

For sites affected by respiratory motion, improved normal tissue sparing can often be achieved by delivering therapeutic radiation only during specified respiratory amplitude or phase "windows." This can be achieved most simply by treating under full inspiration (breath hold) or by gating the beam to treat during selected portions of the respiratory cycle. Whichever respiratory management scheme (breath hold or gated) is chosen for treatment must also be applied during image guidance.

2D radiographic imaging of bony anatomy under breath hold or with respiratory gating is simple and reasonably efficient. However, soft-tissue localization is often important when respiratory motion is a concern, and thus, tomographic imaging is often required. Unfortunately, the long scan duration of CBCT is especially problematic for the imaging of respiration-affected anatomic sites. Breath-hold CBCT scans can usually only be acquired by piecing together multiple partial scans, acquired over several patient breath-hold attempts, with recovery time allotted between each partial scan (Peng et al., 2011). Thus, the total acquisition time for breath-hold CBCT can sometimes be as long as 5 min, impeding the clinical workflow and sometimes introducing artifacts due to inconsistent breath-hold levels across the partial scans.

4D CBCT has recently been developed as a means of reconstructing 3D images of the tumor volume at different respiratory phases to assess tumor motion for gated treatment verification. The retrospective sorting and binning of cone-beam projections based on respiratory phases has been proposed as a method for obtaining phase-resolved 4D CBCT image sets (Sonke et al., 2005; Dietrich et al., 2006; Li et al., 2006, 2007; Leng et al., 2008a,b; Yin et al., 2008; Park et al., 2011). Preliminary studies have shown that 4D CBCT can provide accurate delineation of the internal target volume (ITV) for target localization. However, aliasing artifacts have been reported in 4D CBCT due to the limited number of projections acquired nonuniformly over a full scan angle for each phase (Sonke et al., 2005; Li et al., 2006). Furthermore, the imaging dose from 4D CBCT is significantly higher than that from 3D CBCT. For a 4D CBCT acquisition with 10 phases, 10 times the 3D CBCT imaging dose is required to achieve a complete set of projections for each phase. Thus, for a typical 3D CBCT imaging dose of 1–10 cGy, the accumulated imaging dose from 4D CBCT scans for a hypofractionated treatment (e.g., 4fx for lung stereotactic body radiotherapy) could sum up to as much as 4 Gy, which is clinically significant. Moreover, because accurate target localization requires the imaging volume to be much larger than the treatment volume (for assessment of the tumor position in relation to the adjacent organs), there is a large volume of healthy tissue surrounding the tumor that receives the large imaging dose. The accumulated imaging dose over the entire treatment course to the tumor, as well as surrounding normal organs, will not be negligible and may therefore reduce the applicability of 4D CBCT imaging. Furthermore, 4D CBCT requires very long acquisition times to sufficiently populate each of the respiratory phase bins with projection data, which might make the technique unsuitable for daily setup verification in a busy clinical setting.

14.1.2 CONVENTIONAL IGRT METHODS FOR PROSTATE BRACHYTHERAPY

Prostate brachytherapy is widely performed as an effective treatment for prostate cancer. The treatment typically involves permanent implantation of 35–140 radioactive seeds (usually ^{125}I or ^{103}Pd) into the prostate gland using needles to ensure that the entire prostate receives the prescription dose. The quality of the treatment delivery depends on the accuracy of the placement of the implanted seeds, relative to the pretreatment plan. Transrectal ultrasound (TRUS) and fluoroscopy are typically employed independently during the procedure to guide the needle placement and verify the total number of implanted seeds. Several weeks following the procedure, postimplant verification of the individual seed locations is performed via a computed tomography (CT) scan, to evaluate the dose deposition to the prostate and surrounding anatomy, and provide for the possibility of salvage therapy if necessary. As an alternative to a postimplant verification CT, intraoperative localization of the individual seeds via x-ray fluoroscopy and ultrasound has been proposed, to enable the possibility of correcting an implant by adding additional seeds to improve target coverage in real time, while the patient is still under anesthesia on the operation table (Nag et al., 2001; Gong et al., 2002). In this approach, the locations of the implanted seeds are reconstructed from fluoroscopy images and the 3D prostate volume is reconstructed from the ultrasound images. By fusing the fluoroscopy and ultrasound data, the locations of the seeds in the prostate are determined and the actual dose delivered to the prostate is calculated.

Several methods have been developed to localize the seed locations from fluoroscopy images. One widely used method is the three-film approach, which uses only three projection images to reconstruct the seeds (Amols et al., 1981; Biggs et al., 1983; Rosenthal et al., 1983; Altschuler et al., 1997). A major limitation of this method is that seeds are sometimes undetectable when they overlap with other seeds in a projection image. To account for this, a manual search for the missing seeds is required, which is time consuming and limits the suitability of the method for intraoperative usage.

14.2 TOMOSYNTHESIS FOR IGRT

After many years of development for diagnostic imaging, the first decade of the twenty-first century saw tomosynthesis proposed for IGRT in both the brachytherapy (Person, 2001; Tutar et al., 2003) and external beam (Baydush et al., 2005; Godfrey et al., 2006a) radiotherapy settings. To date, brachytherapy digital tomosynthesis (DTS) research has been primarily composed of c-arm-based tomosynthesis for the real-time localization of radioactive seeds implanted into the prostate, allowing for on-the-fly dose calculation in the operating room. The majority of tomosynthesis-based IGRT research, however, has investigated DTS as a fast and low-dose 3D target localization method prior to the delivery of external beam therapy, employing either an MV detector to capture images with the treatment beam itself (Pang et al., 2008) or a separate gantry-mounted system consisting of a kV x-ray tube and flat-panel detector. Other promising investigational developments include the use of arrays of 20 or so cold cathodes to generate rapid kV tomosynthesis images along the beams-eye view (Zhang et al., 2006; Sprenger et al., 2010).

14.2.1 TOMOSYNTHESIS IMAGE ACQUISITION

For external beam target localization, tomosynthesis is acquired by collecting a series of projection images as the treatment accelerator gantry rotates about the patient. Thus, the imaging geometry is, by definition, isocentric and is essentially just a subset of a full CBCT scan.

It has been shown that excellent soft-tissue visibility can be generated from DTS scans acquired over as little as 40° of gantry rotation, using the same exposure per projection image as that used in a conventional full rotation (360°) CBCT scan (Godfrey et al., 2006a,2007). Thus, a DTS scan can be acquired with as little as 1/9 the total exposure of a CBCT acquisition and in less than 10 s duration (a full 360° CBCT requires roughly a minute). This makes tomosynthesis a promising alternative to CBCT for breath-hold localization of soft tissues prior to breath-hold irradiation.

At our institution, projection images are acquired with a kV OBI system (Varian Medical Systems, Palo Alto, CA), which is composed of an x-ray tube and an amorphous silicon (a-Si) flat-panel detector, mounted orthogonal to the treatment beam axis. The x-ray source is located 100 cm from the isocenter, while the isocenter-to-detector distance is 50 cm. Projection radiographs are collected approximately every 0.5° during the scan, using the detector's rapid acquisition, downsampled mode (1024 × 768 pixels, 0.26 mm pixel pitch at isocenter). The gantry is limited to a rotation velocity of 1 rpm (6°/s), as per IEEE standards. A typical DTS scan would thus collect approximately 80 projections over 40° of rotation in roughly 7 s.

Figure 14.3 shows a Varian 2100EX accelerator in our clinic, equipped with an MV detector and a kV imaging system (OBI), which can be rotated around the patient to collect tomographic projection data.

Alternatively, DTS imaging can be performed with the MV treatment beam, employing an EPID (Descovich et al.,

2008; Pang et al., 2008; Mestrovic et al., 2009). One potential advantage of MV tomosynthesis is the generation of slices with high resolution in the beams-eye view, without having to rotate the gantry 90°, a feat which is not possible using an orthogonally mounted imaging system. This is also the motivation behind a proposed treatment-head-mounted cold-cathode array system (Zhang et al., 2006; Sprenger et al., 2010), which could potentially yield real-time slices oriented in the beams-eye view, with enhanced anatomic contrast provided by kV sources.

14.2.2 TOMOSYNTHESIS RECONSTRUCTION

Numerous approaches to tomosynthesis reconstruction have been investigated for the external beam IGRT task. These include techniques such as iterative restoration (Santoro et al., 2010), iterative reconstruction with a total variation (TV) regularization function (Zhang and Wu, 2010), and the Feldkamp (FDK) algorithm (Godfrey et al., 2006a), which is perhaps the most common, as it is fast and already employed in most clinical CBCT systems.

Most of the DTS IGRT research conducted at our institution has been performed using the FDK reconstructions. Sample breath-hold FDK DTS image data from two thoracic subjects are displayed in Figure 14.4 and illustrate the ability of limited-angle DTS imaging to render soft-tissue target anatomy.

All traditional DTS reconstruction methods suffer from the inability to provide acceptable resolution along the plane-to-plane direction, and thus fail to produce quality images in either the orientation parallel to the central projection (e.g., the sagittal view from a scan centered around an AP projection) or in the axial view. Because of this, the utility of traditional DTS is limited where accurate volume information is required, such as for proposed adaptive radiation therapy techniques, which require frequently updated volumetric image data.

A potential solution to the problem of anisotropic resolution from limited-angle scans was recently introduced (Ren et al., 2008). Ren's method relies upon prior 3D image data, such as a CBCT or treatment planning CT scan, to aid limited-angle reconstructions. The estimated volume images are computed by minimizing the bending energy of the deformation field between the two image sets (prior 3D volume and new volume estimate) such that simulated forward projections through the volume estimate match the newly acquired limited-angle projection data. Sample prior image-based reconstructions are displayed in Figure 14.5.

Other IGRT tomosynthesis research has employed graphics processing unit (GPU)-based computing to improve the computational efficiency of DTS reconstruction (Yan et al., 2007, 2008). Studies have indicated that off-loading the reconstruction to even a single GPU can yield a factor of 10–70× acceleration in the reconstruction time compared with conventional central processing unit (CPU)-based reconstruction.

14.2.3 TOMOSYNTHESIS IMAGE QUALITY

14.2.3.1 Resolution characteristics

As is typical of tomosynthesis, DTS planes oriented parallel to the central projection from an isocentric scan exhibit very high in-plane resolution but much lower "depth" resolution in the

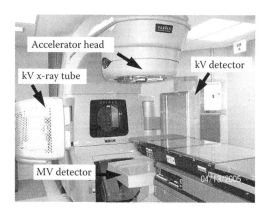

Figure 14.3 Varian 2100EX Clinac with MV EPID and orthogonally mounted kV imaging system (OBI).

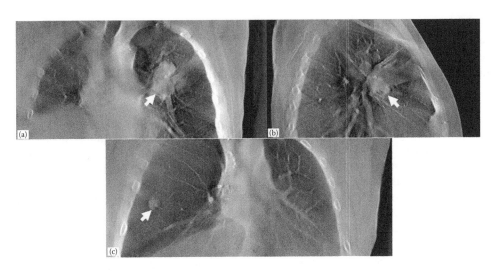

Figure 14.4 On-board 44° DTS slices of thoracic subjects with arrows pointing to visible target malignancies. (a) and (b) Coronal and sagittal slices of a single subject with a large growth in the left lung. (c) Coronal slice from a subject with a small right pulmonary nodule. All three DTS scans were acquired under breath hold. (Reproduced with permission from Godfrey D et al. 2007. *Med Phys* **34**:3374–3384.)

plane-to-plane direction. This can be understood using the Fourier slice theorem, which predicts that a limited-angle tomographic scan leaves a large wedge of Fourier space unsampled (Dobbins and Godfrey, 2003). As a result, information is blurred along the depth dimension and high-quality DTS slices can only be reconstructed in a single orientation (e.g., coronal or sagittal). Quality axial images cannot be rendered at all without the use of a reconstruction algorithm that utilizes prior image information, such as the aforementioned method proposed by Ren et al. (2008).

The reduced depth resolution of tomosynthesis results in reconstructed planes with effectively thick slice profiles. Unlike traditional CT imaging, the DTS slice width cannot be selected during reconstruction, but instead is a function of the scan angle, much like a camera aperture, with larger scan angles resulting in effectively narrower slices. Furthermore, the slice thickness heavily depends on the makeup of the reconstructed object, as the slice profile is thick for low frequencies, but thin for high-frequency image content (Godfrey et al., 2006b). Thus, specifying an effective tomosynthesis voxel size is not straightforward. When an isocentric geometry is employed, matters are complicated further, as the effective slice thickness is then also dependent on location and varies throughout the reconstructed volume. In

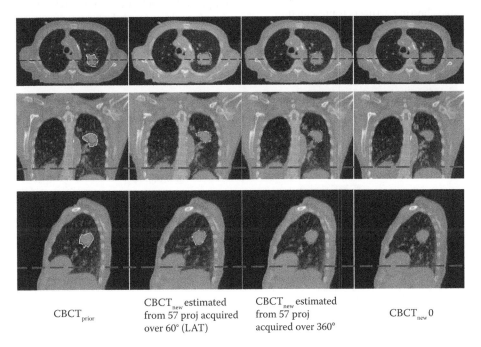

CBCT$_{prior}$ CBCT$_{new}$ estimated from 57 proj acquired over 60° (LAT)   CBCT$_{new}$ estimated from 57 proj acquired over 360°   CBCT$_{new}$ 0

Figure 14.5 Full FDK CBCT (left- and right-most columns) and prior image-based CBCT estimates (middle two columns) of a lung cancer patient from scans acquired on two different days (labeled "prior" and "new"). With only 57 projections spread over either 60° (i.e., tomosynthesis) or 360° (i.e., sparse CBCT), the prior image-based reconstruction approach correctly captures the anatomic changes present in the new CBCT volume, including variation in the diaphragm level (dashed line in coronal and sagittal views) and lung tumor shape and location (contour). (Reproduced from *Int J Radiat Oncol Biol Phys*, **82**, Ren L, Chetty IJ, Zhang J, Jin JY, Wu QJ, Yan H, Brizel DM, Lee WR, Movsas B, and Yin FF, Development and clinical evaluation of a three-dimensional cone-beam computed tomography estimation method using a deformation field map,1584–1593, Copyright 2012, with permission from Elsevier.)

general, however, the point spread function (PSF) at the isocenter in a 40° DTS slice acquired with Varian's OBI has been shown in simulations to have a full-width at half-maximum (FWHM) of approximately 5 mm in the plane-to-plane direction (Blessing et al., 2006).

14.2.3.2 Noise and dose characteristics

Like almost all other forms of imaging, tomosynthesis is governed by the typical trade-off between noise and resolution. For this reason, the thick effective slice profile inherent to DTS results in images that exhibit much less stochastic noise than a full CBCT acquired with the same net exposure. We have found that acceptable gantry-based DTS image noise levels can be attained with the total scan exposures on the order of 1/10th the exposure used for CBCT imaging (Godfrey et al., 2006a), and thus, the typical DTS imaging dose is expected to be on the order of 0.1 cGy. It has been shown both theoretically and experimentally that tomosynthesis noise levels primarily depend on the scan angle and total scan exposure, and not on the number of projection images acquired, provided that the exposure is not so low that baseline detector noise becomes prominent (Godfrey et al., 2009).

14.2.3.3 Artifacts

The poor localization of low-frequency information in the plane-to-plane direction results in subtle blur artifacts that show up in DTS planes some distance away from their actual source, emanating most prominently from high-contrast features such as bony anatomy. These "structured noise" artifacts, which appear to be faint, blurred versions of the actual object structures, are sometimes referred to as "ghosting" artifacts and have the effect of reducing the contrast in reconstructed DTS slices. A sample DTS slice containing visible ghosting artifacts is presented in Figure 14.6. Fortunately, when viewing a stack of DTS slices, it is generally quite easy to distinguish between real structures and very faint ghosting artifacts.

14.2.4 CHOICE OF REFERENCE IMAGE DATA

For DTS-based target localization to be effective, the technique must provide the ability to detect subtle translational and rotational misalignment between DTS images acquired with the OBI, and an original treatment planning (reference) CT

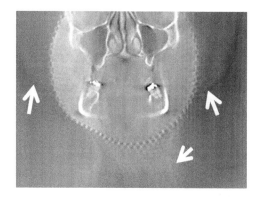

Figure 14.6 Coronal tomosynthesis slice showing anterior facial structures. Faint, low-frequency tomosynthesis "ghost" artifacts (white arrows) from distant structures are easy to distinguish from the actual in-plane anatomy but reduce the contrast in tomosynthesis images.

volume. However, due to the complexity and anisotropy of DTS resolution characteristics, image content is not well matched between DTS and CT volumes. It has been shown that this severely limits the potential precision and accuracy of DTS-to-CT registration (Godfrey et al., 2007).

One effective solution to this problem is to simulate reference DTS (RDTS) data from the treatment planning CT (Godfrey et al., 2006a). This is achieved by mimicking the actual on-board DTS scan geometry to compute a set of forward projections through the CT volume, which are then fed into the DTS reconstruction routine. The final result is a stack of RDTS slices whose image content very closely matches the actual on-board DTS slices, and can be used for precise and accurate registration of the therapy target volume. Figure 14.7 illustrates the process of simulating the projections, also known as digitally reconstructed radiographs (DRR). Figure 14.8 shows coronal CT, RDTS, and DTS slices from a subject being treated for a head-and-neck (H&N) cancer, with arrows highlighting regions where the RDTS exhibits far superior matching of DTS image content than the planning CT.

To quantify the impact of matching the image content between the reference and on-board image volumes, we conducted a study using a volume of interest (VOI) from the spine region of an anthropomorphic chest phantom (Godfrey et al., 2007). Reference CT and on-board DTS scans of the phantom were acquired, and then simulated translations and rotations spanning ±5 mm and 5° were applied to the reference CT data to mimic the initial misalignment of the image volumes. RDTS volumes were also simulated from each of the CT poses. Next, the normalized 3D mutual information (NMI) shared between the reference and an on-board image volume was computed for each "pose" of the reference volume. These values were then plotted and the accuracy was assessed by locating the peaks of the mutual information (MI) curves, while the precision was evaluated by considering the width of the peaks. This was performed for DTS scan angles of 22°, 44°, and 65°.

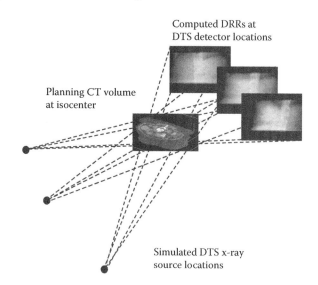

Computed DRRs at DTS detector locations

Planning CT volume at isocenter

Simulated DTS x-ray source locations

Figure 14.7 Forward projection of a planning CT volume to generate cone-beam DRRs, which form the basis for reference DTS reconstructions. (Reproduced from *Int J Radiat Oncol Biol Phys*, **65**, Godfrey D, Yin F, Oldham M et al. Digital tomosynthesis with an on-board tomosynthesis kilovoltage imaging device, 8–15, Copyright 2006a, with permission from Elsevier.)

Clinical applications

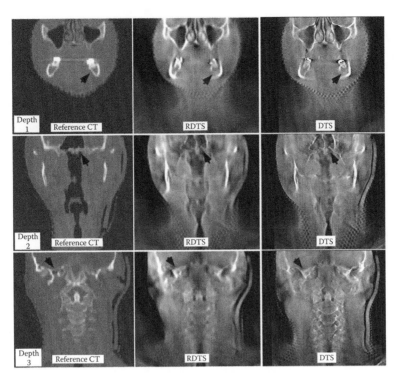

Figure 14.8 Coronal reference CT, 44° RDTS, and 44° on-board DTS slices at three different depths of an H&N subject. The black arrows point to anatomy that shares a similar appearance in RDTS and on-board DTS images, but has a different appearance in the corresponding reference CT slice. (Reproduced with permission from Godfrey D et al. 2007. *Med Phys* **34**:3374–3384.)

Figure 14.9 shows that well-defined NMI peaks were present <1 mm from the absolute truth for all six translations and rotations, when RDTS was used as the reference image set. It can be seen that the peaks were the broadest for translations in the *y* or plane-to-plane direction, but narrowed with increasing scan angle, as one would expect given the knowledge of the DTS depth resolution characteristics. All the other peaks were very sharp, suggesting the potential for excellent image registration precision for the DTS–RDTS pairing.

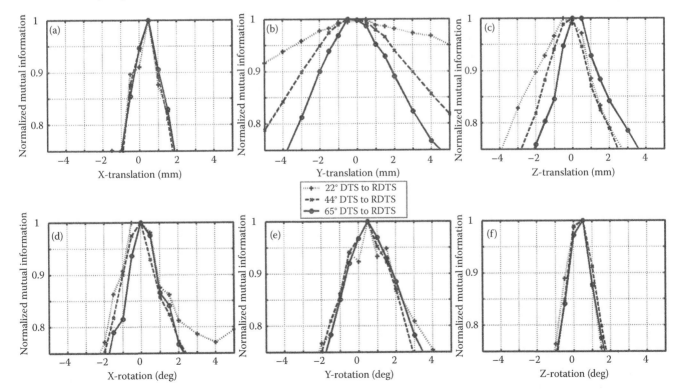

Figure 14.9 Effect of DTS scan angle. DTS to RDTS NMI versus each of the six rigid-body geometric perturbations for 22°, 44°, and 65° DTS scans. (Reproduced with permission from Godfrey D et al. 2007. *Med Phys* **34**:3374–3384.)

Clinical applications

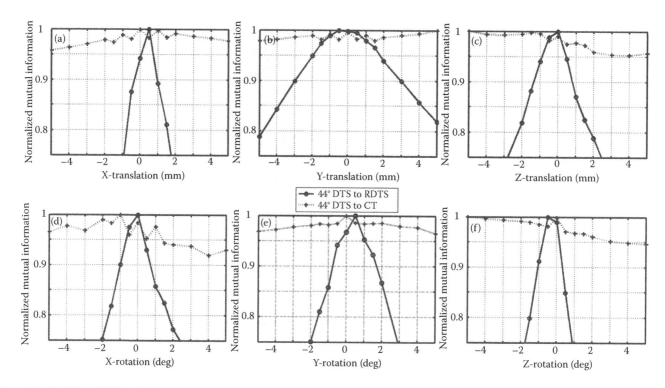

Figure 14.10 DTS to RDTS and DTS to CT NMI as a function of each of the six rigid-body geometric perturbations. (Reproduced with permission from Godfrey D et al. 2007. *Med Phys* **34**:3374–3384.)

On the other hand, the NMI shared between a 44° on-board DTS and a traditional reference CT volume did not result in sharp peaks when the two volumes were correctly registered. Figure 14.10 shows a comparison between the NMI curves obtained by registering a 44° on-board DTS volume to either an RDTS (similar to Figure 14.9) or a reference CT volume. It is clear from the figure that CT reference image data do not provide similar enough image content to allow for accurate registration with a DTS image set even when imaging simple bony anatomy. Thus, we conclude that any DTS target localization technique is likely to require the creation of RDTS image data for registration purposes.

14.3 TOMOSYNTHESIS IGRT APPLICATIONS

14.3.1 3D EBRT LOCALIZATION

14.3.1.1 Head and neck

H&N was the first clinical site studied for DTS-based IGRT application at our institution (Wu et al., 2007). This clinical study evaluated DTS as a daily imaging technique for patient positioning based on bony anatomy and compared the results of DTS-to-RDTS registration with 3D CBCT-to-CT and 2D radiographic localization to evaluate its accuracy. DTS image data were acquired using the standard CBCT imaging acquisition protocol for the H&N region and were reconstructed in the coronal and sagittal orthogonal views, using a 40° scan angle for each. Sixty-five imaging datasets from 10 H&N patients were analyzed. The daily patient positioning variation was retrospectively measured after all the image data were collected. All on-board images (CBCTs and DTS images), as well as their corresponding reference images, were loaded into a dedicated OBI

image evaluation station. To evaluate the potential for localizing with only a single DTS scan, the two DTS datasets (coronal and sagittal views) were treated as independent modalities. The image datasets for each modality were grouped and analyzed together, but information about the positioning variation detected by each modality was not available when performing the localization of the other modalities, to avoid introducing bias.

Table 14.1 summarizes the positioning differences (only translations) between the DTS and CBCT techniques for H&N (Wu et al., 2007). The mean differences between any of the DTS-based and CBCT-based positioning methods were <1.0 mm in all directions, including the out-of-plane direction, and the mean vector differences were on the order of 1.5 mm, with standard deviations of <1 mm, suggesting that either DTS-based method is very effective for the localization of bony H&N anatomy.

Table 14.1 Patient positioning differences (DTS vs. CBCT) for H&N Subjects

	H&N (BONY ANATOMY)	
	SAGITTAL	CORONAL
Vertical	0.7 (0.7)	0.7 (0.7)
Longitudinal	0.6 (0.6)	0.8 (0.7)
Lateral	0.7 (0.6)	0.8 (0.6)
Vector	1.4 (0.8)	1.5 (0.8)

Source: Reproduced from *Int J Radiat Oncol Biol Phys*, **69**, Wu QJ, Godfrey DJ, Wang Z, Zhang J, Zhou S, Yoo S, Brizel DM, and Yin FF. On-board patient positioning for head-and-neck IMRT: Comparing digital tomosynthesis to kilovoltage radiography and cone-beam computed tomography, 598–606, Copyright 2007, with permission from Elsevier.

Note: The standard deviations are shown in parentheses. All values are in mm.

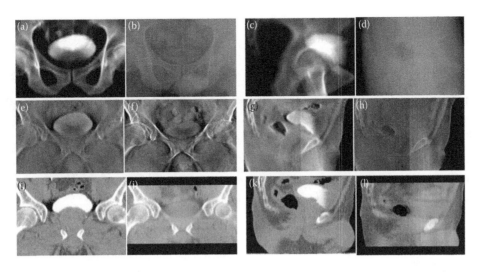

Figure 14.11 Reference and on-board images of a human subject with prostate cancer. (a) Antero-posterior DRR, (b) antero-posterior kV, (c) lateral DRR, (d) lateral kV, (e) coronal RDTS, (f) coronal DTS, (g) sagittal RDTS, (h) sagittal DTS, (i) coronal planning CT, (j) coronal CBCT, (k) sagittal planning CT, and (l) sagittal CBCT. All slice images are through the isocenter plane. (Reproduced from *Int J Radiat Oncol Biol Phys*, **73**, Yoo S, Wu QJ, Godfrey D, Yan H, Ren L, Das S, Lee WR, and Yin FF. Clinical evaluation of positioning verification using digital tomosynthesis and bony anatomy and soft tissues for prostate image-guided radiotherapy, 296–305, Copyright 2009, with permission from Elsevier.)

14.3.1.2 Prostate

A similar experimental design was employed to study DTS-based localization of subjects undergoing external beam prostate radiotherapy, but the prostate study included an examination of soft-tissue-based patient positioning in addition to the registration of bony anatomy (Yoo et al., 2009). Figure 14.11 displays sample reference and the corresponding on-board images from a prostate subject. It can be seen that 2D radiography, in the form of DRRs computed from the treatment planning CT along with matching on-board kV radiographs, captures only bony anatomy information, while both DTS and CBCT provide substantial additional soft-tissue information. The coronal view DTS exhibits superior soft-tissue contrast compared to the sagittal view, as the increased lateral tissue thickness reduces the signal of the projection image in the sagittal direction.

Ninety-two imaging sets from nine prostate patients were analyzed in the study, in the same manner as described above for the H&N study. Table 14.2 summarizes the positioning

differences (only translations) between the DTS and CBCT techniques for prostate patients (Yoo et al., 2009). As in the H&N study, bony anatomy positioning verification with DTS was found to be similar to the results achieved by CBCT. The mean differences between any of the DTS-based and CBCT-based positioning methods were <1 mm in all directions, including the out-of-plane direction, and the mean vector differences were in the range of 1.0–1.5 mm, with standard deviations of <1 mm. The results of matching based on soft tissue for prostate showed a slightly lower correlation than those based on bony anatomy (mean difference of 2–3 mm, with a standard deviation as large as 1.7 mm). In general, the sagittal–DTS method resulted in a larger variation from the CBCT method than the coronal–DTS method for the matching of soft tissues. While coronal–DTS appears to produce results similar enough to CBCT to potentially justify its use for soft-tissue localization, sagittal–DTS alone is likely to be insufficient for soft-tissue-based positioning verification without improvements in the sagittal DTS image quality.

14.3.1.3 Partial breast

Another interesting application of DTS imaging is for localizing accelerated partial breast irradiation (APBI) treatments. In APBI treatments, the irradiated volume is relatively small compared to conventional tangential fields, and hence, there is particular interest in using 3D imaging techniques to guide patient positioning. Recent studies have reported that CBCT imaging can achieve 1–2 mm positioning accuracy for APBI setups (Purdie et al., 2007; Fatunase et al., 2008). However, the current CBCT techniques may not be optimal for APBI treatments in terms of radiation dose, acquisition time, and geometric clearance (Godfrey et al., 2006a; Zhang et al., 2009). Since the isocenter for breast patients is often placed several centimeters lateral to the midline, near the chestwall, the couch often needs to be shifted medially (i.e., back to a centered position) for CBCT data acquisition, to enable a full gantry rotation of 360° without table or patient collisions. The contralateral breast and lung receive the same CBCT x-ray imaging dose (5–8 cGy) as

Table 14.2 **Patient positioning differences (DTS vs. CBCT) for prostate subjects**

	PROSTATE (BONY ANATOMY)		PROSTATE (SOFT TISSUE)	
	SAGITTAL	CORONAL	SAGITTAL	CORONAL
Vertical	0.7 (0.8)	0.5 (0.7)	1.7 (1.7)	1.1 (1.2)
Longitudinal	0.4 (0.6)	0.4 (0.6)	1.0 (0.9)	0.7 (0.6)
Lateral	0.7 (0.6)	0.3 (0.6)	1.5 (1.4)	0.8 (0.7)
Vector	1.4 (0.8)	1.1 (0.8)	3.0 (1.7)	1.9 (1.1)

Source: Reproduced from *Int J Radiat Oncol Biol Phys*, **73**, Yoo S, Wu QJ, Godfrey D, Yan H, Ren L, Das S, Lee WR, and Yin FF. Clinical evaluation of positioning verification using digital tomosynthesis and bony anatomy and soft tissues for prostate image-guided radiotherapy, 296–305, Copyright 2009, with permission from Elsevier.

Note: The standard deviations are shown in parentheses. All values are in mm.

the ipsilateral breast and lung (Kim et al., 2008). To protect the healthy tissue and minimize secondary cancer occurrences, the dose to the contralateral breast and lung should be minimized if possible (Fowble et al., 2001; Hall, 2006; Hall and Brenner, 2008). In DTS acquisition, contrary to CBCT acquisition, the patient can be left in the treatment position (no couch shift is necessary) because only the gantry has to rotate over a small angle. Furthermore, the contralateral breast and lung dose can be minimized because the scanning volume and angle can be selectively limited to the treatment site (Winey et al., 2009).

In a pilot study performed at our institution (Zhang et al., 2009), nine patients receiving external beam partial breast irradiation were evaluated for CBCT and DTS imaging guidance in reducing patient setup variations. Surgical clips were present at the excision bed in six of the nine patients and were used for image registrations when available.

DTS scans were generated along coronal, sagittal, and oblique orientations. Oblique DTS images were reconstructed using the projections between 225° + 22.5° and 225° − 22.5° for the right breast and the projections between 315° + 22.5° and 315° − 22.5° for the left breast, due to the 15-cm detector shift applied in the half-fan CBCT scan mode (enables a larger CBCT field of view). Figures 14.12 and 14.13 show the coronal, sagittal, and oblique CBCT and DTS images for breast cancer subjects with and without implanted surgical clips, respectively. As illustrated in both figures, it can be difficult to identify the postlumpectomy surgical bed of the tumor in one or both of the coronal and sagittal scans because off-plane bone and breast tissue contrast may strongly shade the tumor bed. However, the oblique scan orientation consistently allows the breast tissue, bones, and lung to be well separated, and the soft-tissue contrast of the tumor bed in the oblique view appears sufficient for registration of the DTS images.

For patient positioning, the set of CBCT and DTS images from the first treatment fraction were registered to each subsequent CBCT or DTS image set, acquired throughout the course of treatment, to determine the translational shifts between fractions. For CBCT images, the registration was based on

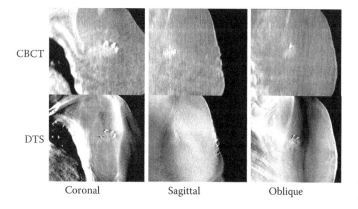

Figure 14.12 Three views of CBCT and DTS for a patient with surgical clips. While the surgical clips were visible in all three DTS scans, the tumor bed was clearly visible only in the coronal and oblique scans. (Reproduced from *Int J Radiat Oncol Biol Phys*, **73**, Zhang J, Wu QJ, Godfrey DJ, Fatunase T, Marks LB, and Yin FF. Comparing digital tomosynthesis to cone-beam CT for position verification in patients undergoing partial breast irradiation, 952–957, Copyright 2009, with permission from Elsevier.)

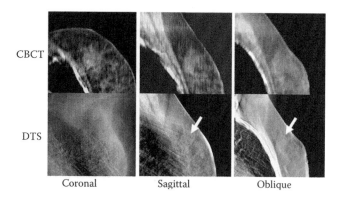

Figure 14.13 Three views of CBCT and DTS for a patient without surgical clips. The tumor bed (indicated by the arrows) is not visible in the coronal scan and is barely visible in the sagittal scan, but is clearly visible in the oblique scan. (Reproduced from *Int J Radiat Oncol Biol Phys*, **73**, Zhang J, Wu QJ, Godfrey DJ, Fatunase T, Marks LB, and Yin FF. Comparing digital tomosynthesis to cone-beam CT for position verification in patients undergoing partial breast irradiation, 952–957, Copyright 2009, with permission from Elsevier.)

the location of the soft-tissue target *seroma* (a postlumpectomy collection of serous fluid marking the spot where the tumor was located) and surgical clips within the breast. For DTS images, the registration was based on the two clips closest to the treatment isocenter when surgical clips were available, and was based on the tumor bed itself when no surgical clips were present. As shown in Figures 14.12 and 14.13, it was difficult to identify the tumor bed in one or both of the coronal and sagittal DTS images in some patients because residual tomosynthesis artifacts from out-of-plane bone and breast tissue added strong shading to the tumor bed. Therefore, only oblique scans were used for patients without clips. Translational shifts of DTS images were compared to the translational shifts of the corresponding CBCT images to evaluate the registration accuracy.

Table 14.3 shows the positioning difference between the DTS and CBCT registrations for patients with (group A) and without (group B) surgical clips. For group A, coronal DTS exhibited smaller root mean square (RMS) error than sagittal and oblique DTS along the lateral and vertical directions. Overall, DTS registrations based on the tumor bed displayed significantly lower accuracy than registrations based on surgical clips. For the oblique DTS, group B had comparatively larger RMS error values in the lateral and longitudinal directions than group A, and also for the vector sum. The RMS error along the vertical direction was not statistically different between the two groups.

Table 14.4 displays the positioning accuracy along the in-plane and out-of-plane directions for each DTS scan. For the six patients with surgical clips (group A), the off-plane accuracy was significantly lower than the in-plane accuracy for all three DTS scans. There was no significant difference between in-plane and off-plane accuracy for the three patients without surgical clips (group B).

14.3.2 4D EBRT LOCALIZATION

4D DTS was recently introduced as a possible alternative to 4D CBCT for faster 4D localization using much less imaging dose (Maurer et al., 2008). In 4D DTS, on-board cone-beam

Table 14.3 **RMS of the positioning difference between DTS and CBCT registration**

	GROUP A: SIX PATIENTS/52 FRACTIONS BASED ON SURGICAL CLIPS			GROUP B: THREE PATIENTS/25 FRACTIONS BASED ON TUMOR BED
	CORONAL	SAGITTAL	OBLIQUE	OBLIQUE
Lateral	0.90	1.67	1.29	2.03
Vertical	1.31	1.28	1.35	1.24
Longitudinal	0.85	1.25	1.10	1.43
Vector	1.80	2.45	2.17	2.78

Source: Reproduced from *Int J Radiat Oncol Biol Phys*, **73**, Zhang J, Wu QJ, Godfrey DJ, Fatunase T, Marks LB, and Yin FF. Comparing digital tomosynthesis to cone-beam CT for position verification in patients undergoing partial breast irradiation, 952–957, Copyright 2009, with permission from Elsevier.
Note: All values are in mm.

projections are acquired only within a limited scan angle for each respiratory phase, and 3D DTS images are reconstructed for each phase. Two methods have been proposed to realize the 4D DTS acquisition (Maurer et al., 2008, 2010). The first is slow gantry rotation, in which the gantry revolution speed is reduced relative to a 3D scan, to allow the continuously acquired projections to adequately sample all the respiratory phase bins. For this method, the optimal gantry rotation speed and projection acquisition frequency need to be determined based on the patient respiratory cycle and phase window to ensure that enough projections are acquired for each respiratory phase, while attempting to minimize the imaging dose. Alternatively, a 4D DTS scan can be performed in a step-and-shoot manner, with the gantry rotation stopped at a fixed number of angular positions, distributed evenly across the desired DTS scan angle. At each position, the projection images are acquired continuously for at least one respiratory period, before the gantry is rotated again. No image acquisition occurs during gantry rotation. Regardless of the selected method, projection images are retrospectively sorted, binned, and reconstructed according to the patient's respiratory phase. Owing to the anisotropic resolution characteristics of tomosynthesis, it may be necessary to perform coronal and sagittal 4D DTS in combination for tracking tumor motion in all three directions (Maurer et al., 2008).

Properly sorting the projection data into correct respiratory bins is a challenge for all the 4D tomographic imaging techniques (DTS, CBCT, and CT). Phase-tracking techniques can be distinguished according to whether they rely upon the tracking of an external marker or, instead, employ an internal surrogate to judge motion. External tracking methods include the detection of an infrared marker placed on a patient's abdomen, as well as pressure belts worn by the patient that convert the expansion of the abdomen directly into a voltage signal. Errors in the temporal correlation between any external signal and the internal target motion will cause errors in phase tracking and consequent image artifacts. As an alternative, internal surrogate tracking may be employed to track the location of implanted fiducial markers or natural anatomical surrogates from 2D projection images, to estimate the respiratory signal. The accuracy of this method relies upon the automatic detection of the surrogate in projection images, and the strength of the correlation between surrogate motion and the motion of the target.

Preliminary studies with a dynamic thorax phantom showed that 60° 4D DTS was able to track the motion of a simple target with submillimeter accuracy (Maurer et al., 2008). A follow-up study using lung cancer patient data also demonstrated the feasibility of using 4D DTS for lung tumor localization (Maurer et al., 2010). The results suggested that the scan angle was far more important in achieving tumor visibility than the number of projections acquired. The scan angle required to visualize the tumor was shown to depend on the tumor location, shape, size, and the surrounding patient anatomy. In the study, three lung tumors of different size and location were examined. All tumors were visible in the coronal view with 40–60° DTS scan angles. The largest tumor was visible in the sagittal view using a 40° scan angle, but the two smaller tumors required 100–120° scan angles for sagittal visualization. Figure 14.14 displays the effect of

Table 14.4 **RMS of the positioning differences for the in-plane and out-of-plane directions**

	GROUP A: SIX PATIENTS/52 FRACTIONS BASED ON SURGICAL CLIPS			GROUP B: THREE PATIENTS/25 FRACTIONS BASED ON TUMOR BED
	CORONAL	SAGITTAL	OBLIQUE	OBLIQUE
In-plane: Longitudinal	0.85	1.28	1.10	1.43
In-plane: Axial	0.90	1.25	0.96	1.75
Off-plane	1.31	1.67	1.60	1.62
p-Value	0.0001	0.01	0.0001	0.5

Source: Reproduced from *Int J Radiat Oncol Biol Phys*, **73**, Zhang J, Wu QJ, Godfrey DJ, Fatunase T, Marks LB, and Yin FF. Comparing digital tomosynthesis to cone-beam CT for position verification in patients undergoing partial breast irradiation, 952–957, Copyright 2009, with permission from Elsevier.
Note: All values are in mm. *P*-values were calculated for the hypothesis that in-plane errors were smaller than out-of-plane errors.

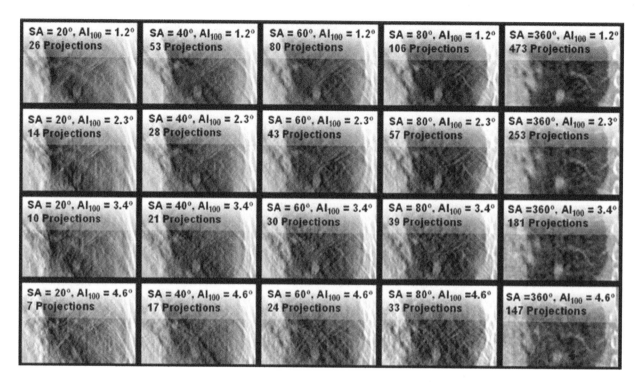

Figure 14.14 Coronal DTS reconstructions of a lung tumor (approximate volume 0.24 cm³) from a single 4D DTS phase, with varying scan angle and number of projections. (Reproduced with permission from Maurer J, Pan T, and Yin FF. 2010. *Med Phys* **37**:921–933.)

varying the scan angle and number of projections on 4D coronal DTS reconstructions of a lung tumor.

Although 4D DTS has been demonstrated to be potentially useful for 4D EBRT localization, several limitations also exist for this technique. Like 3D DTS, 4D DTS does not provide full volumetric information due to the limited scan angle, which may limit its localization accuracy, particularly for low-contrast, soft-tissue targets. Errors in respiratory signal tracking also potentially limit the accuracy of 4D DTS reconstructions, although this is true for 4D CT and 4D CBCT, as well. Finally, the accuracy of phase sorting and binning is affected by the presence of any irregularity in a patient's breathing pattern. This concern can be addressed by employing additional patient motion management, such as an active breathing control device and/or postprocessing of the breathing signal.

14.3.3 TOMOSYNTHESIS FOR PROSTATE BRACHYTHERAPY

Although most tomosynthesis IGRT research has focused on employing the technique for EBRT guidance, tomosynthesis has also been investigated as an alternative to fluoroscopy for discerning seed positions in real time during prostate implants, thus improving the efficacy of real-time dose calculation in the operating room. For this endeavor, DTS has the particular advantage of being able to distinctly render seeds that would otherwise appear superimposed in a traditional radiograph. Tutar et al. developed a novel brachytherapy-specific DTS reconstruction technique with built-in blur compensation, which they refer to as the selective backprojection method (SBM) (Tutar et al., 2003). In the SBM, seeds are segmented in each projection image to generate seed-only images that are then used for DTS reconstruction. In a preliminary study with a tissue-equivalent ultrasound phantom containing 61 radiographically visible dummy ^{125}I seeds, SBM was

able to localize the implanted seeds with submillimeter accuracy using only seven projections distributed over a 30° scan angle. Lee et al. developed a different tomosynthesis-based seed localization method, utilizing modified distance map images (Lee et al., 2008). In an initial trial with two patients, this method enabled researchers to localize all 60 seeds with submillimeter accuracy, using as few as three projections acquired over a 10° scan angle, and was robust to C-arm pose errors. Although preliminary studies suggest that tomosynthesis may be valuable for intraoperative prostate seed localization, the application is still being investigated, and more comprehensive clinical evaluations need to be performed before its clinical value can be fully determined.

14.4 FUTURE DIRECTIONS

14.4.1 DUAL-SOURCE TOMOSYNTHESIS

Dual-source imaging has been a topic of recent interest for IGRT applications (Yin et al., 2009). Owing to the complementary properties of kV and MV beams, Yin et al. proposed to use an MV/kV aggregated imaging technique to improve the imaging efficiency while reducing CBCT reconstruction artifacts caused by high-density materials (Yin et al., 2005; Zhang et al., 2007). Massachusetts General Hospital in collaboration with Varian Medical Systems, Inc. has developed the integrated radiotherapy imaging system (IRIS), which consists of two gantry-mounted diagnostic kV x-ray tubes and two flat-panel a-Si detectors, for a dual kV source approach (Berbeco et al., 2004). Kamino et al. at Mitsubishi Heavy Industries, Ltd. also recently developed a dual-detector and dual-tube system mounted to a 4D-IGRT therapy system (MHI-TM2000) (Kamino et al., 2006). In addition to performing radiographic imaging and CBCT, these dual-source imaging systems can also be employed to simultaneously acquire coronal and sagittal DTS images, which would reduce the image acquisition time and minimize the motion artifacts associated with two independent scans. As in dual-source CBCT, combining MV and kV beams for DTS imaging may potentially reduce metal and scatter artifacts that are sometimes present in pure kV DTS images. Further comprehensive investigations are needed to determine the value of dual-source DTS imaging for IGRT applications.

14.4.2 FUNCTIONAL TOMOSYNTHESIS

Functional imaging modalities such as positron emission tomography (PET) and single photon emission computed tomography (SPECT) track radiotracer distribution in the body to identify physiological processes—such as angiogenesis, apoptosis, hormone receptor status, hypoxia, and proliferation—that have important implications in cancer management. Meanwhile, all currently available in-room imaging guidance tools only provide anatomical information for target localization, which may not be sufficient for localizing low-contrast tumors or for targeting subclinical disease.

Perhaps, the most important IGRT advances still to come involve the development of in-treatment-room functional imaging techniques that might allow treatment to be guided by imaged physiology, rather than just anatomy. A novel on-board SPECT system has recently been proposed for the localization of a biological target prior to radiation therapy (Roper et al., 2009, 2010). In addition to potentially increasing the contrast-to-noise ratio of targeted tumors, the additional information provided by SPECT might ultimately enable the delivery of biologically conformal radiation therapy that can reasonably be expected to enhance the therapeutic success (Ling et al., 2000). Initial simulation studies have demonstrated that on-board SPECT is capable of localizing a tumor to within 2 mm, given a 180° scan angle and a 4-min scan duration (Roper et al., 2009, 2010). It is easy to envision tomosynthesis as a means to potentially reduce the scan time and therefore minimize motion artifacts in SPECT reconstructions, which are a limiting factor for SPECT-based IGRT localization accuracy.

14.5 DISCUSSION

Radiation oncology applications of tomosynthesis are currently at somewhat of a crossroad. Although DTS has been investigated for many image guidance tasks and has been shown in studies to be a potentially valuable alternative to 2D radiographic imaging or fully 3D CBCT, the modality has yet to gain a foothold for any single application and currently, there are no commercial implementations of DTS-based IGRT. Further investigation of DTS applications by many research groups is still under way, many of which are supported by industrial collaborations, and it would be a surprise if the technique did not formally find its way into the clinic at some point in the relatively near future. We believe that a particularly strong case can be made right now for using tomosynthesis to guide the treatment of sites affected by respiratory motion, as breath-hold CBCT is cumbersome and requires multiple full inspirations to piece together a single scan, and 4D CBCT may be too time and dose intensive for frequent clinical use. Furthermore, given the recent focused efforts to reduce the

imaging dose in radiation oncology, tomosynthesis-based IGRT may represent an effective trade-off between dose and resolution for daily IGRT needs.

REFERENCES

Altschuler MD and Kassaee A. 1997. Automated matching of corresponding seed images of three simulator radiographs to allow 3D triangulation of implanted seeds. *Phys Med Bio* **42**:293–302.

Amols HI and Rosen II. 1981. A three-film technique for reconstruction of radioactive seed implants. *Med Phys* **8**:210–214.

Baydush A, Godfrey D, Oldham M et al. 2005. Initial application of digital tomosynthesis with on-board imaging in radiation oncology. *Proc SPIE Phys Med Imaging* **5745**:1300–1305.

Berbeco RI, Jiang SB, Sharp GC, Chen GT, Mostafavi H, and Shirato H. 2004. Integrated radiotherapy imaging system (IRIS): Design considerations of tumour tracking with linac gantry-mounted diagnostic x-ray systems with flat-panel detectors. *Phys Med Bio* **49**:243–255.

Biggs PJ and Kelley DM. 1983. Geometric reconstruction of seed implants using a three-film technique. *Med Phys* **10**:701–704.

Blessing M, Godfrey D, Lour F, and Yin F. 2006. Analysis of the point spread function of isocentric digital tomosynthesis (DTS). *Med Phys* **33**:2266.

Descovich M, Morin O, Aubry JF, Aubin M, Chen J, Bani-Hashemi A, and Pouliot J. 2008. Characteristics of megavoltage cone-beam digital tomosynthesis. *Med Phys* **35**:1310–1316.

Dietrich L, Jetter S, Tucking T, Nill S, and Oelfke U. 2006. Linac-integrated 4D cone beam CT: First experimental results. *Phys Med Bio* **51**:2939–2952.

Dobbins III JT and Godfrey DJ. 2003. Digital x-ray tomosynthesis: Current state of the art and clinical potential. *Phys Med Bio* **48**:R65–R106.

Fatunase T, Wang Z, Yoo S, Hubbs JL, Prosnitz RG, Yin FF, and Marks LB. 2008. Assessment of the residual error in soft tissue setup in patients undergoing partial breast irradiation: Results of a prospective study using cone-beam computed tomography. *Int J Radiat Oncol Biol Phys* **70**:1025–1034.

Fowble B, Hanlon A, Freedman G, Nicolaou N, and Anderson P. 2001. Second cancers after conservative surgery and radiation for stages I–II breast cancer: Identifying a subset of women at increased risk. *Int J Radiat Oncol Biol Phys* **51**:679–690.

Godfrey D, Yin F, Oldham M et al. 2006a. Digital tomosynthesis with an on-board tomosynthesis kilovoltage imaging device. *Int J Radiat Oncol Biol Phys* **65**:8–15.

Godfrey DJ, McAdams HP, and Dobbins JT III. 2006b. Optimization of the matrix inversion tomosynthesis (MITS) impulse response and modulation transfer function characteristics for chest imaging. *Med Phys* **33**:665–667.

Godfrey DJ, McAdams HP, and Dobbins JT III. 2009. Stochastic noise characteristics in matrix inversion tomosynthesis (MITS). *Med Phys* **36**:1521–1532.

Godfrey D, Ren L, Yan H et al. 2007. Evaluation of three types of reference image data for external beam radiotherapy target localization using digital tomosynthesis (DTS). *Med Phys* **34**:3374–3384.

Gong L, Cho PS, Han BH, Wallner KE, Sutlief SG, Pathak SD, Haynor DR, and Kim Y. 2002. Ultrasonography and fluoroscopic fusion for prostate brachytherapy dosimetry. *Int J Radiat Oncol Biol Phys* **54**:1322–1330.

Hall EJ. 2006. Intensity-modulated radiation therapy, protons, and the risk of second cancers. *Int J Radiat Oncol Biol Phys* **65**:1–7.

Hall EJ and Brenner DJ. 2008. Cancer risks from diagnostic radiology. *Br J Radiol* **81**:362–378.

Kamino Y, Takayama K, Kokubo M, Narita Y, Hirai E, Kawawda N, Mizowaki T, Nagata Y, Nishidai T, and Hiraoka M. 2006. Development of a four-dimensional image-guided radiotherapy system with a gimbaled x-ray head. *Int J Radiat Oncol Biol Phys* **66**:271–278.

Lee J, Liu X, Jain AK, Prince JL, and Fichtinger G. 2008. Tomosynthesis-based radioactive seed localization in prostate brachytherapy using modified distance map images. *Proc IEEE Int Symp Biomed Imaging* 680–683.

Leng S, Tang J, Zambelli J, Nett B, Tolakanahallie R, and Chen GH. 2008a. High temporal resolution and streak-free four-dimensional cone-beam computed tomography. *Phys Med Bio* **53**:5653–5673.

Leng S, Zambelli J, Tolakanahalli R, Nett B, Munro P, Star-Lack J, Paliwal B, and Chen GH. 2008b. Streaking artifacts reduction in four-dimensional cone-beam computed tomography. *Med Phys* **35**:4649–4659.

Li T, Xing L, Munro, P, McGuinness C, Chao M, Yang Y, Loo B, and Koong A. 2006. Four-dimensional cone-beam computed tomography using an on-board imager. *Med Phys* **33**:3825–3833.

Li T, Koong A, and Xing L. 2007. Enhanced 4D cone-beam CT with inter-phase motion model. *Med Phys* **34**:3688–3695.

Ling CC, Humm J, Larson S, Amols H, Fuks Z, Leibel S, and Koutcher JA. 2000. Towards multidimensional radiotherapy (MD-CRT): Biological imaging and biological conformality. *Int J Radiat Oncol Biol Phys* **47**:551–560.

Maurer J, Godfrey D, Wang Z, and Yin FF. 2008. On-board four-dimensional digital tomosynthesis: First experimental results. *Med Phys* **35**:3574–3583.

Maurer J, Pan T, and Yin FF. 2010. Slow gantry rotation acquisition technique for on-board four-dimensional digital tomosynthesis. *Med Phys* **37**:921–933.

Mestrovic A, Nichol A, Clark B et al. 2009. Integration of on-line imaging, plan adaptation and radiation delivery: Proof of concept using digital tomosynthesis. *Phys Med Biol* **54**:3803.

Nag S, Ciezki JP, Cormack R, Doggett S, DeWyngaert K, Edmundson GK, Stock RG, Stone NN, Yu Y, and Zelefsky MJ. 2001. Intraoperative planning and evaluation of permanent prostate brachytherapy: Report of the American Brachytherapy Society. *Int J Radiat Oncol Biol Phys* **51**:1422–1430.

Pang G, Bani-Hashemi A, Au P, O'Brien PF, Rowlands JA, Morton G, Lim T, Cheung P, and Loblaw A. 2008. Megavoltage cone beam digital tomosynthesis (MV-CBDT) for image-guided radiotherapy: A clinical investigational system. *Phys Med Biol* **53**:999–1013.

Park JC, Park SH, Kim JH, Yoon SM, Kim SS, Kim JS, Liu Z, Watkins T, and Song WY. 2011. Four-dimensional cone-beam computed tomography and digital tomosynthesis reconstructions using respiratory signals extracted from transcutaneously inserted metal markers for liver SBRT. *Med Phys* **38**:1028–1036.

Peng Y, Vedam S, Chang JY, Gao S, Sadagopan R, Bues M, and Balter P. 2011. Implementation of feedback-guided voluntary breath-hold gating for cone beam CT-based stereotactic body radiotherapy. *Int J Radiat Oncol Biol Phys* **80**:909–917.

Purdie TG, Bissonnette JP, Franks K, Bezjak A, Payne D, Sie F, Sharpe MB, and Jaffray DA. 2007. Cone-beam computed tomography for on-line image guidance of lung stereotactic radiotherapy: Localization, verification, and intrafraction tumor position. *Int J Radiat Oncol Biol Phys* **68**:243–252.

Ren L, Zhang J, Thongphiew D et al. 2008. A novel digital tomosynthesis (DTS) reconstruction method using a deformation field map. *Med Phys* **35**:3110.

Ren L, Chetty IJ, Zhang J, Jin JY, Wu QJ, Yan H, Brizel DM, Lee WR, Movsas B, and Yin FF. 2012. Development and clinical evaluation of a three-dimensional cone-beam computed tomography estimation method using a deformation field map. *Int J Radiat Oncol Biol Phys* **82**:1584–1593.

Roper J, Bowsher J, and Yin FF. 2009. On-board SPECT for localizing functional targets: A simulation study. *Med Phys* **36**:1727–1735.

Roper J, Bowsher J, and Yin FF. 2010. Direction-dependent localization errors in SPECT images. *Med Phys* **37**:4886–4896.

Rosenthal MS and Nath R. 1983. An automatic seed identification technique for interstitial implants using three isocentric radiographs. *Med Phys* **10**:475–479.

Santoro J, Kriminski S, Lovelock DM et al. 2010. Evaluation of respiration-correlated digital tomosynthesis in lung. *Med Phys* **37**:1237–1245.

Sonke JJ, Zijp L, Remeijer P, and Van Herk M. 2005. Respiratory correlated cone beam CT. *Med Phys* **32**:1176–1186.

Sprenger F, Calderon-Colon X, Cheng Y et al. 2010. Distributed source x-ray tube technology for tomosynthesis imaging. *Proc SPIE Phys Med Imaging* **7622**:76225M-1–76225M-8.

Tutar IB, Managuli R, Shamdasani V, Cho PS, Pathak SD, and Kim Y. 2003. Tomosynthesis-based localization of radioactive seeds in prostate brachytherapy. *Med Phys* **30**:3135–3142.

Winey B, Zygmanski P, and Lyatskaya Y. 2009. Evaluation of radiation dose delivered by cone beam CT and tomosynthesis employed for setup of external breast irradiation. *Med Phys* **36**:164–173.

Wu QJ, Godfrey DJ, Wang Z, Zhang J, Zhou S, Yoo S, Brizel DM, and Yin FF. 2007. On-board patient positioning for head-and-neck IMRT: Comparing digital tomosynthesis to kilovoltage radiography and cone-beam computed tomography. *Int J Radiat Oncol Biol Phys* **69**:598–606.

Yan H, Ren L, Godfrey DJ et al. 2007. Accelerating reconstruction of reference digital tomosynthesis using graphics hardware. *Med Phys* **34**:3768–3776.

Yan H, Godfrey DJ, and Yin F. 2008. Fast reconstruction of digital tomosynthesis using on-board images. *Med Phys* **35**:2162.

Yin FF, Guan H, and Lu W. 2005. A technique for on-board CT reconstruction using both kilovoltage and megavoltage beam projections for 3D treatment verification. *Med Phys* **32**:2819–2826.

Yin FF, Das S, Kirkpatrick J, Oldham M, Wang Z, and Zhou SM. 2006. Physics and imaging for targeting of oligometastases. *Sem Rad Onc* **16**:85–101.

Yin FF, Wang Z, Yoo S, Wu QJ, Kirkpatrick J, Larrier N, Meyer J, Willett CG, and Marks LB. 2008. Integration of cone-beam CT in stereotactic body radiation therapy. *Technol Cancer Res Treat* **7**:133–140.

Yin FF, Wong J, Balter J, and Benedict S. 2009. The role of in-room kV x-ray imaging for patient setup and target localization. *Rep AAPM Task Group* 104.

Yoo S, Wu QJ, Godfrey D, Yan H, Ren L, Das S, Lee WR, and Yin FF. 2009. Clinical evaluation of positioning verification using digital tomosynthesis and bony anatomy and soft tissues for prostate image-guided radiotherapy. *Int J Radiat Oncol Biol Phys* **73**:296–305.

Zhang J, Godfrey D, and Yin FF. 2006. A study of megavoltage beam tomosynthesis. *Med Phys* **33**:1982.

Zhang J, Yang G, Lee YZ et al. 2006. A multi-beam x-ray imaging system based on carbon nanotube field emitters. *Proc SPIE Phys Med Imaging* **6142**:614204–1:614204–8.

Zhang J and Yin FF. 2007. Minimizing image noise in on-board CT reconstruction using both kilovoltage and megavoltage beam projections. *Med Phys* **34**:3665–3673.

Zhang J, Wu QJ, Godfrey DJ, Fatunase T, Marks LB, and Yin FF. 2009. Comparing digital tomosynthesis to cone-beam CT for position verification in patients undergoing partial breast irradiation. *Int J Radiat Oncol Biol Phys* **73**:952–957.

Zhang J and Yu C. 2010. A novel solid-angle tomosynthesis (SAT) scanning scheme. *Med Phys* **37**:4186–4192.

Clinical applications

Future developments in breast tomosynthesis

Martin J. Yaffe, Roberta Jong, and James G. Mainprize

Contents

15.1 STATUS AND LIMITATIONS OF CURRENT TOMOSYNTHESIS

Systems for digital breast tomosynthesis (DBT) are now being used for clinical work in several parts of the world, including Europe, Asia, the United States, and Canada. Early findings from trials suggest that tomosynthesis may replace digital mammography as the standard tool for screening because of its improved sensitivity and specificity. The role of breast computed tomography (CT) versus DBT is still unclear; however, because of the true isotropic three-dimensional (3D) nature of CT, but with inferior x–y spatial resolution to DBT, CT may be found very useful for diagnostic applications, particularly with the use of contrast media. The technology of DBT and CT is still quite young and there is ample scope for further optimization and development. Areas for possible improvement of DBT are discussed in the previous chapters and include the choice of angular range and increment for acquisition of projections, the distribution of radiation dose among the projections (dose budgeting), the reconstruction algorithm, image processing, detector technology, and the overall user interface. The potential benefits that are worth pursuing and would facilitate the acceptance of tomosynthesis are dose reduction from current

levels and improved efficiency for the radiologist in working with the image sets. There are also opportunities for the extraction of additional useful information from the examination by the use of contrast media, density measurement, and computer-assisted diagnostic techniques. Finally, there may be value in systems that provide fusion of tomosynthesis with other complementary breast imaging modalities.

15.2 CLINICAL PERSPECTIVE

The clinical role of tomosynthesis is evolving. There are many questions to be resolved. Should it be used for screening or diagnosis or both? Is it a stand-alone technology to replace current mammography or an ancillary examination? Is the usual two-dimensional (2D) mammogram necessary? Will one tomosynthesis view—the mediolateral oblique—be sufficient or is the craniocaudal view also required? Will tomosynthesis in one or two projections with a synthesized 2D image in the same or other projection (see Section 15.4.3 below) be the optimum? Clinical trials and experience will help clarify these issues.

For radiologist workflow, tomosynthesis needs integration of all components from the acquisition, image processing, storage, retrieval, and display. Meeting Integrating the Healthcare Enterprise (IHE) and digital imaging and communications in medicine standard (DICOM) requirements will assist in this process. The efficient review of the large number of reconstructed images will be assisted by computer-aided detection (CAD) algorithms, especially for calcifications that might otherwise be overlooked. The ability to segment and rotate these in an maximum intensity projection (MIP) view would be helpful to assess their spatial relationship. Combining images into slabs of varying thickness decreases the number of images needed to review. After an initial overview, it would be advantageous to be able to skip from one CAD-marked image to another. For recalled patients, it would be useful to be able to readily display only annotated images. As in CT for other body areas, different default window width and level presentations or possibly different reconstruction algorithms for soft tissue or calcifications may prove to be useful.

Digital tomosynthesis data also lends itself to other potential uses such as volume measurements of lesions and quantification of the amount of fibroglandular tissue in the breast, which is a known risk factor for developing breast cancer. Image fusion to data from other modalities such as ultrasound, contrast-enhanced mammography, nuclear medicine (e.g., sestamibi, fluorodeoxyglucose (FDG), or fluorodeoxythymidine (FLT)) studies, or optical imaging may lead to valuable anatomical and functional information correlates.

15.3 DIGITAL MAMMOGRAPHY, BREAST TOMOSYNTHESIS, OR BREAST CT?

Full-field digital mammography has essentially supplanted conventional film mammography in North America and Europe and is making inroads in other regions. For example, in the United States, 87% (10887/12456) of Mammographic Quality Standards Act (MQSA)-accredited units were digital (including computed radiography) in 2012 (US Food and Drug Administration 2012). From the results of the Digital Mammography Imaging Screening Trial (DMIST), digital mammography was shown to be better than film for younger women (<50 years), for dense breasts, and/or for premenopausal women (Pisano et al. 2005). Preliminary results from studies of screening with DBT suggest that improved sensitivity and specificity and reduced call back rates may be possible (Philpotts et al. 2012). Two large screening studies, Malmö and the Oslo Tomosynthesis Screening Trials (U.S. National Institutes of Health 2012a,b) are currently enrolling women to compare the accuracy of tomosynthesis and digital mammography on a single-vendor DBT system. A larger expanded study evaluating DBT performance on multiple vendors, the Tomosynthesis Mammographic Imaging Screening Trial (TMIST), has been proposed to evaluate DBT in a North American screening context (ACRIN 2012). If these studies are successful, it is likely that DBT will supplant FFDM. However, concerns exist regarding the increased amount of data (number of images to be interpreted in an examination) and new interpretation skills that need to be learned (Section 15.2). Furthermore, most DBT protocols also include conventional 2D digital mammograms obtained on the same unit. This increases dose and interpretation time.

As an alternative to DBT, dedicated breast CT is a system used to obtain full 3D CT datasets of the breast (Boone et al. 2001). With these systems, imaging is carried out with the patient lying prone with one breast suspended into a portal to allow for a circular orbit of the x-ray tube and detector around the breast for a cone-beam CT acquisition. Careful optimization of the x-ray spectrum and use of low-noise, rapid readout detectors can be used to reduce the radiation dose such that a breast CT exam is equivalent in dose to that of 1–2 digital mammograms. The resolution characteristics are quite different. Breast CT systems generally produce isotropic voxels of side lengths 200–300 μm, whereas DBT can have in-plane (x–y) resolutions ranging from 50 to 140 μm and a nominal cross-plane resolution of 0.5–1 mm (Boone et al. 2001).

For breast CT, chest wall and axillary tail coverage is somewhat problematic with a pendent breast geometry, although a table top with a gentle swale helps improve the coverage. A design that employed an x-ray tube and detector motion with a tilted orbit or a saddle orbit such as that proposed by Madhav et al. (2009) would likely improve coverage.

One of the chief concerns is the appearance of clusters of small microcalcifications, which may be lost in CT versus DBT. This arises partly from the relatively coarse CT detector resolution, but also from the image noise related to the restricted radiation dose used in an exam (Chen et al. 2008). When clusters of microcalcifications are viewed in serial slice images, the radiologist may be inhibited in their ability to perceive signs of malignancy based on cluster arrangement for both DBT and breast CT. Slab view imaging may assist the radiologist in recognizing suspicious clusters for DBT (Diekmann et al. 2009). A similar approach could be used for breast CT, or other more advanced 3D visualization techniques, provided the signal-to-noise ratio is sufficient (Karellas and Vedantham 2008).

Reading breast CT datasets may be burdensome. The number of slices can be staggering. Following an example given by Glick (2007), for a breast with a 15-cm chest-wall-to-nipple distance, the image set could contain as many as 750 transverse slices if reconstructed with 200 μm isotropic voxels, which could be viewed along the other two primary axes, or any arbitrary axis for that matter. It is likely that CAD or new methods for visualization of 3D data may be very important to maintain efficient workflow (Glick 2007).

Image appearance for CT is quite different than FFDM likely requiring additional training (O'Connell et al. 2010). Although many of the key mammographic indicators of disease (e.g., spiculated margins, architectural distortion) are conserved in DBT, additional training for DBT interpretation will be required to achieve proficiency (Smith et al. 2008). DBT fits well with hanging protocols for mammography. On the other hand, breast CT may have a natural match to breast MRI hanging protocols (Glick 2007).

Because DBT is essentially an "add-on" to many mammography systems, DBT will likely see much more rapid adoption in the clinics. If efficacy is proven by the large clinical trials, DBT may become a standard screening tool. Because breast CT is quite different in operation and review, there will probably be a slower uptake, at least in the immediate future. Breast CT will appeal to women who dislike compression and it might be a very accessible alternative to MRI.

15.4 ACQUISITION PARAMETERS AND IMAGE RECONSTRUCTION

15.4.1 ANGULAR RANGE AND INCREMENT FOR ACQUISITION OF PROJECTIONS

As described in Chapter 2, the acquisition geometry has an impact on reconstructed image quality. Two competing phenomena affect image quality. Generally speaking, wider angles lead to improved detectability of masses (Reiser and Nishikawa 2010; Sechopoulos and Ghetti 2009). However, detectability of fine detail structures (e.g., microcalcifications) can suffer from insufficiently fine angular sampling (Reiser and Nishikawa 2010; Sechopoulos and Ghetti 2009). At low doses (higher relative quantum noise), increasing the number of projections beyond a certain threshold may not significantly improve image quality, because an inadequate number of quanta are used to record each projection. The result is a trade-off between angular range and number of projections, resulting in a compromise between detectability of masses and calcifications. The use of model observers, as described in Chapter 10, will be instrumental in determining the impact of choice of geometric parameters on image quality and lesion detectability.

15.4.2 RECONSTRUCTION ALGORITHM

There are several approaches to reconstructing tomosynthesis volumes (Chapters 7 and 8). Certainly, these approaches can be optimized to yield high-quality images. An area for continuing improvement is in the reduction of image noise and artifact appearance. Of particular interest are specialized iterative algorithms that incorporate prior information and/or constraints into the reconstruction process to reduce the noise level. Noise-tolerant methods like total variation minimization (Sidky et al. 2009; Velikina et al. 2007) and penalized maximum-likelihood estimation (Das et al. 2011) incorporate constraints that monitor image variance or roughness measures to reduce fluctuations in the reconstructed images due to image noise. Generally, these methods require more computational overhead, but even with moderate desktop computers, reconstruction times are reasonable and can be easily accelerated with dedicated hardware.

15.4.3 IMAGE PROCESSING

Projection images or reconstructed slices may be processed to improve image quality, for example, by employing pre- or postreconstruction filters (Das et al. 2011; Wang et al. 2012) to reduce noise or artifacts. New image views may be created such as maximum intensity projection or average "slab" views for better visualization of clustered microcalcifications (Diekmann et al. 2009). One interesting development is the "synthesized mammogram" that is a digitally reconstructed radiograph created by reprojecting the tomosynthesis dataset to create an image that appears very similar to a conventional mammogram (Gur et al. 2011). Because tomosynthesis datasets are generally very noisy, special care is required to filter or constrain the projection data to produce a mammogram with sufficient signal to noise ratio (SNR). If found to be practical and acceptable to radiologists, the synthesized mammogram may eliminate the need for additional conventional mammographic views, currently suggested for some screening protocols.

15.5 DETECTOR TECHNOLOGY

Digital x-ray detector technology has matured in recent years. Active matrix flat-panel imagers (AMFPI) are largely either indirect conversion systems with a cesium iodide structured phosphor coupled to a light-sensitive flat-panel array or a direct conversion system that employs an x-ray-absorbing selenium photoconductor coupled to a flat-panel readout. Further details of detector technologies for tomosynthesis are presented in Chapter 3.

Generally, first-generation DBT systems simply used a commercial-grade digital mammography detector. Electronic noise levels, which were low enough to provide excellent performance for digital mammography, become of much greater concern for tomosynthesis, especially when considered in emerging techniques such as contrast-enhanced tomosynthesis (Section 15.9.1) in which multiple datasets may be acquired in a single exam. Newer detectors have been developed that have significantly reduced electronic noise and increased detective quantum efficiency (DQE) (Ghetti et al. 2008), and as the flat-panel imager technology improves, these noise levels will be reduced further.

There is a definite trend to higher kilovoltages to increase the transmission of x-rays through the breast, providing increased image signal-to-noise ratios (Smith et al. 2008; Varjonen et al. 2008). Also, contrast-enhanced and spectral imaging need much higher kilovoltages than standard mammography. Performance will be ultimately limited by the quantum efficiency of the x-ray absorption layer in the detector, which drops with increasing energy. It is likely then that new DBT detectors will have thicker absorption layers to meet these needs.

A recently developed alternative to the AMPFI are arrays using a complementary metal oxide semiconductor (CMOS) read-out technology (Konstantinidis et al. 2012). These systems would generally be constructed of smaller (e.g., 2 × 2) subunits tiled together to create a full-field detector. Like the indirect conversion AMPFI systems, the x-ray absorber is generally CsI deposited either directly on the detector or on a fiber optic plate. CMOS technology can accomodate much more complex electronic circuitry than AMPFI systems, allowing the use of "active pixel sensors." Active pixel sensors have on-element individual gain amplification for every detector element, resulting in very low noise read-out. Proof-of-principle full-field prototype DBT systems with detector element size of 74.8×74.8 μm have been demonstrated (Choi et al. 2012; Naday et al. 2010).

Another detector technology that has been adapted to DBT employs photon counting (see Chapter 3). An example is the MicroDose system from Philips (originally Sectra) that uses a multislit silicon strip detector that can achieve photon counting rates in excess of 200 kHz (counts per second) per detector element (Åslund et al. 2007). Photon counting techniques have a performance advantage by nearly eliminating electronic noise in the detector. This can be very beneficial for tomosynthesis because of the low amount of radiation used to acquire each projection view. There is also potential for multienergy spectral imaging with a photon counting system, which may be useful for contrast-enhanced imaging (see Section 15.9.1). A prototype system has been described (Fredenberg et al. 2009) that uses two energy bins for photon counting. By acquiring both high- and low-energy projections simultaneously, it would nearly eliminate motion artifacts and maintain short imaging times. Likely improvements to existing detector technology will include increased numbers of energy channels, and improved charge sharing correction and high-count rate correction (Åslund et al. 2010).

There are alternative detector materials that show promise for mammographic applications, such as CdZnTe and CdTe, used as either integrating (Mainprize et al. 2002) or photon counting detectors (Iwanczyk et al. 2011), but to date, none have been demonstrated in a practical full-field image acquisition system that would be suitable for breast tomosynthesis. Systems based on "avalanche gain" to boost the signal over the noise floor may be especially appealing for low-dose projection acquistion and a number of approaches have been proposed (Reznik et al. 2009). Again, these systems are in the early development phase.

15.6   USER INTERFACE AND DISPLAY

A screening mammography exam generally consists of four images (1 CC and 1 MLO view for each breast) plus prior mammograms. For tomosynthesis reviews, there may be one or two tomosynthesis views (20–80 slices/view) and possibly an additional one, or two, 2D mammographic images for each breast. Even without including prior exams (mammography or tomosynthesis), the radiologist may be required to view hundreds of images for each patient. To maintain efficiency, accuracy, and throughput, the radiologist must have the usual digital viewing tools plus rapid image processing, display and retrieval of priors, and individualized hanging protocols.

Typically, the tomosynthesis slices are viewed as a ciné sequence and with an interface that permits the radiologist to manually scroll through the slices. There should be automatic variable ciné speeds set for the individual radiologist preference. The scrolling interface device must be ergonomic. The radiologist may require alternate views or different reconstruction settings for visualizing microcalcifications and convenient access to CAD analyses that can be toggled on or off should be available.

These new tools require new hardware. For example, medical displays suitable for tomosynthesis must conform to mammographic standards, but also have the added requirement of a faster temporal response to allow rapid image update and ciné loop review. A reader analysis of one vendor's monitors suggested that improved temporal response improved radiologists' performance (ROC) by 6–10% (Marchessoux et al. 2011). In that study, there appeared to be improved performance at 25 fps (compared to 50 fps). Others (Lång et al. 2011) found no statistically significant changes in performance between 9 and 25 fps, although the time required to review a case was minimized at 14 fps update rate.

Because of the size of the data files and the highly asymmetric nature of reconstructions, some have suggested on-demand reconstruction of regions of interest that might permit zooming or tilting of the view plane (Kuo et al. 2011). Better resolution and reduced artifacts are achieved because the resliced image is not limited by the fixed grid of a conventionally reconstructed dataset. On-demand reconstructions are computer resource intensive; nevertheless, Kuo et al. were able to demonstrate user interactive updates near 15 fps.

One of the most common concerns of radiologists is the possible loss of conspicuity of microcalcifications in a tomosynthesis dataset, or rather the loss of the ability to perceive clusters of microcalcifications that may span several slices. The radiologist may see only one or two microcalcifications on a given slice and not recognize that these calcifications are part of a more ominous cluster. This may be overcome by the use of slab views that combine image information from several slices into a much thicker slice (3–10 mm typically) (Poplack et al. 2007) that captures more of the extent of the clusters.

15.7   DOSE REDUCTION IN TOMOSYNTHESIS

If DBT is used as a screening tool, it will likely be governed by radiation dose limitations similar to the situation with digital mammography. What these dose limits should be is an open question. Historically, the dose employed for analog mammography was related to the technical characteristics of the particular technology, first the sensitivity of film and later that of xeroradiography and various generations of screen-film technology. Sensitivity depended on the quantum efficiency of the x-ray absorption element and any additional amplification (e.g., provided by converting x-rays to light in a phosphor). The quantum efficiency in turn depends on the thickness of the x-ray absorption element. This has practical limits because if the absorber is too thick there will be unacceptable blur due to spreading of the signal in traversing that element. Additionally, in film imaging, the gradient of the characteristic curve of the film, which strongly affects image contrast, is dependent on the level of exposure to the film. So the amount of radiation incident on the detector, which is directly proportional to radiation dose to the breast, is largely determined by the need to expose the film to the point where the gradient is optimal. In digital imaging techniques such as tomosysnthesis, image brightness and contrast are adjustable, independent of the radiation level, so this constraint is removed.

All x-ray imaging is governed by x-ray absorption statistics, which follow a Poisson distribution, causing a random fluctuation of signal, that is, noise. This can be described in simple terms by the standard deviation about the mean signal and this is given by $N^{1/2}$, where N is the mean number of x-ray quanta used by the detector over a defined area to form the image. This results in a signal-to-noise ratio, which, in simple analysis, is also related to $N^{1/2}$. It is this fundamental phenomenon, not the sensitivity and contrast considerations described above, that should set the level of exposure used to form an image.

As the tomosynthesis slice images are reconstructed from conventional x-ray projections, the source of the noise is identical to that in 2D imaging; however, the noise in the measured projections propagates into the slice images through the reconstruction algorithm and the final image noise is likely to be dependent on factors additional to dose such as the number of projections, the electronic noise of the detectors, and the various filters employed in the reconstruction algorithm. Each of the multiple acquired views will contribute a certain level of electronic detector noise independent of that due to x-ray Poisson effects. Therefore, the image is likely to contain a higher electronic noise level than that of a 2D mammogram. To avoid domination by such effects requires a certain basic level of x-ray exposure for each projection and this may be more than that resulting by dividing the dose used for digital mammography by the number of projections. This may cause the total dose to exceed that used in 2D mammography. Is this acceptable? Possibly, if the DBT exam provides greater sensitivity and/or specificity of breast cancer detection or a clearer picture of extent of disease.

Most early implementations of breast tomosynthesis used identical x-ray technique factors for all projections with a goal that when integrated over all projections the mean glandular dose to the breast would be similar to that for digital mammography. This yields a dose per projection that is nominally equal, although the dose could vary by somewhat less than 20% with projection angle for the CC view and slightly more for MLO views due to the breast shape and tissue distribution (Chapter 4) (Sechopoulos et al. 2007). When averaged over all projections, the dose from a tomosynthesis set is generally within 10% of an equivalent view mammogram with the same total technique factors (Dance et al. 2011).

Because the dose delivered for a given choice of technique factors is similar to that in digital mammography, some of the same approaches to dose reduction that are used for digital mammography are applicable in breast tomosynthesis. For example, anode-filter combinations that provide a more penetrating beam such as W/Rh, W/Al, and W/Ag have been proposed for both digital mammography and breast tomosynthesis. One analysis has suggested that for a W/Rh spectrum, the dose reduction over an equivalent Mo/Mo spectrum for digital mammography can range between 9% and 60% while maintaining the same SDNR for both masses and calcifications (Ranger et al. 2010). Already, most tomosynthesis systems use more energetic x-ray beams than in digital mammography, achieved through filtration or higher kV to help reduce dose (Feng and Sechopoulos 2012).

Dose reduction can be achieved through the use of noise-reducing reconstruction schemes. Certainly, many of the iterative techniques (Chapter 8) have achieved better reconstructed image quality compared to filtered backprojection. Nonlinear methods such as penalized maximum likelihood (Das and Gifford 2011; Das et al. 2009b), and total variation (Sidky et al. 2009; Velikina et al. 2007) achieve lesion conspicuity that is better than that obtained with filtered backprojection (FBP). For example, Das et al. (2009b) showed that for a microcalcification phantom, a penalized maximum-likelihood algorithm achieved better SNR at 1.5 mGy compared to conventional FBP at 4 mGy despite using an optimized Butterworth filter (<0.25 cycles/pixel) to maximize SNR for the detection of microcalcifications for the FBP reconstruction.

15.8 DISTRIBUTION OF RADIATION DOSE AMONG PROJECTIONS

Most approaches have, until recently, apportioned the x-ray exposure equally over all projections. However, it might be beneficial to distribute the dose unequally across projections. Image SNR may be improved if the reconstruction relies on information from certain projections more than from others. Hu and Zhao (2011) used a Fourier-based task analysis of the detection of Gaussian objects in uniform backgrounds to evaluate increasing the dose for the central projections while reducing the dose at higher angles as compared to a scheme using uniform dose at the same total exposure. They showed that for small objects (<200 μm full-width half maximum (FWHM)), the angular-dependent dose profile provided better results than for a uniform dose profile. For larger objects (>1 mm FWHM), the uniform dose profile appeared to work better. Detectability was highly dependent on the choice of reconstruction filters as well as the dose profile used.

Another motivation for using different angular dose profiles may be to produce some higher-quality projections. For example, the central projection of the tomosynthesis set provides identical positioning to that of a conventional mammogram. A significant dose saving can be achieved if the central projection serves such a dual role if a conventional mammogram is required. To maintain good image quality for the mammogram, as much as one-half of the dose of the total acquired image set might be used in the central projection, with the remainder distributed across all other projections. Nishikawa et al. (2007) suggested that a high-dose central projection could be used directly by radiologists to improve the speed and accuracy of microcalcification detection while maintaining high mass detection rates in the DBT reconstruction. However, others have shown equivalence (Vecchio et al. 2011) or inferior performance (Das et al. 2009a) for the detection of microcalcifications (or specks simulating microcalcifications) with the aid of a higher-dose central projection. Further investigation is warranted to determine whether a higher-dose central projection will have clinical benefit.

15.9 QUANTITATIVE IMAGING

15.9.1 CONTRAST-ENHANCED TOMOSYNTHESIS

Tomosynthesis offers the potential for improved lesion conspicuity by removing background structures. However, the contrast between healthy fibroglandular tissue and cancer is fundamentally limited by the x-ray attenuation properties of breast tissues. As shown in Figure 15.1, the differences in the linear attenuation coefficients measured from samples of invasive ductal carcinoma and that measured from normal fibroglandular tissue are small (Johns and Yaffe 1987). One possible solution to increase image contrast of a lesion is by the use of an exogenous x-ray contrast agent.

Tumors that grow larger than about 1 mm in diameter require additional oxygen and nutrient supplies causing them to recruit new blood vessels by a complex series of molecular signals (Folkman 2000). These neo-angiogenic vessels tend to be poorly formed, disorganized, thin-walled, and leaky, allowing intravenous contrast agents to pool in the extravascular space of tumors (Fukumura and Jain 2008). In the 1980s, it was demonstrated that angiogenesis could potentially be an indicator for breast cancer, as they showed that malignant lesions enhance more strongly than benign lesions following injection of an iodinated contrast medium, when imaged with a body CT

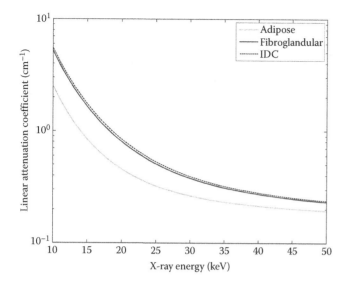

Figure 15.1 A plot of linear attenuation coefficients versus x-ray energy as measured by Johns and Yaffe (1987) for breast adipose tissue (dotted line), fibroglandular tissue (solid), and invasive ductal carcinoma (IDC, dashed). The similarity between the curves between IDC and healthy fibroglandular tissue results in very low image contrast.

scanner (Chang et al. 1982) or by digital subtraction angiography (Watt et al. 1986).

The use of contrast agents for imaging breast lesions increased markedly following the introduction of Gd-based contrast agents for breast magnetic resonance imaging (Heywang et al. 1986; Kaiser 1985). Interestingly, it was found that the pharmacokinetics of the contrast agent over time in dynamic contrast-enhanced MRI (DCE-MRI) as measured by signal enhancement and wash-out could be used to differentiate between malignant and benign lesions (Buadu et al. 1996). The classic indicator for malignant lesions was an early, rapid uptake of the contrast agent followed by a fast washout, whereas a slower steadily increasing uptake of contrast agent was usually associated with benign lesions (Heywang et al. 1989; Mussurakis et al. 1995). Breast DCE-MRI has a very high sensitivity, reported between 88% and 100%; however, this technique can suffer from low specificity of 30–70% (Zakhireh et al. 2008).

It has been shown in a number of small pilot studies that contrast-enhanced digital mammography (CEDM) has the potential to yield similar information to that of breast DCE-MRI (Diekmann et al. 2005; Dromain et al. 2006; Jong et al. 2003; Lewin et al. 2003). The radiation dose of a temporal CEDM exam is similar to a conventional digital mammography exam (Diekmann et al. 2005; Dromain et al. 2006). In small reader studies, the sensitivity and specificity of CEDM read with mammography was equivalent or improved compared to mammography alone (Diekmann et al. 2011; Dromain et al. 2011). This suggests that CEDM may be useful as an adjunct to mammography. Further, Diekmann et al. (2011) also showed that CEDM is effective in the evaluation of the dense breast where specificity was increased from 0.35 to 0.59. The use of contrast enhancement with digital breast tomosynthesis (CE-DBT) has been proposed to address the limitation of tissue superposition in CEDM, which can cause the iodine in normal tissue to mask the

appearance and kinetics of the contrast uptake in a lesion (Chen et al. 2007).

Similar to CEDM, the imaging protocols for CE-DBT are either temporal subtraction or dual-energy techniques. Temporal subtraction involves the acquisition of an initial "mask" DBT dataset prior to injection (Carton et al. 2010; Chen et al. 2007). Standard iodinated, nonionic, small-molecule contrast agents are then administered by intravenous injection, typically via a catheter into the antecubital vein contralateral to the breast of concern, at a dosage of about 1–1.5 mL/kg body weight, over a period of between 30 and 60 s, usually with a power injector. Tomosynthesis datasets are acquired at several time points following the injection over a period of several minutes. Reconstruction can be performed with any of the techniques described in the previous chapters. A volume difference image is created by subtracting the mask reconstruction from the reconstructions at each time point, leaving behind the iodine signal. Alternatively, the mask projection data (log normalized) may be subtracted from the postinjection projections for each time point and then reconstructed. For linear reconstruction algorithms such as filtered back projection, this order is not important (within numerical error). For other algorithms, there may be a benefit to choosing to subtract in the projection domain or in the reconstructed volume. Some algorithms (especially those that have nonnegativity criteria) may not work well when using difference projections.

An early pilot study by Chen et al. used a modified digital mammography system, which required manual positioning of the gantry arm for every projection and approximately 30 s between projections for image processing and transfer. Studies from 13 women with BI-RADS category 4 or 5 lesions using temporal subtraction contrast-enhanced tomosynthesis were acquired with nine projection images over a 50° angular range (Chen et al. 2007). Compared to mammography, CE-DBT was found in qualitative assessments to provide better lesion localization, better lesion margin visualization, and better assessment of morphological features of the lesion. Similar to breast MRI, lesion vascular assessment could be performed with CE-DBT.

Because of the large number of images and long scan times required for temporal subtraction CE-DBT compared to CEDM, patient motion is an important limitation (Chen et al. 2007). Subjects may be seated to help reduce motion and light breast compression is used during the acquisition. There is concern that breast compression may cause alterations in blood flow and hence contrast kinetics, but it must be maintained throughout the procedure to maintain image registration.

A second approach for CE-DBT is to employ a dual-energy technique. Dual-energy CE-DBT involves acquiring a high-energy and a low-energy projection pair. A weighted log subtraction of the projections suppresses the noniodine signal, and tomosynthesis reconstruction can be performed by any suitable algorithm. Imaging protocols may be at a single or at multiple time points following injection. Dual-energy CE-DBT can mitigate motion artifacts when the time interval between the corresponding high- and low-energy projections is very short (Carton et al. 2010; Schmitzberger et al. 2011). Furthermore, the breast need not be held under compression during injection, or between successive time points, allowing for a more natural blood flow within the breast.

Carton et al. (2010) noted that dual-energy CE-DBT reconstructions have high relative noise, possibly requiring additional image processing. In their work, a total variation noise reduction algorithm was used to reduce the appearance of noise without introducing significant additional blurring.

Schmitzberger et al. (2011) describe a dual-energy DBT system that uses a photon counting detector. By energy discrimination, photons are segregated to form a low-energy (below the k-edge of iodine) or a high-energy (above the k-edge) image during a single exposure. Thus, simultaneous acquisition of the high- and low-energy projection sets is achieved. This reduces the chance for motion or other temporal artifacts to occur. Because the photon counting system has no added electronic noise, the resulting images have improved SNR over those of conventional integrating detectors. This is particularly of importance with dual-energy DBT, which requires twice the number of projection images per time point. Schmitzberger et al. described a pilot study of 10 women who had a single lesion and were scheduled for biopsy. A dual-energy or "spectral imaging" approach for tomosynthesis was used in which images were acquired at two time points post injection: 120 and 480 s. Reconstruction was performed with an iterative maximum-likelihood (convex) Lange–Fessler algorithm for each of the low- and high-energy datasets (Dahlman et al. 2011; Schmitzberger et al. 2011). Weighted subtraction of the resulting 3-mm slices was performed yielding the iodine image volume. Example CE-DBT slices from a spectral imaging exam are shown in Figure 15.2 for a 69-year-old woman with a 25-mm ductal carcinoma *in situ* in the left breast. Results of the pilot study were encouraging, with all four radiologists agreeing on the level of suspicion (BIRADS 4 or 5) for 7 of 10 lesions with spectral imaging. Washout was seen on all those malignant lesions that exhibited enhancement.

In all systems to date, the total scan time for acquisition of all projections is relatively long (~15–60 s). Because the contrast agent concentrations are changing dynamically, each projection image will capture a slightly different iodine concentration.

The reconstruction will, therefore, be based on the combination of projections with inconsistent iodine concentrations. Hill et al. (2008) examined the effect of long scan times when undergoing different contrast enhancement profiles (e.g., rapid wash in and wash out, increasing, and plateau) for temporal CE-DBT. This was evaluated using a series of projection images acquired at several static iodine concentrations and resorting the images to simulate various enhancement rates. They were able to demonstrate that iodine concentrations in the reconstructed images were equivalent to the temporal mean of the enhancement profile and that the signals could be converted to quantitative iodine concentrations provided that the calibration was performed as a function of lesion size. In similar work, Hill et al. (2012) showed that for dual-energy CE-DBT, near-simultaneous acquisition of corresponding projections at low and high energy yielded the best image quality. On the other hand, for a real iodine enhancement rate of 0.1 mg/mL/s, if two separate tube sweeps were used to acquire the low-energy projection set followed by the high-energy set (15 s per scan, with no delay between scans), then the apparent iodine concentration in the reconstructed images would drop by as much as 1 mg/mL during the wash in phase. To reduce temporal effects such as motion artifacts and enhancement rate-related effects, it has been suggested that CE-DBT should be acquired with fewer projections over a narrower angular span to shorten acquisition times.

15.9.2 BREAST DENSITY INFORMATION FROM TOMOSYNTHESIS

Breast density is a known indicator of breast cancer risk, with odds ratios in excess of 4× between the densest and least dense breast categories (Boyd et al. 1995, 2007). Breast density is often measured as a percent area measure of projected dense areas relative to total projected breast area on mammograms. Percent density may be estimated by human readers or by semiautomatic threshold techniques (Byng et al. 1996). More recently, breast density has been estimated volumetrically from mammograms by estimating the volume of fibroglandular tissue in the breast consistent with producing the transmission profiles seen in a mammogram (Alonzo-Proulx et al. 2010; Jeffreys et al. 2006; Pawluczyk et al. 2003; van Engeland et al. 2006).

Area percent density can be obtained from the central projection image from a DBT exam with good correlation to percent density measurements on a standard mammogram despite lower doses used in acquiring the central projection image (Bakic et al. 2009). Because DBT provides a slice stack that yields quasi-3D information about tissue composition through the breast, an alternative would be to attempt to measure volumetric breast density.

Alternative measures of breast density examine the parenchymal patterns or texture in an effort to extract other biomarkers for breast cancer risk. In a small preliminary study, measures of texture in DBT slices such as contrast, homogeneity, and "energy" showed significant correlations with percent breast density in corresponding mammograms (Kontos et al. 2009). When several different texture measures were combined as a signature, principal component analysis revealed a significant trend ($p < 0.003$) with breast density categories with an $R^2 = 0.21$.

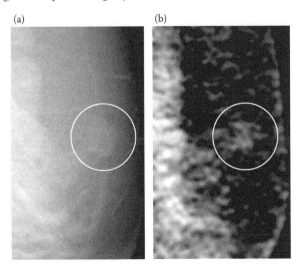

(a)   (b)

Figure 15.2 (**See insert.**) Examples of zoomed CE-DBT images of a 25-mm ductal carcinoma *in situ* lesion (circled) for a 69-year-old woman. Total energy image (a) and energy subtraction image (b) from a photon counting spectral imaging system at 120 s following contrast agent injection. Slice thickness was 3 mm. (Images courtesy of Dr. Florian Schmitzberger.)

These results suggest that an automatic analysis of parenchymal texture may be useful as a biomarker. However, this study is only very preliminary and a proper case–control risk study has yet to be performed with this approach.

15.10 COMPUTER-ASSISTED DIAGNOSTIC TECHNIQUES

CAD has been shown to improve cancer detection in several radiological applications, including mammography. A CAD system could possibly act as a prescreening reader to alleviate radiologists' workloads, or as a second reader to improve detection rates. Although breast tomosynthesis offers the promise of improved lesion conspicuity by virtue of eliminating obscuring tissue structures, it also offers markedly increased numbers of images (generally 20–80 slices per view per breast) that a radiologist must review. As the workload for the radiologists increases, the possibility increases that a small lesion will be missed. It seems natural, then, to apply CAD to tomosynthesis, especially, as mentioned above, to improve the efficiency of identifying microcalcifications.

The techniques for CAD for tomosynthesis are generally similar to that for mammography usually with an image preprocessing step, segmentation, feature extraction, and decision analysis. Tomosynthesis datasets can be analyzed by CAD on the raw projection images, the 3D reconstructed slices, or a combination of the two (Chan et al. 2008).

By applying CAD on the projection images, mature validated algorithms from mammography can be employed. This approach also has the advantage that it can be applied to many systems regardless of reconstruction algorithm. Several groups have investigated variations on this approach (Chan et al. 2005; Reiser et al. 2006; Singh et al. 2008). In one implementation, likelihood maps are generated for each projection and by means of a simple backprojection technique, these likelihood maps are projected back into the reconstruction volume. For objects seen in many views, the likelihood scores are reinforced by overlapping rays, increasing the 3D likelihood score for a mass candidate in the volume. Segmentation of the object is achieved by a likelihood threshold.

One potential drawback from this approach is that the noise characteristics are significantly different from a mammogram. If any step in the 2D CAD process is sensitive to noise, this may yield inferior results in terms of reduced sensitivity or an increased number of false positives. For example, Chan et al. (2008) noted that when operating at 90% sensitivity in their tests, this produced on average approximately four false positives in mass detection per breast.

Applying CAD to the reconstructed slices requires the development of new algorithms tailored to tomosynthesis and most likely specifically tailored to the reconstruction algorithm used. Here, mass candidates may be segmented in 3D, but because of the highly asymmetric resolution, feature analysis is likely to be performed only "in plane."

In a third approach, the 3D likelihoods are combined with the 2D likelihoods. Chan et al. (2008) first normalized and then averaged the two responses to provide a fused estimate. They pointed out, however, that the optimal combination of the 2D and 3D estimates may involve something other than a simple average, perhaps by a linear or nonlinear classifier. In their comparison of the three approaches, the 3D method was better than the 2D method, and the combined method provided the best performance.

15.11 FUSION WITH OTHER MODALITIES

Simultaneous or coregistered imaging with tomosynthesis and other imaging modalities is an emerging area of research. Tomosynthesis provides a reasonably cost-effective means of acquiring quasi-3D anatomic information, which is complementary to that obtained from other modalities that are more sensitive to physiological responses. Williams et al. (2010) describe a prototype dual-modality tomosynthesis system that has a gamma camera mounted on the same pivot as the x-ray tube and detector above the compression paddle. The gamma camera can be translated out of view for transmission tomosynthesis and returned for molecular breast imaging tomosynthesis (MBIT). In a pilot study of 17 volunteers scheduled for breast biopsy, the women received 25 mCi of ^{99m}Tc sestamibi. Mild breast compression was used to minimize motion and permit coregistration between the DBT and MBIT volumes. Estimated radiation dose to the breast was 2 mGy from the nuclear medicine and approximately 3–4 mGy for DBT imaging. Note, however, that because the radiopharmaceutical for MBIT is administered systemically (injection), radiation dose is also delivered to other radio-sensitive organs (Hendrick 2010). Imaging time for DBT was 30 s and 11 min for MBI. An example of the fused images is shown in Figure 15.3. Seven malignant and 14 benign lesions were detected. Combined DBT and MBI reading exhibited high sensitivity (85%) and excellent specificity (100%). These results suggest the feasibility of a dual-modality breast tomosynthesis system and its role in tumor detection, localization, and characterization.

Another fusion system is a combined breast tomosynthesis and ultrasound breast imaging system that uses a sonolucent compression paddle to permit a robotic arm to scan a conventional ultrasound transducer in raster fashion across the compressed breast (Wu et al. 2003). A 3D x-ray dataset and 3D ultrasound dataset are acquired and registered. Because of the nonisotropic nature of both modalities, there are preferred viewing planes for both DBT and the 3D automated US dataset (AUS). In this case, the preferred plane for DBT is parallel to the detector (body axial), which is perpendicular to the AUS viewing plane (body sagittal). One approach to viewing would be to view both modalities in one plane with suboptimal image quality for one of the modalities. Alternatively, the visualization approach of Goodsitt et al. (2008) provides one view for each modality in its highest-resolution plane while providing a synchronized navigation system with a special "volume of interest" block shown on both views to allow the reader to maintain localization in both views. An example of the reviewing software is shown in Figure 15.4.

Fang et al. (2009, 2011) have demonstrated the use of a combined optical and breast tomosynthesis system. Near-infrared lasers are used in the tomographic optical breast imaging system,

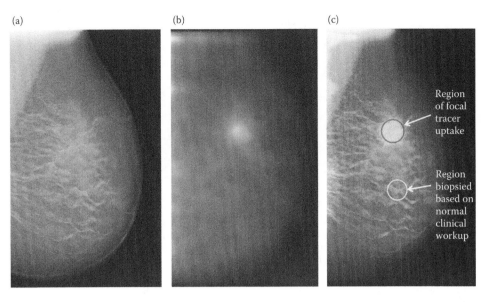

Figure 15.3 (**See insert.**) Examples of fused DBT and MBI images for a 52-year-old woman. (a) DBT slice at 1 mm thick. (b) Corresponding gamma tomosynthesis image at the same depth. (c) Merged sections from (a) and (b). The enhancing region was biopsy-confirmed ductal carcinoma *in situ* (arrow). (Images courtesy of Dr. Mark Williams, University of Virginia.)

which consists of a series of optical source fibers mounted on one side of the compressed breast and a series of optical detectors (probe) mounted in an array with 5 mm spacing on the opposite surface of the breast. The probe is incorporated into a modified compression paddle, with detachable components to remove them from the x-ray field during radiographic imaging. The tomosynthesis dataset is registered to the optical source/detector locations and the reconstructed breast surface is used to define the breast boundaries for the optical reconstruction algorithms and thereby constrain the reconstruction. The images of optical

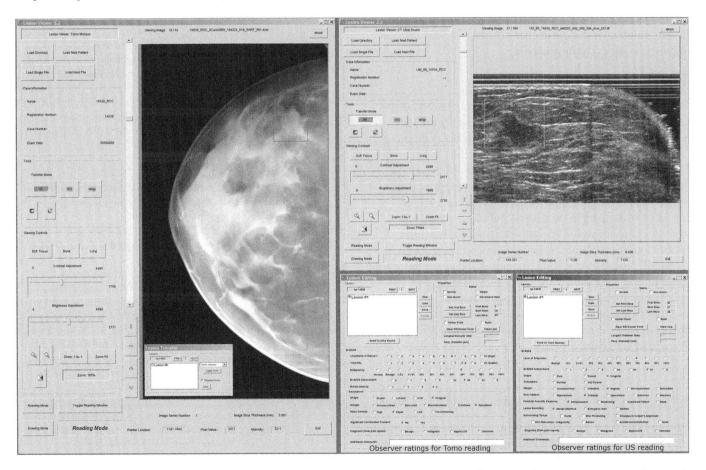

Figure 15.4 Screen capture of GUI display for dual-modality (tomosynthesis and 3D ultrasound) observer studies. (Figure courtesy of Dr. Mitch Goodsitt, University of Michigan. Copyrighted display developed by CAD group at the University of Michigan, Department of Radiology, all rights reserved.)

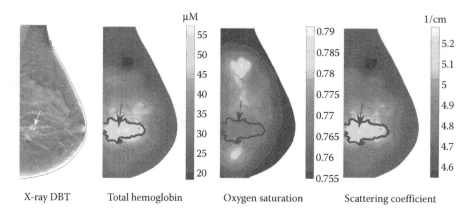

X-ray DBT   Total hemoglobin   Oxygen saturation   Scattering coefficient

Figure 15.5 (**See insert.**) Registered DBT slice (far left) and optical images for (left to right) total hemoglobin concentration, blood oxygen saturation, a scattering coefficient for a 45-year-old woman with a 2.50-cm invasive ductal carcinoma (arrow on DBT image and outlined in optical images). (Figure courtesy of Dr. Qianqian Fang, Massachussetts General Hospital.)

absorption and scattering at multiple wavelengths, were used to calculate the oxy- and deoxygenated hemoglobin concentration, and oxygen saturation in the breast. The coregistered images can be viewed together to determine regions of correlation. In a pilot study ($n = 189$), the normalized total hemoglobin concentration could be used to differentiate solid benign lesions, malignant masses, and cysts. Low oxygen saturation was also indicative of cysts. The functional images from the optical data were well correlated with structural DBT overlays. Example coregistered images are shown in Figure 15.5 for a woman diagnosed with invasive ductal carcinoma. Fang et al. (2011) suggest that in the future, the tomosynthesis data may be used as spatial priors to improve the optical reconstructions.

15.12 SUMMARY

DBT is an emerging modality with the potential to transform how breast cancer is managed. In particular, it may yield improvement of both the sensitivity and specificity of breast cancer detection. Early clinical study results have been promising. If the proposed much larger multicenter trials support these findings, tomosynthesis could, in a few years, supplant digital mammography as the primary tool for screening. There is considerable diversity in the designs and image acquisition strategies implemented in the current first generation of DBT systems and the optimum imaging technique has yet to be identified. New reconstruction techniques and image processing methods may allow dose reduction or provide additional information to the radiologist. Contrast-enhanced imaging and fusion imaging modalities may add useful functional physiological information to the radiologist's array of tools for differential diagnosis.

REFERENCES

ACRIN. 2012. *ACRIN 2009 Fall Meeting. ACRIN.* Vol. 104. Pentagon City, Arlington, Virginia.

Alonzo-Proulx, O., N. Packard, J. M. Boone, A. Al-Mayah, K. K. Brock, S. Z. Shen, and M. J. Yaffe. 2010. Validation of a method for measuring the volumetric breast density from digital mammograms. *Physics in Medicine and Biology* 55 (11) (June): 3027–3044.

Åslund, M., B. Cederström, M. Lundqvist, and M. Danielsson. 2007. Physical characterization of a scanning photon counting digital mammography system based on Si-Strip detectors. *Medical Physics* 34 (6): 1918–1925.

Åslund, M., E. Fredenberg, M. Telman, and M. Danielsson. 2010. Detectors for the future of x-ray imaging. *Radiation Protection Dosimetry* 139 (1–3): 327–333.

Bakic, P. R., A. Carton, D. Kontos, C. Zhang, A. B. Troxel, and A. D. A. Maidment. 2009. Breast percent density: Estimation on digital mammograms and central tomosynthesis projections. *Radiology* 252 (1): 40–49.

Boone, J. M., T. R. Nelson, K. K. Lindfors, and J. A. Seibert. 2001. Dedicated breast CT: Radiation dose and image quality evaluation. *Radiology* 221 (3): 657–667.

Boyd, N. F., J. W. Byng, R. A. Jong, E. K. Fishell, L. E. Little, A. B. Miller, G. A. Lockwood, D. L. Tritchler, and M. J. Yaffe. 1995. Quantitative classification of mammographic densities and breast cancer risk: Results from the canadian national breast screening study. *JNCI* 87: 670–675.

Boyd, N. F., H. Guo, L. J. Martin, L. Sun, J. Stone, E. Fishell, R. A. Jong et al. 2007. Mammographic density and the risk and detection of breast cancer. *New England Journal of Medicine* 356 (3) (January): 227–236.

Buadu, L. D., J. Murakami, S. Murayama, N. Hashiguchi, S. Sakai, K. Masuda, S. Toyoshima, S. Kuroki, and S. Ohno. 1996. Breast lesions: Correlation of contrast medium enhancement patterns on MR images with histopathologic findings and tumor angiogenesis. *Radiology* 200 (3) (September): 639–649.

Byng, J. W., N. F. Boyd, E. Fishell, R. A. Jong, and M. J. Yaffe. 1996. Automated analysis of mammographic densities. *Physics in Medicine and Biology* 41 (5) (May): 909–923.

Carton, A-K, S. C. Gavenonis, J. A. Currivan, E. F. Conant, M. D. Schnall, and A. D. A. Maidment. 2010. Dual-energy contrast-enhanced digital breast tomosynthesis—A feasibility study. *The British Journal of Radiology* 83 (988) (April): 344–350.

Chan, H-P, J. Wei, B. Sahiner, E. A. Rafferty, T. Wu, M. A. Roubidoux, R. H. Moore, D. B. Kopans, L. M. Hadjiiski, and M. A. Helvie. 2005. Computer-aided detection system for breast masses on digital tomosynthesis mammograms: Preliminary experience. *Radiology* 237 (3): 1075.

Chan, H-P, J. Wei, Y. Zhang, M. a. Helvie, R. H. Moore, B. Sahiner, L. Hadjiiski, and D. B. Kopans. 2008. Computer-aided detection of masses in digital tomosynthesis mammography: Comparison of three approaches. *Medical Physics* 35 (9): 4087–4095.

Chang, C. H., D. E. Nesbit, D. R. Fisher, S. L. Fritz, S. J. Dwyer, A. W. Templeton, F. Lin, and W. R. Jewell. 1982. Computed tomographic mammography using a conventional body scanner. *American Journal of Roentgenology* 138 (3) (March): 553–558.

Chen, L., C. C. Shaw, M. C. Altunbas, C.-J. Lai, and X. Liu. 2008. Spatial resolution properties in cone beam CT: A simulation study. *Medical Physics* 35 (2): 724–734.

Chen, S. C., A.-K. Carton, M. Albert, E. F. Conant, M. D. Schnall, and A. D. A. Maidment. 2007. Initial clinical experience with contrast-enhanced digital breast tomosynthesis. *Academic Radiology* 14 (2): 229–238.

Choi, J.-G., H.-S. Park, Y.-s. Kim, Y.-W. Choi, T.-H. Ham, and H.-J. Kim. 2012. Characterization of prototype full-field breast tomosynthesis by using a CMOS array coupled with a columnar CsI(Tl) scintillator. *Journal of the Korean Physical Society* 60 (3) (February 14): 521–526.

Dahlman, N., E. F., M. Åslund, M. Lundqvist, F. Diekmann, and M. Danielsson. 2011. Evaluation of photon-counting spectral breast tomosynthesis. *Proceedings of SPIE* 7961: 796114-1-10.

Dance, D. R., K. C. Young, and R. E. van Engen. 2011. Estimation of mean glandular dose for breast tomosynthesis: Factors for use with the UK, European and IAEA breast dosimetry protocols. *Physics in Medicine and Biology* 56 (2) (January 21): 453–471.

Das, M. and H. C. Gifford. 2011. Comparison of model-observer and human-observer performance for breast tomosynthesis: Effect of reconstruction and acquisition parameters. *Proceedings of SPIE* 7961: 796118-1–796118-9.

Das, M., H. C. Gifford, J. M. O'Connor, and S. J. Glick. 2011. Penalized maximum likelihood reconstruction for improved microcalcification detection in breast tomosynthesis. *IEEE Transactions on Medical Imaging* 30 (4) (April): 904–914.

Das, M., H. C. Gifford, J. M. O'Connor, and S. J. Glick. 2009a. Evaluation of a variable dose acquisition technique for microcalcification and mass detection in digital breast tomosynthesis. *Medical Physics* 36 (6): 1976–1984.

Das, M. H. Gifford, M. O'Connor, and S. J. Glick. 2009b. Dose reduction in digital breast tomosynthesis using a penalized maximum likelihood reconstruction. *Proceedings of SPIE* 7258: 725855-1–725855-8.

Diekmann, F., S. Diekmann, F. Jeunehomme, S. Muller, B. Hamm, and U. Bick. 2005. Digital mammography using iodine-based contrast media: Initial clinical experience with ynamic contrast edium Enhancement. *Investigative Radiology* 40 (7): 397–404.

Diekmann, F., M. Freyer, S. Diekmann, E. M. Fallenberg, T. Fischer, U. Bick, and A. Pöllinger. 2011. Evaluation of contrast-enhanced digital mammography. *European Journal of Radiology* 78 (1) (April): 112–121.

Diekmann, F., H. Meyer, S. Diekmann, S. Puong, S. Muller, U. Bick, and P. Rogalla. 2009. Thick slices from tomosynthesis data sets: Phantom study for the evaluation of different algorithms. *Journal of Digital Imaging* 22 (5) (October): 519–526.

Dromain, C., C. Balleyguier, S. Muller, M.-C. Mathieu, F. Rochard, P. Opolon, and R. Sigal. 2006. Evaluation of tumor angiogenesis of breast carcinoma using contrast-enhanced digital mammography. *American Journal of Roentgenology* 187 (5): 528–537.

Dromain, C., F. Thibault, S. Muller, F. Rimareix, S. Delaloge, A. Tardivon, and C. Balleyguier. 2011. Dual-energy contrast-enhanced digital mammography: Initial clinical results. *European Radiology* 21 (3) (March): 565–574.

Fang, Q., S. A. Carp, J. Selb, G. Boverman, Q. Zhang, D. B. Kopans, R. H. Moore, E. L. Miller, D. H. Brooks, and D. A. Boas. 2009. Combined optical imaging and mammography of the healthy breast: Optical contrast derived from breast structure and ompression. *IEEE Transactions on Medical Imaging* 28 (1) (January): 30–42.

Fang, Q., J. Selb, S. A. Carp, G. Boverman, E. L. Miller, D. H. Brooks, R. H. Moore, D. B. Kopans, and D. A. Boas. 2011. Combined optical and x-ray tomosynthesis breast imaging. *Radiology* 258 (1): 89–97.

Feng, S. S. J. and I. Sechopoulos. 2012. Clinical digital breast tomosynthesis system: Dosimetric characterization. *Radiology* 263 (1): 35–42.

Folkman, J. 2000. Incipient angiogenesis. *Journal of the National Cancer Institute* 92 (2) (January): 94–95.

Fredenberg, E., M. Lundqvist, M. Åslund, M. Hemmendorff, B. Cederström, and M. Danielsson. 2009. A photon-counting detector for dual-energy breast tomosynthesis. *Proceedings of SPIE* 7258: 72581 J–72581 J-11.

Fukumura, D. and R. K. Jain. 2008. Imaging angiogenesis and the microenvironment. *APMIS: Acta Pathologica, Microbiologica, Et Immunologica Scandinavica* 116 (7–8): 695–715.

Ghetti, C., A. Borrini, O. Ortenzia, R. Rossi, and P. L. Ordóñez. 2008. Physical characteristics of GE senographe essential and DS digital mammography detectors. *Medical Physics* 35 (2): 456–463.

Glick, S. J. 2007. Breast CT. *Annual Review of Biomedical Engineering* 9 (January): 501–526.

Goodsitt, M. M., H.-P. Chan, L. Hadjiiski, G. L. Lecarpentier, and P. L. Carson. 2008. Automated registration of volumes of interest for a combined x-ray tomosynthesis and ultrasound breast imaging system. *LNCS* 5116: 463–468.

Gur, D., M. L. Zuley, M. I. Anello, G. Y. Rathfon, D. M. Chough, M. a. Ganott, C. M. Hakim, L. Wallace, A. Lu, and A. I Bandos. 2011. Dose reduction in digital breast tomosynthesis (DBT) screening using synthetically reconstructed projection images an observer performance study. *Academic Radiology* 19 (2) (November 17): 166–171.

Hendrick, R. E. 2010. Radiation doses and cancer risks from breast imaging studies. *Radiology* 257 (1): 246–253.

Heywang, S. H., D. Hahn, H. Schmidt, I. Krischke, W. Eiermann, R. Bassermann, and J. Lissner. 1986. MR imaging of the breast using gadolinium-DTPA. *Journal of Computer Assisted Tomography* 10 (2): 199–204.

Heywang, S. H., A. Wolf, E. Pruss, T. Hilbertz, W. Eiermann, and W. Permanetter. 1989. MR imaging of the breast with Gd-DTPA: Use and limitations. *Radiology* 171 (1) (April): 95–103.

Hill, M., J. Mainprize, S. Puong, A.-K. Carton, R. Iordache, S. Muller, and M. Yaffe. 2011. SU-F-BRA-05 Dual-energy contrast-enhanced breast tomosynthesis: Signal response to tissue contrast uptake kinetics. *Medical Physics* 38 (6): 3701.

Hill, M. L., J. G. Mainprize, S. Puong, A.-K. Carton, R. Iordache, S. Muller, and M. J. Yaffe. 2012. Impact of image acquisition timing on image quality for dual energy contrast-enhanced breast tomosynthesis. *Proceedings of SPIE* 8313: 831308.

Hill, M. L., J. G. Mainprize, and M. J. Yaffe. 2008. Sensitivity of contrast-enhanced digital breast tomosynthesis to changes in iodine concentration during Acquisition. *LNCS* 5116: 643–650.

Hu, Y.-H. and W. Zhao. 2011. The effect of angular dose distribution on the detection of microcalcifications in digital breast tomosynthesis. *Medical Physics* 38 (5): 2455–2466.

Iwanczyk, J. S., E. Nygard, J. C. Wessel, N. Malakhov, G. Wawrzyniak, N. E. Hartsough, T. Gandhi, and W. C. Barber. 2011. Optimization of room-temperature semiconductor detectors for energy-resolved x-ray imaging. *2011 IEEE Nuclear Science Symposium Conference Record* (23–29 October): 4745–4750.

Jeffreys, M., R. Warren, R. Highnam, and G. D. Smith. 2006. Initial experiences of using an automated volumetric measure of breast density: The standard mammogram form. *BJR* 79 (941) (May): 378–382.

Johns, P. C. and M. J. Yaffe. 1987. X-ray characterisation of normal and neoplastic breast tissues. *Physics in Medicine and Biology* 32 (6): 675–695.

Jong, R. A., M. J. Yaffe, M. Skarpathiotakis, R. S. Shumak, N. M. Danjoux, and A. Gunesekara. 2003. Contrast-enhanced digital mammography: Initial clinical experience. *Radiology* 228: 842–850.

Kaiser, W. 1985. MRI of the female breast. First clinical results. *Archives Internationales De Physiologie Et De Biochimie* 93 (5) (December): 67–76.

Karellas, A. and S. Vedantham. 2008. Breast cancer imaging: A perspective for the next decade. *Medical Physics* 35 (11): 4878–4897.

Konstantinidis, A. C., M. B. Szafraniec, R. D. Speller, and A. Olivo. 2012. The dexela 2923 CMOS x-ray detector: A flat panel detector based on CMOS active pixel sensors for medical imaging applications. *Nuclear Instruments and Methods in Physics Research Section A* 689 (October): 12–21.

Kontos, D., P. R. Bakic, A.-K. Carton, and A. B. Troxel. 2009. Parenchymal texture analysis in digital breast tomosynthesis for breast cancer risk estimation: A preliminary study. *Academic Radiology* 16 (3): 283–298.

Kuo, J., P. A. Ringer, S. G. Fallows, P. R. Bakic, A. D. A. Maidment, and S. Ng. 2011. Dynamic reconstruction and rendering of 3D tomosynthesis images. *Physics* 7961: 796116–796116-11.

Lång, K., S. Zackrisson, K. Holmqvist, M. Nystrom, I. Andersson, D. Förnvik, A. Tingberg, and P. Timberg. 2011. Optimizing viewing procedures of breast tomosynthesis image volumes using eye tracking combined with a free response human observer study. *Breast* 7966: 796602–796602-11.

Lewin, J. M., P. K. Isaacs, V. Vance, and F. J. Larke. 2003. Dual-energy contrast-enhanced digital subtraction mammography: Feasibility. *Radiology* 229 (1) (October): 261–268.

Madhav, P., D. J. Crotty, R. L. McKinley, and M. P. Tornai. 2009. Evaluation of tilted cone-beam CT orbits in the development of a dedicated hybrid mammotomograph. *Physics in Medicine and Biology* 54 (12) (June 21): 3659–3676.

Mainprize, J. G., N. L. Ford, S. Yin, E. E. Gordon, W. J. Hamilton, T. O. Tümer, and M. J. Yaffe. 2002. A CdZnTe slot-scanned detector for digital mammography. *Medical Physics* 29 (12): 2767–2781.

Marchessoux, C., N. Vivien, A. Kumcu, and T. Kimpe. 2011. Validation of a new digital breast tomosynthesis medical display. *Proceedings of SPIE* 7966: 79660R-1–79660R-12.

Mussurakis, S., D. L. Buckley, S. J. Bowsley, P. J. Carleton, J. N. Fox, L. W. Turnbull, and A. Horsman. 1995. Dynamic contrast-enhanced magnetic resonance imaging of the breast combined with pharmacokinetic analysis of gadolinium-DTPA uptake in the diagnosis of local recurrence of early stage breast carcinoma. *Investigative Radiology* 30 (11) (November): 650–662.

Naday, S., E. Bullard, S. Gunn, J. E. Brodrick, E. O. O'Tuairisg, A. McArthur, H. Amin, M. B. Williams, P. G. Judy, and A. Konstantinidis. 2010. Optimised breast tomosynthesis with a novel CMOS flat panel detector. *LNCS* 6136: 428–435.

Nishikawa, R. M., I. Reiser, P. Seifi, and C. J. Vyborny. 2007. A new approach to digital breast tomosynthesis for breast cancer screening. *Proceedings of SPIE* 6510: 65103C–65103C-8.

O'Connell, A., D. L. Conover, Y. Zhang, P. Seifert, W. Logan-young, C.-F. L. Lin, L. Sahler, and R. Ning. 2010. Cone-beam CT for breast imaging: Radiation dose, breast coverage, and image quality. *American Journal of Roentgenology* 195 (2): 496–509.

Pawluczyk, O., B. J. Augustine, M. J. Yaffe, D. Rico, J. Yang, G. E. Mawdsley, and N. F. Boyd. 2003. A volumetric method for estimation of breast density on digitized screen-film mammograms. *Medical Physics* 30 (3): 352–364.

Philpotts, L., M. Raghu, M. Durand, R. Hooley, R. Vashi, L. Horvath, J. Geisel, and R. Butler. 2012. Initial experience with digital breast tomosynthesis in screening mammography. In *AJR*. Vol. 198. Vancouver, B.C.

Pisano, E. D., C. Gatsonis, E. Hendrick, M. Yaffe, J. K. Baum, S. Acharyya, E. F. Conant et al. 2005. Diagnostic performance of digital versus film mammography for breast-cancer screening. *New England Journal of Medicine* 353 (17): 1773–1783.

Poplack, S. P., T. D. Tosteson, C. A. Kogel, and H. M. Nagy. 2007. Digital breast tomosynthesis: Initial experience in 98 women with abnormal digital screening mammography. *American Journal of Roentgenology* 189 (3): 616–623.

Ranger, N. T., J. Y. Lo, and E. Samei. 2010. A technique optimization protocol and the potential for dose reduction in digital mammography. *Medical Physics* 37 (3): 962–969.

Reiser, I. and R. M. Nishikawa. 2010. Task-based assessment of breast tomosynthesis: Effect of acquisition parameters and quantum noise. *Medical Physics* 37 (4): 1591–1600.

Reiser, I., R. M. Nishikawa, M. L. Giger, T. Wu, E. A. Rafferty, R. Moore, and D. B. Kopans. 2006. Computerized mass detection for digital breast tomosynthesis directly from the projection images. *Medical Physics* 33 (2): 482–491.

Reznik, A., W. Zhao, Y. Ohkawa, K. Tanioka, and J. A. Rowlands. 2009. Applications of avalanche multiplication in amorphous selenium to flat panel detectors for medical applications. *Journal of Materials Science: Materials in Electronics* 20 (November 6): S63–S67.

Schmitzberger, F. F., E. M. Fallenberg, R. Lawaczeck, M. Hemmendorff, E. Moa, M. Danielsson, U. Bick et al. 2011. Development of low-dose photon-counting contrast-enhanced tomosynthesis with spectral imaging. *Radiology* 259 (2): 558–564.

Sechopoulos, I. and C. Ghetti. 2009. Optimization of the acquisition geometry in digital tomosynthesis of the breast. *Medical Physics* 36 (4): 1199–1207.

Sechopoulos, I. S. Suryanarayanan, S. Vedantham, C. D'Orsi, and A. Karellas. 2007. Computation of the glandular radiation dose in digital tomosynthesis of the breast. *Medical Physics* 34 (1): 221–232.

Sidky, E. Y., X. Pan, I. S. Reiser, R. M. Nishikawa, R. H. Moore, and D. B. Kopans. 2009. Enhanced imaging of microcalcifications in digital breast tomosynthesis through improved image-reconstruction algorithms. *Medical Physics* 36 (11): 4920–4932.

Singh, S., G. D. Tourassi, J. A. Baker, E. Samei, and J. Y. Lo. 2008. Automated breast mass detection in 3D reconstructed tomosynthesis volumes: A featureless approach. *Medical Physics* 35 (8): 3626–3636.

Smith, A. P., E. A. Rafferty, and L. Niklason. 2008. Clinical performance of breast tomosynthesis as a function of radiologist experience. *LNCS* 5116: 61–66.

U.S. National Institutes of Health. 2012a. Malmö Breast Tomosynthesis Screening Trial (MBTST) NCT01091545.

U.S. National Institutes of Health. 2012b. Tomosynthesis in the Oslo Breast Cancer Screening Program (DBT) NCT01248546.

US Food and Drug Administration. 2012. MQSA National Statistics. *Radiation-Emitting Products: MQSA National Statistics.*

Varjonen, M., P. Strömmer, and P. Oy. 2008. Optimizing the target-filter combination in digital mammography in the sense of image quality and average glandular dose. *LNCS* 5116: 570–576.

van Engeland, S., P. R. Snoeren, H. Huisman, C. Boetes, and N. Karssemeijer. 2006. Volumetric breast density estimation from full-field digital mammograms. *IEEE Transactions on Medical Imaging* 25 (3) (March): 273–282.

Vecchio, S., A. Albanese, P. Vignoli, and A. Taibi. 2011. A novel approach to digital breast tomosynthesis for simultaneous acquisition of 2D and 3D images. *European Radiology* 21 (6) (June): 1207–1213.

Velikina, J., S. Leng, and G.-H. Chen. 2007. Limited view angle tomographic image reconstruction via total variation minimization. *Proceedings of SPIE* 6510: 651020-1–12.

Wang, X., J. G. Mainprize, G. Wu, and M. J. Yaffe. 2012. Wiener filter for filtered back projection in digital breast tomosynthesis. *Proceedings of SPIE* 8313: 83134Z-1–83134Z-8.

Watt, C. A., L. V. Ackerman, J. P. Windham, P. C. Shetty, M. W. Burke, M. J. Flynn, C. Grodinsky, G. Fine, and S. J. Wilderman. 1986.

Breast lesions: Differential diagnosis using digital subtraction angiography. *Radiology* 159 (1): 39–42.

Williams, M. B., P. G. Judy, and S. Gunn. 2010. Dual-modality breast tomosynthesis. *Radiology* 255 (1): 191–198.

Wu, T., A. Stewart, M. Stanton, T. McCauley, W. Phillips, D. B. Kopans, R. H. Moore et al. 2003. Tomographic mammography using a limited number of low-dose cone-beam projection images. *Medical Physics* 30: 365–380.

Zakhireh, J., R. Gomez, and L. Esserman. 2008. Converting evidence to practice: A guide for the clinical application of MRI for the screening and management of breast cancer. *European Journal of Cancer* 44 (18) (December): 2742–2752.

Index